THE MOBILITY AND DIFFUSION

OF IONS IN GASES

WILEY SERIES IN PLASMA PHYSICS

SANBORN C. BROWN ADVISORY EDITOR
RESEARCH LABORATORY OF ELECTRONICS
MASSACHUSETTS INSTITUTE OF TECHNOLOGY

McDANIEL AND MASON · THE MOBILITY AND DIFFUSION OF IONS IN
    GASES
GILARDINI · LOW ENERGY ELECTRON COLLISIONS IN GASES
MITCHNER AND KRUGER · PARTIALLY IONIZED GASES
TIDMAN AND KRALL · SHOCK WAVES IN COLLISIONLESS PLASMAS
NASSER · FUNDAMENTALS OF GASEOUS IONIZATION IN PLASMA
    ELECTRONICS
LICHTENBERG · PHASE-SPACE DYNAMICS OF PARTICLES
BROWN · INTRODUCTION TO ELECTRICAL DISCHARGES IN GASES
BEKEFI · RADIATION PROCESSES IN PLASMAS
MACDONALD · MICROWAVE BREAKDOWN IN GASES
HEALD AND WHARTON · PLASMA DIAGNOSTICS WITH MICROWAVES
MCDANIEL · COLLISION PHENOMENA IN IONIZED GASES

THE MOBILITY AND DIFFUSION

OF IONS IN GASES

EARL W. McDANIEL

EDWARD A. MASON

JOHN WILEY & SONS, NEW YORK · LONDON · SYDNEY · TORONTO

PHYSICS

Library of Congress Cataloging in Publication Data

McDaniel, Earl Wadsworth, 1926–
    The mobility and diffusion of ions in gases.

    (Wiley series in plasma physics)
    "A Wiley-Interscience publication."
    Includes bibliographies.
    1. Ionic mobility. 2. Diffusion. I. Mason,
Edward Allen, 1926–   joint author. II. Title.

QC717.M18   530.4′3   72–13414
ISBN 0-471-58387-1

Printed in the United States of America

10 9 8 7 6 5 4 3 2 1

PREFACE

Research on the mobility and diffusion of ions in gases began more than 75 years ago, and there is now a substantial volume of literature on this subject. Experimentation in this field is difficult, however, and despite the ingenuity and skill of the early workers few mobility data obtained before about 1960 are of more than historical interest today. Apparently no reliable direct measurements of ionic diffusion coefficients were made until the 1960's. As a result the experimental literature in this area is confusing, and a newcomer to the field is hard put to know where to start to read. Also, persons wishing to utilize ionic mobility and diffusion data in various applications may have considerable difficulty in selecting reliable numbers for their use. On the theoretical side there are fewer misleading papers in the literature, but no general exposition of the subject, dealing with all of its aspects, is available.

We have attempted in this book to prepare a statement of what is known today about both the experimental and theoretical aspects of the mobility and diffusion of ions in gases. We have made no effort to describe the historical development of the subject systematically—that has been done by Loeb, Tyndall, Massey, and Thomson and Thomson in books that we cite. As a result we have failed to give proper credit to many workers who played an important role in advancing the field to its present state, but

v

we believe that we have been able to paint a less confusing picture. We exclude all results in which the interactions of ions with other ions are important. The long-range Coulomb interaction requires the adoption of new theoretical techniques to avoid divergences, and this subject is more appropriately to be considered as plasma physics.

The audience for whom this book was written includes scientists and engineers working in this area, graduate students entering the field, and researchers in gaseous electronics, atomic collisions, the atmospheric sciences, and the kinetic theory of gases who have an interest in the motion of ions through gases. We assume on the part of the reader a knowledge of such basic things as the Maxwellian velocity distribution, the difference between elastic and inelastic collisions, the fundamentals of scattering theory, and the nature of a tensor quantity. On the other hand, we do not assume any detailed knowledge of transport processes—it seems desirable to build from the ground level here.

Chapter 1 is an introduction that relates mainly to the experimental portion of the book. In it we introduce a number of definitions and attempt to give a qualitative picture of the physical processes discussed in detail later. In addition, equations are derived that describe the geometrical spread of ions diffusing in gases under different conditions. These equations are of particular interest with respect to the experiments in Chapter 4.

Chapters 2 and 3 deal with experimental techniques for measuring mobilities and diffusion coefficients in drift tubes. Here emphasis is placed on apparatus in current use. A great deal of attention is devoted to the effects of chemical reactions between ions and molecules of the gas through which they are moving and to other general considerations that affect the success of experiments. Chapter 4 is concerned with afterglow techniques.

Chapter 5 is the major theoretical section of the book. Here we have attempted to include more of the historical background than in the experimental section. We give first some qualitative arguments and an elementary theoretical discussion to bring out the basic physics involved before proceeding to the more elaborate mathematical theories. Work on the theory of ion motions in gases has often gone along an apparently different path from that followed in the conventional kinetic theory of gases, and we have tried to give a unified account in which the connections would be clear. We have also tried to point out the gaps remaining in the subject.

Chapter 6 is a summary of how the theoretical results of Chapter 5 can be applied to the calculation of mobilities, especially in the face of meager initial information. A large fraction of this chapter is of necessity devoted to the theory and determination of ion-neutral interaction potentials. Information of use in such calculations is collected in the tables of Appendices I and II at the end.

Chapter 7 contains a collection of experimental data on ionic mobilities and diffusion that we consider to be useful at present.

Since we are concerned primarily with unclustered ions, the suggestion was made that this book should be titled "Naked Came the Ion". It was with some reluctance that we finally decided against this suggestion.

We wish to thank the U.S. Air Force Office of Scientific Research, the Office of Naval Research, and the National Aeronautics and Space Administration for their support of our research on the subject of this book through the years. We are also indebted to J. E. Boyd, Vernon Crawford, J. R. Stevenson, Hong-sup Hahn, P. W. Langhoff, J. H. Whealton, and Mrs. S. Whitney for their comments and their help. Finally, we wish to express our appreciation to the reviewers of our manuscript, Professor A. Dalgarno and Dr. D. Edelson, and to the staff of John Wiley and Sons for their cooperation.

EARL W. McDANIEL
EDWARD A. MASON

September 1972
Atlanta, Georgia
Providence, Rhode Island

CONTENTS

CHAPTER 1   INTRODUCTION, 1

1-1   Qualitative Description of the Motion of Slow Ions
     in Gases, 1
1-2   The Parameters E/N and E/p, 4
1-3   General Facts about Mobilities and Diffusion
     Coefficients, 5
1-4   Ion-Ion Interactions and the Effect of Space Charge, 6
1-5   The Importance of Data on Ionic Mobilities and
     Diffusion Coefficients, 8
1-6   The Differences in Behavior of Ions and Electrons, 9
1-7   The Spreading of a Cloud of Ions by Diffusion
     Through an Unbounded Gas, 10
1-8   The Spreading of an Ion Cloud during Its Drift in
     an Electric Field, 12
1-9   The Diffusion Equation, 13
1-10 Boundary Conditions, 14
1-11 Solution of the Time-Independent Diffusion Equation
     for Various Geometries, 16
1-12 Ambipolar Diffusion, 24
     References, 27

CHAPTER 2   THE MEASUREMENT OF DRIFT VELOCITIES AND
            LONGITUDINAL DIFFUSION COEFFICIENTS, 29

2-1   General Considerations in Drift-Tube Experiments, 30
2-2   Basic Aspects of Drift Velocity Measurements, 35
2-3   The Determination of Longitudinal Diffusion
      Coefficients, 44
2-4   The Determination of Reaction Rate Coefficients from
      Arrival-Time Spectra, 46
2-5   Description of Drift Tubes, 50
2-6   The Mathematical Analysis of Ionic Motion in Drift
      Tubes, 75
      References, 82

CHAPTER 3   THE MEASUREMENT OF TRANSVERSE
            DIFFUSION COEFFICIENTS, 85

3-1   The Attenuation Method, 85
3-2   The Townsend Method, 92
      References, 98

CHAPTER 4   AFTERGLOW TECHNIQUES, 99

4-1   The Technique of Lineberger and Puckett, 100
4-2   The Technique of Smith and His Colleagues, 107
4-3   Microwave Techniques, 114
      References, 116

CHAPTER 5   KINETIC THEORY OF DIFFUSION
            AND MOBILITY, 118

5-1   Definitions and General Results, 118
5-2   Elementary Theories and Qualitative Arguments, 120
5-3   Low-Field Theory, 136
5-4   Medium-Field Theory, 165
5-5   High-Field Theory, 187
5-6   Connection Formulas, 204
5-7   Resonant Charge Transfer, 209
5-8   Ion Transfer, 229
      References, 232

CHAPTER 6   INTERACTION POTENTIALS AND MOBILITIES, 236

6-1   Mobilities from Interaction Potentials, 236
6-2   Theory of Ion-Atom and Ion-Molecule Interactions, 248
6-3   Determination of Interaction Potentials, 257
6-4   Estimation of Mobilities from Meager Data, 260
       References, 263

CHAPTER 7   EXPERIMENTAL DATA ON MOBILITIES AND
            DIFFUSION COEFFICIENTS, 266

7-1   The Mobility of Ions in Pure Gases at or Near Room
       Temperature; the Mobility of Ions in Vapors, 267
7-2   The Mobility of Ions in Mixtures of Gases—
       Blanc's Law, 304
7-3   The Variation of Ionic Mobilities with Gas
       Temperature, 306
7-4   The Diffusion of Ions in Gases, 313

APPENDIX I   TABLES OF TRANSPORT CROSS SECTIONS AND
             COLLISION INTEGRALS, 327
APPENDIX II TABLES OF PROPERTIES USEFUL IN THE ESTIMATION
             OF ION-NEUTRAL INTERACTION ENERGIES, 343

       AUTHOR INDEX, 361

       SUBJECT INDEX, 367

1

INTRODUCTION

1-1. QUALITATIVE DESCRIPTION OF THE MOTION OF SLOW IONS IN GASES

Consider a localized collection of ions of a single type in a gas of uniform temperature and total pressure and suppose that the number density n of the ions is low enough to ignore the Coulomb forces of repulsion. As is well known, the ions will become dispersed through the gas by the process of diffusion,* in which there is a net spatial transport of the ions produced by the gradient in their relative concentration. The diffusive flow takes place in the direction opposite that of this gradient and the flow rate is directly proportional to its magnitude. The constant of proportionality is called the (scalar) diffusion coefficient and is denoted by the symbol D. Thus the ionic flux density $\mathbf{J}$ is given by Fick's law of diffusion (Fick, 1855),

$$\mathbf{J} = -D \, \nabla n. \qquad (1\text{-}1\text{-}1)$$

The magnitude of $\mathbf{J}$ equals the number of ions flowing in unit time through unit area normal to the direction of flow. The minus sign indicates that the flow occurs in the direction of decreasing concentration; D is a joint property of the ions and the gas through which they are diffusing and (1-1-1) shows

---

* Good general references to the subject of diffusion are Crank (1956) and Hirschfelder et al. (1964).

it to be a measure of the transparency of the gas to the diffusing particles. Since the velocity of the diffusive flow $\mathbf{v}$ is given by

$$\mathbf{J} = n\mathbf{v}, \qquad (1\text{-}1\text{-}2)$$

we may also write Fick's law in the form

$$\mathbf{v} = -\frac{D}{n}\nabla n. \qquad (1\text{-}1\text{-}3)$$

The diffusive flow continues until inequalities in composition have been eliminated by interdiffusion of the ions and gas molecules. This type of flow is separate and distinct from the flow that would result from a nonuniformity of the total pressure.

If a weak uniform electric field is now applied throughout the gas, a steady flow of the ions along the field lines will develop, superimposed on the much faster random motion that leads to diffusion. The velocity of the center of mass of the ion cloud, or equivalently the average velocity of the ions, is called the drift velocity $\mathbf{v_d}$, and this velocity is directly proportional to the electric field intensity $\mathbf{E}$, provided that the field is kept weak. Thus

$$\mathbf{v_d} = K\mathbf{E}, \qquad (1\text{-}1\text{-}4)$$

where the constant of proportionality K is called the (scalar) mobility of the ions; K, like D, is a joint property of the ions and the gas through which the motion occurs.

A simple relation, known as the Einstein equation,* exists between the weak-field mobility and diffusion coefficient. This equation is exact in the limit of vanishing electric field and ion concentration and states that

$$K = \frac{eD}{kT}. \qquad (1\text{-}1\text{-}5)$$

Here e is the ionic charge, k is the Boltzmann constant, and T is the gas temperature. If K is expressed in the usual units, square centimeters per volt per second, D in the usual units, square centimeters per second, and T in degrees Kelvin, we have

$$K = 1.1605 \times 10^4 \frac{D}{T}. \qquad (1\text{-}1\text{-}6)$$

---

* This relationship is derived in Chapter 5 (Section 5-1-A). A simple derivation also appears in McDaniel (1964a). Equation 1-1-5 was first established for ions in gases by Townsend in 1899 (Townsend, 1899). It is frequently referred to as the Nernst-Townsend relation because it had been previously derived (in a different context) by Nernst (1888). Einstein's derivation was published in a paper dealing with the theory of Brownian motion (Einstein, 1905).

(A factor of 299.79 comes into play here because the electrostatic unit of mobility is square centimeters per statvolt per second and 1 statvolt equals 299.79 V.) It is not surprising that K should be directly proportional to D here, since both quantities are a measure of the ease with which the ions can flow through the gas. It is important to point out that (1-1-5) is valid only when the electric field is so weak that the ions are close to being in thermal equilibrium with the gas molecules, that is to say, when "low-field" conditions obtain. Under these conditions the ionic velocity distribution is very nearly Maxwellian. The ionic motion is largely the random thermal motion produced by the heat energy of the gas, with a small drift component superimposed in the direction of the applied field.

The constant, steady-state drift velocity of the ions given by (1-1-4) is achieved as a balance between the accelerations in the field direction between collisions with gas molecules and the decelerations that occur during collisions. Since the ionic mass is usually comparable to the molecular mass, only a few collisions are normally required for the ions to attain a steady-state condition after the electric field is applied.

If the electric field intensity is now raised to a level at which the ions acquire an average energy appreciably in excess of the thermal energy of the gas molecules, a number of complications develop. The thermal energy becomes less important, but two large components of motion are produced by the drift field: a directed component along the field lines and a random component representing energy acquired from the drift field but converted into random form by collisions with molecules. The mobility K appearing in (1-1-4) is no longer a constant in general but will usually depend on the ratio of the electric field intensity to the gas number density E/N, which is the parameter that determines the average ionic energy gained from the field in steady-state drift above the energy associated with the thermal motion. In addition, the energy distribution of the ions becomes distinctly non-Maxwellian and cannot be calculated accurately by existing theory. Furthermore, the diffusion now takes place transverse to the field direction at a rate different from that of the diffusion in the direction of the electric field, so that the diffusion coefficient becomes a tensor rather than a scalar. The diffusion coefficient tensor* has the form

$$\mathbf{D} = \begin{vmatrix} D_T & 0 & 0 \\ 0 & D_T & 0 \\ 0 & 0 & D_L \end{vmatrix}, \tag{1-1-7}$$

where D_T is the (scalar) transverse diffusion coefficient that describes the rate of diffusion in directions perpendicular to $\mathbf{E}$ and D_L is the (scalar)

* The manner in which this tensor is used in calculations is indicated in Section 2-6-B.

longitudinal diffusion coefficient characterizing diffusion in the field direction (Wannier, 1953). In the "intermediate-field" and "high-field" regions* described here the Einstein equation (1-1-5) no longer applies.

It may be of interest to note that the presence of a magnetic field also renders an ionized gas anisotropic and makes both the mobility and the diffusion coefficient assume a tensor form (McDaniel, 1964b). In this case, however, **D** is not in the form indicated by (1-1-7), since two of the off-diagonal components are now different from zero.

1-2. THE PARAMETERS E/N AND E/p

We now attempt to make plausible the statement in Section 1-1 that E/N is the parameter that determines the average ionic energy acquired from the electric field, that is, the "field energy." The electric force on an ion of charge e is eE and the resulting acceleration is eE/m, where m is the mass of the ion. We make the crude assumption that when an ion undergoes a collision it loses, on the average, all the energy it acquired from the field during the preceding free path. Then, if τ denotes the collision period, or mean free time, the velocity acquired just before a collision is $eE\tau/m$. Since $\tau \sim 1/N$, the energy obtained between collisions from the field is thus seen to be proportional to $(E/N)^2$. Rigorous calculations also show E/N to be the parameter that determines the field energy of the ions.

Although E/N is the more fundamental quantity, until recently most experimentalists have reported the results of their measurements in terms of E/p, where p is the gas pressure or, in terms of E/p_0, where p_0 is the "reduced pressure," normalized to 0°C. The reduced pressure is defined by the equation

$$p_0 = \frac{273.16}{T} \, p, \tag{1-2-1}$$

where T is the absolute temperature at which the measurement was made. This convention has not been wholly satisfactory, since p_0 has also been used to represent normalizations to temperatures other than 0°C. If we use the parameter E/N, however, there is no ambiguity in comparing experimental results. The conversion relations are

$$\frac{E}{N} = (1.0354 \times T \times 10^{-2})\left(\frac{E}{p}\right) \tag{1-2-2}$$

or

$$\frac{E}{N} = 2.828 \times \frac{E}{p_0}, \tag{1-2-3}$$

* More precise meaning is given to the terms "intermediate field" and "high field" in Chapter 5.

where E/N is in units of 10^{-17} V-cm^2, T is in degrees Kelvin, and E/p or E/p_0 is in volts per centimeter per Torr. Huxley, Crompton, and Elford (1966) have suggested that the units of E/N be denoted by the "Townsend," or "Td," where 1 Td $= 10^{-17}$ V-cm^2, and this designation is attaining widespread usage. Both E/N and E/p are used in this book.

The field energy is negligible compared with the thermal energy if

$$\left(\frac{M}{m} + \frac{m}{M}\right)eE\lambda \ll kT, \tag{1-2-4}$$

where M and m are the molecular and ionic masses, respectively, and $eE\lambda$ is the energy gained by an ion in moving a mean free path λ in the field direction. The factor involving the masses accounts for the ability of the ions to store the acquired energy over many collisions if the masses are significantly different. Using the ideal gas law $NkT = p$ and the relationship $\lambda = 1/NQ$, where N is the molecular number density and Q is the ion-molecule collision cross section, we may express the foregoing inequality as $(M/m + m/M)eE \ll pQ$. Taking a singly charged ion moving through the parent gas and making the reasonable assumption that $Q = 50 \times 10^{-16}$ cm^2, we find that the field energy is much less than the thermal energy if $E/p \ll 5 \times 10^{-6}$ (statvolt/cm) per (dyne/cm^2) ≈ 2 V/cm-Torr. The electric field is said to be "low" when this criterion is satisfied and "high" when the inequality is reversed. It should be noted that a given field in a gas of given density may change from "low" to "high" if the gas temperature is lowered sufficiently.

From what has been said here it may be inferred that the mobility K will be constant, independent of E/p, provided that $E/p \lesssim 2$ V/cm-Torr, so that the ionic energy is close to thermal. Actually, theory predicts that K will also be constant at higher ionic energies, provided that the collision frequency does not depend on the energy of the ions. Usually, however, the mobility begins to vary with E/p at the upper end of the low-field region. Under such conditions the concept of mobility loses some of its convenience, but the phenomenological definition of the mobility as the ratio of v_d to E is still useful in comparing experimental data and is used here.

1-3. GENERAL FACTS ABOUT MOBILITIES AND DIFFUSION COEFFICIENTS

The mobility of a given ionic species in a given gas is inversely proportional to the number density of the molecules but relatively insensitive to small changes (a few degrees Kelvin) in the gas temperature if the number density is held constant. To facilitate the comparison and use of data a measured mobility K is usually converted to a "reduced mobility," K_0, defined by the equation

$$K_0 = \frac{p}{760} \frac{273.16}{T} K = \frac{p_0}{760} K, \qquad (1\text{-}3\text{-}1)$$

where p is the gas pressure in Torr and T is the gas temperature in degrees Kelvin at which the mobility K was obtained.* Under the standard conditions of pressure and temperature (760 Torr and 0°C) the gas number density is $2.69 \times 10^{19}/cm^3$. It must be emphasized that the use of (1-3-1) merely provides a standardization or normalization with respect to the molecular number density; the temperature to which the reduced mobility actually refers is the temperature of the gas during the measurement. For ions of atmospheric interest in atmospheric gases the reduced mobility is of the order of several square centimeters per volt per second. In the modern literature, when a single value is quoted as "the mobility" of an ion in a gas, the value cited is the reduced mobility extrapolated to zero field strength.

Ions of the atmospheric gases in their parent gases have diffusion coefficients of the order of 50 cm^2/sec at 1 Torr pressure and low E/N. The diffusion coefficients usually increase dramatically as E/N is raised above the low-field region. As expected, D is found to vary inversely with the number density of the gas, and diffusion coefficient data are usually presented in the form of the product DN.

Methods of measuring mobilities and diffusion coefficients are described in Chapters 2, 3, and 4 of this volume. The calculation of these quantities from kinetic theory is discussed in Chapters 5 and 6. Data are collected in Chapter 7.

1-4. ION-ION INTERACTIONS AND THE EFFECT OF SPACE CHARGE

Under the usual conditions more than one ionic species is present at a given time in an ionized gas. If the density of ionization is low, each species of ions may be considered separately, and the ions of each type drift and diffuse through the gas without interacting appreciably with one another or with members of the other species. For reasons discussed in Chapter 2 this condition must obtain in ion drift experiments if reliable measurements

---

* All existing mobility theories predict that the reduced mobility K_0 should be independent of the gas-number density N except at pressures so high that three-body elastic interactions would be expected to become significant. Elford (1971), however, has reported the observation of a slight dependence of K_0 on N for K^+ ions in He, Ne, Ar, H_2, and N_2 at pressures of the order of 5 Torr. McDaniel (1972), Gatland (1972), and Thomson et al. (1973) have discussed Elford's observations in terms of the clustering of molecules about the K^+ ions and the lack of application of diffusion corrections.

are to be made of the diffusion coefficients (and even the mobilities in certain instances), and it is of interest to establish a criterion for this condition to be satisfied (Wannier, 1953).

We consider the space charge effect produced mainly by widely separated ions, and we find that its magnitude depends on the dimensions of the apparatus. In one dimension Poisson's equation, $\nabla^2 V = -4\pi\rho$, may be written in the form $\partial E/\partial x = 4\pi n e$, where ρ is the charge density, e, the ionic charge, and n, the ionic number density. The criterion for negligible space charge distortion of an applied field E_0 is then

$$n \ll \frac{E_0}{4\pi e L}, \qquad (1\text{-}4\text{-}1)$$

where L is the relevant dimension of the apparatus. If the drift-field intensity E_0 is 2 V/cm and the drift distance L is 10 cm as typical values for mobility and diffusion measurements at low E/N, this inequality predicts significant space charge distortion of the applied field at ion densities of the order of $10^5/\text{cm}^3$, a value consistent with experimental evidence discussed in Chapter 2. (It may be of interest to point out that in most measurements of mobilities and diffusion coefficients the gas pressure is in the range 10^{-1} to 10 Torr, corresponding to a neutral gas-number density of 10^{15} to $10^{17}/\text{cm}^3$.) Larger ion densities may be tolerated in measurements at high field and in apparatus of smaller dimensions.

We may also consider a second effect of ion-ion interactions produced by random fluctuations of the ionic-number density that may alter a velocity distribution derived on the assumption of a low ionic density. Neighboring ions are more important than remote ions in this connection, since their relative positions can fluctuate more rapidly. The basis of the disturbance is the randomly fluctuating Coulomb force that produces mutual scattering. This random force may be neglected if it is unable to produce a significant deflection in a single mean free path. Since the magnitude of the force is of the order $e^2/d^2 = e^2 n^{2/3}$, where d is the mean ionic spacing, the effect is small if $e^2 n^{2/3} \lambda \ll$ (mean ion energy). In the low-field region the thermal energy dominates the field energy and the inequality becomes

$$e^2 n^{2/3} \ll pQ. \qquad (1\text{-}4\text{-}2)$$

At high field the relative importance of the thermal and field energies is reversed and the criterion becomes

$$e^2 n^{2/3} \ll eE\left(\frac{M}{m} + \frac{m}{M}\right). \qquad (1\text{-}4\text{-}3)$$

If a pressure of 1 Torr and a collision cross section of 50×10^{-16} cm^2 are assumed, (1-4-2) yields $n \ll 10^{11}$ ions/cm^3; (1-4-3) gives similar results.

Throughout most of this book the assumption is made that the ionic-number density is low enough to neglect all ion-ion interactions. This assumption greatly simplifies the mathematical treatment of the ionic motion, for the equation for the velocity distribution function is then linear instead of quadratic (Wannier, 1953).

1-5. THE IMPORTANCE OF DATA ON IONIC MOBILITIES AND DIFFUSION COEFFICIENTS

Data on ionic mobilities and diffusion coefficients are of both theoretical and practical interest. First of all experimental values of these quantities, and particularly their dependence on E/N and the gas temperature, can provide information about ion-molecule interaction potentials at greater separation distances than are accessible in beam-scattering experiments, as explained in Chapter 6. Second, mobilities are required for the calculation of ion-ion recombination coefficients (Flannery, 1972) and the rate of dispersion of ions in a gas due to mutual repulsion (McDaniel, 1964c). Ionic transport data are also required for the proper analysis of various experiments on chemical reactions between ions and molecules (see Sections 2-4 and 3-1 for discussions of experiments performed with drift tubes). In addition, knowledge of the mobility of an ionic species in a given gas as a function of E/N permits the estimation of the average ionic energy as a function of this parameter (see the second footnote in Section 2-1). Finally, information on both mobilities and diffusion is required for a quantitative understanding of electrical discharges in gases and various atmospheric phenomena.

To date little has been done on the extraction of collision cross sections from experimental data on ionic mobilities and diffusion coefficients. By contrast a considerable amount of information on electron cross sections has been derived from electron swarm experiments (Phelps, 1968; Bederson and Kieffer, 1971; Huxley and Crompton, 1973). It appears likely that with further development of the theory of ionic transport it will become possible to obtain cross sections for elastic, inelastic, and reactive scattering of ions as functions of the impact energy.

The transport properties of ions in gases have been studied experimentally since shortly after the discovery of X-rays in 1895 and theoretically since 1903. The first measurements were performed by Thomson, Rutherford, and Townsend at the Cavendish Laboratory of Cambridge University late in the nineteenth century (Thomson and Thomson, 1969). Surveys of the history of experimentation in this field appear in Loeb (1960) and Massey (1971) and are not repeated here; the calculations, however, are reviewed in Chapter 5.

Meaningful direct measurements of ionic diffusion coefficients were first made only in the 1960's and comparatively few data are available. On the other hand, some reliable mobility measurements were made as long ago as the 1930's and a large amount of good data is now on hand. Recent work has indicated, however, that most of the old data and even some of the newer results are either incorrect or refer to ions whose identities were not known. The main reason for this is that in most cases the drifting ions can undergo chemical reactions with the molecules of the gases through which they are moving and thereby change their identities. Techniques for obtaining reliable results, many of them developed only recently, are discussed in Chapters 2, 3, and 4.

1-6. THE DIFFERENCES IN BEHAVIOR OF IONS AND ELECTRONS

It is appropriate at this time to discuss the differences in behavior of ions and electrons in regard to their drift and diffusion in gases. As might be expected, electrons usually have much higher drift velocities and diffusion coefficients (by orders of magnitude) than ions under given conditions in a given gas. Because of their small mass, electrons are accelerated rapidly by an electric field, and they lose little energy in elastic collisions with molecules (a fraction of the order of m_e/M, where m_e and M are the electronic and molecular masses, respectively.) Therefore electrons can acquire kinetic energy from an electric field faster than ions, and they can store this energy between collisions to a much greater degree until they reach energies at which inelastic collisions become important. Even with only a weak electric field imposed on the gas through which the electrons are moving, the average electronic energy may be far in excess of the thermal value associated with the gas molecules. Furthermore, the electronic energy distribution is not close to Maxwellian except at extremely low values of E/N.

Other differences between electrons and ions develop in connection with their collision cross sections (Massey and Burhop, 1969; Massey, 1969; Massey, 1971; Massey, Burhop, and Gilbody, 1973; Hasted 1972). Electronic excitation of atoms and molecules is frequently an important factor in electron collisions even for impact energies of less than 10 eV, and in molecular gases the onset of vibrational and rotational excitation occurs at energies far below 1 eV (McDaniel, 1964d; Phelps, 1968; Moiseiwitsch and Smith, 1968). These energies are often attained by electrons in situations of common interest. The laboratory-frame thresholds for the corresponding modes of excitation by ions are higher than those for electrons and the excitation cross section curves peak at energies considerably above these thresholds (Thomas, 1972). Therefore ions have insufficient energy to produce much excitation under the usual gas kinetic conditions. These considerations tend to make the analysis of electronic motion in gases more difficult than the analysis of ionic motion. A compensating factor is operative, however, because of the relatively small mass of the electron. Since $m_e/M \ll 1$ in any gas, it is possible to make approximations in the analysis of electronic motion that are not valid in the ionic case. These approximations greatly simplify the mathematics and make it possible to calculate accurately the velocity distribution and transport properties of electrons in many gases at high E/N (Huxley and Crompton, 1973). This is not so with ions, as shown in Chapter 5.

On the experimental side certain other important differences appear. Electrons may be produced much more simply than ions by thermionic emission from filaments, by photoemission from surfaces, or by beta decay of radioactive isotopes. Ionic production usually requires the use of much more elaborate apparatus: electron bombardment or photoionization ion sources or an electrical discharge. Furthermore, in an electron-swarm experiment the electronic component of the charge carriers may be easily separated from any ionic component that may be present and no mass analysis is required to interpret the data. In ionic drift and diffusion experiments, on the other hand, mass analysis of the ions is usually essential if unambiguous results are to be obtained. This requirement involves a great complication of the apparatus.

A final difference between electron and ion experiments relates to the effects of impurities in the gas being studied. Molecular impurities in an atomic gas can hold the average electronic energy well below the level that would be attained in the pure gas because electrons can lose large fractions of their energy by exciting the rotational and vibrational levels of the molecules. The electronic velocity distribution can be seriously altered in the process. In ionic experiments, however, impurities have little effect on the average ionic energy and velocity distribution. The complication that may develop instead is the production of impurity ions by the reaction of the ions of the main gas with impurity molecules. This is frequently a matter of serious concern.

1-7. THE SPREADING OF A CLOUD OF IONS BY DIFFUSION THROUGH AN UNBOUNDED GAS

In this section and in Sections 1-8 through 1-11 we concern ourselves with matters related to the spatial dispersion of ions by diffusion through a gas at thermal energy under conditions of low ionization density in which the ions interact only with the gas molecules and not with other ions or electrons. It is assumed that the ions are incapable of undergoing chemical reactions with the gas molecules. The results to be obtained are useful in the analysis of various practical problems and certain types of experiment of interest to us in the present book. In Section 1-12 we consider ambipolar diffusion, which takes place when ions and electrons are both present at high number density, and in Section 2-6 of Chapter 2 we treat ion-molecule reactions.

First consider a number of ions, S, located at the origin of a one-dimensional coordinate system. If the ions are released at $t = 0$ and allowed to diffuse through a field-free gas filling all space at uniform pressure, the one-dimensional number density of the ions at distance x from the origin at time t is

$$n = \frac{S}{\sqrt{4\pi Dt}} e^{-x^2/4Dt}, \tag{1-7-1}$$

where D is the coefficient characterizing the diffusive motion of the ions through the gas.* This equation, as well as (1-1-5), is known as the Einstein relation. At any instant of time a plot of n as a function of x has the shape of a Gaussian error curve. The curve becomes progressively flatter as time elapses. The mean and root-mean-square displacements of the ions from the origin may be calculated from the distribution function in (1-7-1). The results are

$$|\bar{x}| = \frac{1}{S} \int_{-\infty}^{\infty} |x| n \, dx = \frac{2}{S} \int_{0}^{\infty} xn \, dx = \left(\frac{4Dt}{\pi}\right)^{1/2} \tag{1-7-2}$$

and

$$\sqrt{\bar{x^2}} = \left(\frac{1}{S} \int_{-\infty}^{\infty} x^2 n \, dx\right)^{1/2} = \sqrt{2Dt}. \tag{1-7-3}$$

In three dimensions the ionic number density at radius r and time t is

$$n = \frac{S}{(4\pi Dt)^{3/2}} e^{-r^2/4Dt}. \tag{1-7-4}$$

The mean and root-mean-square displacements are

$$\bar{r} = \left(\frac{16Dt}{\pi}\right)^{1/2} \tag{1-7-5}$$

and

$$\sqrt{\bar{r^2}} = \sqrt{6Dt}, \tag{1-7-6}$$

* Here S is assumed to be small enough that the total pressure will remain essentially constant throughout space.

respectively. In two dimensions

$$\sqrt{\overline{r^2}} = \sqrt{4Dt}. \tag{1-7-7}$$

The equations derived above are useful in the estimation of the average lifetime τ of ions against collision with the walls of a containing vessel. The expressions for the mean displacement indicate that

$$\tau \approx \frac{d^2}{D}, \tag{1-7-8}$$

where d is the relevant dimension of the container. The discussion in Section 1-11 permits the calculation of more accurate values of τ for various geometries. These results are

(a) for an infinitely long rectangular tube of width a and depth b

$$\tau = \left[D\pi^2 \left(\frac{1}{a^2} + \frac{1}{b^2} \right) \right]^{-1}; \tag{1-7-9}$$

(b) for an infinitely long cylinder of radius r_0

$$\tau = \frac{1}{D} \left(\frac{r_0}{2.405} \right)^2; \tag{1-7-10}$$

(c) for a sphere of radius r_0

$$\tau = \frac{1}{D} \left(\frac{r_0}{\pi} \right)^2. \tag{1-7-11}$$

To illustrate the use of this concept we may use (1-7-10) to calculate the lifetime of an ion originating on the axis of a tube of 1-cm radius containing nitrogen at 1 Torr pressure and room temperature. Taking the diffusion coefficient to be 50 cm^2/sec, we find τ to be about 3×10^{-3} sec. The total distance traveled during this time is equal to $\bar{v}\tau \approx 160$ cm, where $\bar{v}$ is the mean thermal velocity.

1-8. THE SPREADING OF AN ION CLOUD DURING ITS DRIFT IN AN ELECTRIC FIELD

It is also of interest to determine the extent of the diffusive spreading of a cloud of ions as it drifts through a gas under the influence of a weak electric field. Let L be the distance of drift during time t, v_d, the drift velocity, E, the electric-field intensity, and V, the potential difference between the extremities of the drift path. The mean displacement of the ions from the center of mass of the moving ion cloud is given by (1-7-2), whereas L is, of course, related to the drift time by the equation $L = v_d t$. Thus

$$\frac{|\bar{x}|}{L} = \left(\frac{4D}{\pi v_d L}\right)^{1/2}. \tag{1-8-1}$$

If we assume a temperature of 0°C, (1-1-6) shows that $D = K/42.465$. Then, using the relationships $K = v_d/E$ and $E = V/L$, we find that

$$\frac{|\bar{x}|}{L} = \frac{0.173}{\sqrt{V}}. \tag{1-8-2}$$

The ratio of the spread of the ion cloud to the drift distance is thus independent of the diffusion coefficient and mobility and depends only on the total voltage drop experienced by the ions. It should be emphasized that only diffusion effects were considered in the foregoing development. Dispersion due to mutual Coulomb repulsion of the ions was neglected.

1-9. THE DIFFUSION EQUATION

Let us now consider an ensemble of ions diffusing through an infinite medium that contains no sources or sinks. By definition of the particle flux density $\mathbf{J}$ the net leakage outward through an arbitrarily shaped, imaginary closed surface within the medium is $\int \mathbf{J} \cdot d\mathbf{A}$. Gauss's law shows that this leakage may also be expressed by the integral $\int \mathbf{\nabla} \cdot \mathbf{J} \, dv$, where the integration is performed over the volume bounded by the surface A. Then, if n denotes the number density of the ions, it follows that

$$\int \frac{\partial n}{\partial t} dv = -\int \mathbf{\nabla} \cdot \mathbf{J} \, dv$$

or

$$\int \left(\frac{\partial n}{\partial t} + \mathbf{\nabla} \cdot \mathbf{J}\right) dv = 0.$$

Since the choice of A was arbitrary, the integrand must itself vanish. Thus

$$\frac{\partial n}{\partial t} + \mathbf{\nabla} \cdot \mathbf{J} = 0, \tag{1-9-1}$$

which is known as the equation of continuity.

According to Fick's law of diffusion,

$$\mathbf{J} = -D\mathbf{\nabla} n. \tag{1-1-1}$$

Therefore

$$\mathbf{V} \cdot \mathbf{J} = -\mathbf{V} \cdot (D \mathbf{V} n) \tag{1-9-2}$$

and the equation of continuity gives

$$\frac{\partial n}{\partial t} = \mathbf{V} \cdot (D \mathbf{V} n), \tag{1-9-3}$$

which is known as the time-dependent diffusion equation or Fick's second law. Note that (1-9-3) allows for the possible dependence of D on position [by its dependence on composition (Crank, 1956)]. In Section 2-6 we use an extension of this equation that allows for the drift of the ions in an electric field, chemical reactions, and the tensor nature of the diffusion coefficient at high E/N.

We are now in position to verify the form of the distribution functions given in (1-7-1) and (1-7-4). Direct substitution into (1-9-3) shows that both functions satisfy the diffusion equation.

Let us now suppose that some steady-state distribution of ions $n_0(x, y, z)$ has been established within a gas-filled container. To maintain steady-state conditions ions must be continuously supplied within the gas to replenish losses to the walls by diffusion. This replenishment can be accomplished by continuously ionizing the gas with X-rays or microwaves. Now imagine the ionization source to be abruptly turned off at $t = 0$. If we assume that D is independent of position and separate variables in (1-9-3) by writing

$$n(x, y, z, t) = n_0(x, y, z)T(t), \tag{1-9-4}$$

we obtain an equation for $T(t)$ whose solution is

$$T(t) = e^{-t/\tau}, \tag{1-9-5}$$

where τ is a time constant describing the decay. Equation 1-9-3 thus yields the time-independent diffusion equation

$$\nabla^2 n_0 + \frac{n_0}{D\tau} = 0. \tag{1-9-6}$$

The solution of (1-9-6) for $n_0(x, y, z)$ is an eigenvalue problem whose solution depends on the geometry of the container and the appropriate boundary conditions.

1-10. BOUNDARY CONDITIONS

Since the diffusion equation is a second-order differential equation, its general solution will contain two arbitrary constants of integration. In the solution of a specific problem the values of these constants are determined by boundary conditions and other physical considerations.

For ions diffusing in a gas-filled container the conditions usually imposed stipulate that the ionic number density must be finite everywhere within the gas and vanish "at" the walls of the container. If this statement is interpreted as meaning that the inward ionic flux density must be zero at the walls so that no ions will be reflected back into the gas on impact, diffusion theory actually requires that the number density vary near the wall in such a way that linear extrapolation would cause it to vanish at a definite distance, d, *beyond* the wall (Fig. 1-10-1). A rather lengthy calculation (McDaniel, 1964e) shows that for a plane boundary

$$d = \tfrac{2}{3}\lambda, \qquad\qquad (1\text{-}10\text{-}1)$$

where λ is the mean free path for elastic scattering of the ions in the gas; d is usually called the linear extrapolation distance and is a measure of the

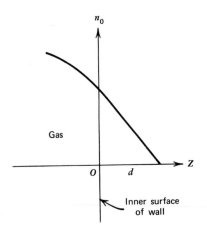

FIG. 1-10-1. Linear extrapolation of the ionic number density n_0 past the physical boundary to obtain the extrapolation distance d. This procedure is dictated by the assumption that no reflection occurs at the wall; that is, the ions will become neutralized at the wall on impact if the wall is conducting or stick to the wall if it is an insulator.

extent by which the dimensions of the container are augmented for the mathematical treatment of the diffusion problem. Slightly different values of d are appropriate for different geometries.

In the study of ions and electrons diffusing in gases the extrapolation distance is negligibly small compared with the container dimensions and ignored in calculations; that is to say, the boundary condition applied is that the ionic or electronic number density go to zero at the inner surface of the containing vessel. In the diffusion of neutrons, on the other hand, d is frequently of significant size and must be taken into account.

1-11. SOLUTION OF THE TIME-INDEPENDENT DIFFUSION EQUATION FOR VARIOUS GEOMETRIES

Here we consider the steady-state diffusion of ions of a single species through a gas of uniform temperature and pressure filling containers of various shapes. In each case the ionic number density is negligible in comparison to the number density of the molecules. No electric field is assumed to be present, so the average ionic energy has the thermal value. Under the conditions described here D will be independent of position. We neglect the extrapolation distance and require n_0 to vanish at the geometrical boundaries of the containers.

A. INFINITE PARALLEL PLATES. As the first example, consider a one-dimensional cavity whose walls are the infinite plane parallel plates shown in Fig. 1-11-1. In this simple case the diffusion equation (1-9-6) becomes

$$\frac{d^2 n_0(x)}{dx^2} + \frac{n_0(x)}{D\tau} = 0. \tag{1-11-1}$$

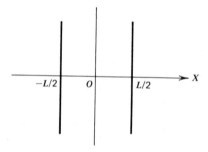

FIG. 1-11-1. A one-dimensional cavity with plane parallel walls.

Since $D\tau$ is positive, the solution of (1-11-1) is

$$n_0(x) = A \cos \frac{x}{\sqrt{D\tau}} + B \sin \frac{x}{\sqrt{D\tau}}, \tag{1-11-2}$$

where A and B are constants of integration that must be determined from the boundary conditions and from the requirement that we impose for symmetry about the midplane. If the width of the cavity is L and the origin of the coordinate system is located at the midplane, the boundary conditions are $n_0(x) = 0$ when $x = \pm L/2$.

The symmetry requirement makes $B = 0$ and the boundary conditions force τ to assume one of the infinite number of values τ_k $(k = 1, 2, 3, \ldots)$ that satisfy the equation

$$\cos \frac{L}{2\sqrt{D\tau_k}} = 0 \quad \text{or} \quad \frac{L}{2\sqrt{D\tau_k}} = (2k-1)\frac{\pi}{2}. \tag{1-11-3}$$

Now define a quantity Λ_k that represents the characteristic diffusion length for the kth mode of diffusion:

$$\Lambda_k^2 = D\tau_k = \left(\frac{1}{2k-1}\frac{L}{\pi}\right)^2. \tag{1-11-4}$$

The diffusion length is useful in describing the shape of a cavity in the diffusion process. The solution for the kth mode can then be written

$$n_0(x)_k = A_k \cos \frac{x}{\Lambda_k}. \tag{1-11-5}$$

The function $\cos(x/\Lambda_k)$ assumes negative values in certain regions within the cavity for all modes of diffusion except the lowest, or fundamental, mode corresponding to $k = 1$. Therefore, if we consider each solution singly, we must discard all but the fundamental mode on physical grounds, since the ionic number density can never be negative. Since, however, the diffusion equation is linear, the total solution consists of an infinite number of modes, many of which may be excited simultaneously. Any sum of these modes is then a possible solution, provided the constants A_k have values that prevent the number density from becoming negative. The use of an ionization source that provides uniform ionization throughout the cavity will ensure that the fundamental mode predominates, but in many other experimental arrangements higher modes must be considered.

After the ionization source is abruptly turned off, say at $t = 0$, each diffusion mode decays out with its own characteristic time constant τ_k. The total solution of the time-dependent diffusion problem is thus given by

$$n(x, t) = \sum_{k=1}^{\infty} A_k \left(\cos \frac{x}{\Lambda_k}\right) e^{-t/\tau_k}. \tag{1-11-6}$$

Equation 1-11-3 shows that

$$\frac{\tau_1}{\tau_k} = (2k-1)^2, \tag{1-11-7}$$

$\tau_1/\tau_2 = 9$, $\tau_1/\tau_3 = 25$, $\tau_1/\tau_4 = 49$, etc. Consequently, if higher modes are initially present, they will decay out much faster than the fundamental mode and only this mode will be observable after a time comparable with τ_1. This fact obviously simplifies the analysis of experiments.

B. RECTANGULAR PARALLELEPIPED. The next case to be treated is that of a cavity in the form of a rectangular parallelepiped (Fig. 1-11-2). Take the

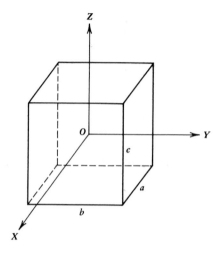

FIG. 1-11-2. A cavity with the shape of a rectangular parallelepiped.

origin of rectangular Cartesian coordinates at the center of the cavity, whose x, y, and z dimensions are a, b, and c, respectively. The time-independent diffusion equation is now

$$\frac{\partial^2 n_0}{\partial x^2} + \frac{\partial^2 n_0}{\partial y^2} + \frac{\partial^2 n_0}{\partial z^2} + \frac{n_0}{D\tau} = 0 \qquad (1\text{-}11\text{-}8)$$

with the boundary conditions that $n_0 = 0$ when $x = \pm a/2$, $y = \pm b/2$, and $z = \pm c/2$. Expressing $n_0(x, y, z)$ as the product of three functions, each of which is a function of only one coordinate,

$$n_0(x, y, z) = X(x)Y(y)Z(z), \qquad (1\text{-}11\text{-}9)$$

we may separate the variables in the diffusion equation and obtain

$$\frac{1}{X}\frac{d^2X}{dx^2} + \frac{1}{Y}\frac{d^2Y}{dy^2} + \frac{1}{Z}\frac{d^2Z}{dz^2} + \frac{1}{D\tau} = 0. \qquad (1\text{-}11\text{-}10)$$

Since $D\tau$ is a constant and each of the first three terms is a function of only one variable, we may equate each of them to a separate constant:

$$\frac{1}{X}\frac{d^2X}{dx^2} = -\alpha^2, \qquad \frac{1}{Y}\frac{d^2Y}{dy^2} = -\beta^2, \qquad \frac{1}{Z}\frac{d^2Z}{dz^2} = -\gamma^2. \quad (1\text{-}11\text{-}11)$$

Equation 1-11-10 shows that

$$\alpha^2 + \beta^2 + \gamma^2 = \frac{1}{D\tau}. \qquad (1\text{-}11\text{-}12)$$

Since there is no essential difference between the x, y, and z directions in this problem and since $\alpha^2 + \beta^2 + \gamma^2$ equals a positive quantity, it follows that α^2, β^2, and γ^2 separately must be positive. Using the boundary conditions and symmetry requirements, we see that the solutions of (1-11-11) are

$$X_i = A_i \cos \frac{(2i-1)\pi x}{a}, \qquad Y_j = B_j \cos \frac{(2j-1)\pi y}{b}, \qquad Z_k = C_k \frac{(2k-1)\pi z}{c},$$

$$(1\text{-}11\text{-}13)$$

where i, j, and k each may assume any positive integral values. The total solution to the time-dependent problem then has the form

$$n(x, y, z, t) = \sum_{i, j, k=1}^{\infty} G_{ijk} \cos \frac{(2i-1)\pi x}{a} \cos \frac{(2j-1)\pi y}{b}$$

$$\times \cos \frac{(2k-1)\pi z}{c} e^{-t/\tau_{ijk}}, \qquad (1\text{-}11\text{-}14)$$

where the three arbitrary constants have been lumped into G_{ijk}. Specification of the mode of diffusion now requires three indices, and corresponding to this triad of indices and this mode of diffusion is a time constant τ_{ijk} given by

$$\frac{1}{\tau_{ijk}} = D\pi^2 \left[\left(\frac{2i-1}{a}\right)^2 + \left(\frac{2j-1}{b}\right)^2 + \left(\frac{2k-1}{c}\right)^2 \right]. \qquad (1\text{-}11\text{-}15)$$

The diffusion length is now given by

$$\Lambda_{ijk}^2 = D\tau_{ijk}. \qquad (1\text{-}11\text{-}16)$$

If the cavity is cubical, $a = b = c$ and

$$\frac{\tau_{111}}{\tau_{211}} = 3.67, \qquad \frac{\tau_{111}}{\tau_{311}} = 9, \qquad \frac{\tau_{111}}{\tau_{411}} = 17.$$

Here the higher modes persist longer in relation to the fundamental mode than in the one-dimensional case and thus their effect is enhanced.

C. SPHERICAL CAVITY. Now consider a spherical cavity of radius r_0 (Fig. 1-11-3). For spherical geometry the diffusion equation is

$$\frac{\partial^2 n_0}{\partial r^2} + \frac{2}{r}\frac{\partial n_0}{\partial r} + \frac{1}{r^2 \sin\theta}\frac{\partial}{\partial\theta}\left(\sin\theta\frac{\partial n_0}{\partial\theta}\right) + \frac{1}{r^2 \sin^2\theta}\frac{\partial^2 n_0}{\partial\phi^2} + \frac{n_0}{D\tau} = 0,$$

$$(1\text{-}11\text{-}17)$$

but since there is no preferred direction here we reduce (1-11-17) to

$$\frac{\partial^2 n_0}{dr^2} + \frac{2}{r}\frac{dn_0}{dr} + \frac{n_0}{D\tau} = 0, \qquad (1\text{-}11\text{-}18)$$

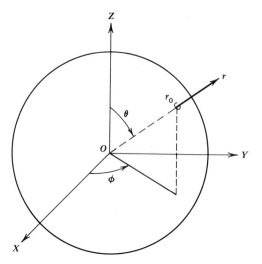

FIG. 1-11-3. A spherical cavity.

discarding in the process all but the fundamental angular mode. To solve this equation easily we put $n_0 = u/r$. We then obtain

$$\frac{d^2u}{dr^2} + \frac{u}{D\tau} = 0 \qquad (1\text{-}11\text{-}19)$$

whose solution is

$$u = A \cos \frac{r}{\sqrt{D\tau}} + B \sin \frac{r}{\sqrt{D\tau}}, \qquad (1\text{-}11\text{-}20)$$

since $D\tau$ is positive. Thus the solution for n_0 has the form

$$n_0 = \frac{A}{r} \cos \frac{r}{\sqrt{D\tau}} + \frac{B}{r} \sin \frac{r}{\sqrt{D\tau}}, \qquad (1\text{-}11\text{-}21)$$

where evidently A must be zero in order that n_0 may remain finite at the origin. The final time-dependent solution is then

$$n(r, t) = \sum_{k=0}^{\infty} \frac{B_k}{r} \sin \frac{r}{\sqrt{D\tau_k}} e^{-t/\tau_k}, \qquad (1\text{-}11\text{-}22)$$

where

$$\frac{r_0}{\sqrt{D\tau_k}} = k\pi \qquad (k = 0, 1, 2, 3, \ldots). \qquad (1\text{-}11\text{-}23)$$

The diffusion length Λ_k is given by the equation

$$\Lambda_k^2 = D\tau_k = \left(\frac{r_0}{\pi k}\right)^2.$$ (1-11-24)

D. CYLINDRICAL CAVITY. As a final example let us treat a cavity in the form of a right circular cylinder of radius r_0 and height H (Fig. 1-11-4).

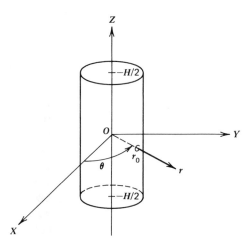

FIG. 1-11-4. A cylindrical cavity.

If we assume symmetry about the axis, there is no dependence on the azimuth angle θ, and the diffusion equation

$$\frac{\partial^2 n_0}{\partial r^2} + \frac{1}{r}\frac{\partial n_0}{\partial r} + \frac{1}{r^2}\frac{\partial^2 n_0}{\partial \theta^2} + \frac{\partial^2 n_0}{\partial z^2} + \frac{n_0}{D\tau} = 0$$ (1-11-25)

reduces to

$$\frac{\partial^2 n_0}{\partial r^2} + \frac{1}{r}\frac{\partial n_0}{\partial r} + \frac{\partial^2 n_0}{\partial z^2} + \frac{n_0}{D\tau} = 0$$ (1-11-26)

(for the fundamental angular mode). We separate variables by writing

$$n_0(r, z) = R(r)Z(z)$$ (1-11-27)

and obtain

$$\frac{1}{R}\left(\frac{d^2 R}{dr^2} + \frac{1}{r}\frac{dR}{dr}\right) + \frac{1}{Z}\frac{d^2 Z}{dz^2} + \frac{1}{D\tau} = 0.$$ (1-11-28)

The first term depends only on r and the second only on z, and since $D\tau$ is a constant each of these terms must be equal to a constant. Set

$$\frac{1}{R}\left(\frac{d^2R}{dr^2} + \frac{1}{r}\frac{dR}{dr}\right) = -\alpha^2 \tag{1-11-29}$$

and

$$\frac{1}{Z}\frac{d^2Z}{dz^2} = -\beta^2 \tag{1-11-30}$$

so that

$$\alpha^2 + \beta^2 = \frac{1}{D\tau}. \tag{1-11-31}$$

We must now determine whether α^2 and β^2 are positive or negative.

In solving the r equation (1-11-29) it is convenient to make the substitution $r = u/\alpha$. We then obtain the equation

$$u^2\frac{d^2R}{du^2} + u\frac{dR}{du} + u^2R = 0. \tag{1-11-32}$$

Now the general Bessel equation of order n is

$$x^2\frac{d^2y}{dx^2} + x\frac{dy}{dx} + (x^2 - n^2)y = 0, \tag{1-11-33}$$

where $(x^2 - n^2)$ is a positive quantity, whereas the modified Bessel equation of order n is

$$x^2\frac{d^2y}{dx^2} + x\frac{dy}{dx} - (x^2 - n^2)y = 0. \tag{1-11-34}$$

We see that (1-11-32) is a Bessel equation of order zero, unmodified if α^2 is positive, modified if α^2 is negative. In the first instance the general solution is

$$R = AJ_0(u) + BY_0(u), \tag{1-11-35}$$

where J_0 and Y_0 are the Bessel functions of the first and second kinds, respectively, of order zero (McLachlan, 1934). In the second instance the solution is

$$R = A'I_0(u) + B'K_0(u), \tag{1-11-36}$$

where I_0 and K_0 are modified Bessel functions of the first and second kinds, respectively, of order zero. By reference to Fig. 1-11-5 we see that the only satisfactory solution is J_0 and the only possible solution for the r part of our diffusion problem is

$$R(r) = AJ_0(u) = AJ_0(\alpha r). \tag{1-11-37}$$

Thus α^2 is required to be positive.

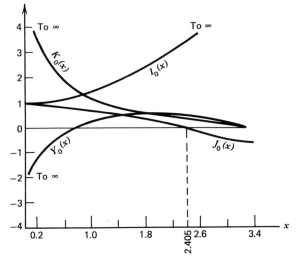

FIG. 1-11-5. Zero-order Bessel functions.

By applying the boundary condition that n_0 must vanish at $r = r_0$ we see that $R(r_0) = AJ_0(\alpha r_0) = 0$. The first zero of J_0 occurs at $\alpha r = 2.405$, so that, for the fundamental mode, $\alpha_1 = 2.405/r_0$ and $R(r)$ is given by

$$R(r) = AJ_0\left(\frac{2.405r}{r_0}\right). \tag{1-11-38}$$

Turning now to the z equation

$$\frac{d^2Z}{dz^2} + \beta^2 Z = 0, \tag{1-11-39}$$

we observe that if β^2 is positive the solutions will be $\cos \beta z$ and $\sin \beta z$, whereas if β^2 is negative the solutions are $\cosh \beta z$ and $\sinh \beta z$. Figure 1-11-6 shows that the hyperbolic functions are unacceptable because they would provide a greater ionic number density at the top than at the center of the cavity. The $\sin \beta z$ solution must be discarded because it is an odd function of z. The only remaining possibility is the cosine solution and β^2 must be positive. The solution of the z equation is then

$$Z(z) = C \cos \beta z. \tag{1-11-40}$$

The boundary conditions that n_0 must vanish at $z = \pm H/2$ require that $\beta_1 = \pi/H$ for the fundamental mode.

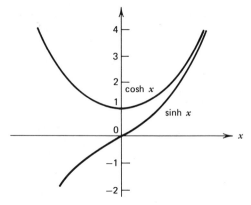

FIG. 1-11-6. The hyperbolic sine and cosine.

The time-dependent solution for the lowest mode of diffusion can now be written

$$n(r, z, t) = G_{11}J_0\left(\frac{2.405r}{r_0}\right)\left(\cos\frac{\pi z}{H}\right)e^{-t/\tau_{11}}, \qquad (1\text{-}11\text{-}41)$$

where

$$\frac{1}{\Lambda_{11}^2} = \frac{1}{D\tau_{11}} = \left(\frac{2.405}{r_0}\right)^2 + \left(\frac{\pi}{H}\right)^2 \qquad (1\text{-}11\text{-}42)$$

The total solution which contains the radial higher modes as well as the fundamental is

$$n(r, z, t) = \sum_{i,\,j=1}^{\infty} G_{ij} J_0(\alpha_i r)\left[\cos\frac{(2j-1)\pi z}{H}\right]e^{-t/\tau_{ij}}. \qquad (1\text{-}11\text{-}43)$$

The diffusion length is given by

$$\frac{1}{\Lambda_{ij}^2} = \frac{1}{D\tau_{ij}} = \alpha_i^2 + \left[\frac{(2j-1)\pi}{H}\right]^2, \qquad (1\text{-}11\text{-}44)$$

where $\alpha_i r_0$ is the ith root of J_0.

1-12. AMBIPOLAR DIFFUSION

We now wish to consider a gas-filled cavity in which both electrons and positive ions are diffusing toward the walls. Usually in studies of such cavities the interaction between the negative and positive particles can be neglected below ionization densities of about 10^7 to $10^8/\text{cm}^3$, but above this level space charge effects produced by the Coulomb forces between the electrons and positive ions become important and must be taken into account.

It may be shown that the number density of electrons in a highly ionized gas must approximate the number density of positive ions at each point, provided we are not within about 1 debye length* of a boundary (McDaniel, 1964f). Any deviation from charge equality produces electrical forces that oppose the charge separation and tend to restore the balance. Because their diffusion coefficient is much higher than that of the ions, the electrons attempt to diffuse more rapidly than the ions toward regions of lower concentration but their motion is impeded by the restraining space charge field thereby created. This same field has the opposite effect on the ions and causes them to diffuse at a faster rate than they would in the absence of the electrons. Both species of charged particles consequently diffuse with the same velocity, and since there is now no difference in the flow of the particles of opposite sign the diffusion is called "ambipolar." The concept of ambipolar diffusion was introduced by Schottky in 1924 in an analysis of the positive column of the glow discharge (Schottky, 1924).

We now derive an expression for the coefficient of ambipolar diffusion. Let n represent the common number density of the electrons and positive ions and v_a, the velocity of ambipolar diffusion. We assume that the gas pressure is high enough for the particles to make frequent collisions. The mobility concept will then be assumed to apply, not only for the ions but for the electrons as well. Let E denote the intensity of the electric field established by the charge separation. Since the velocity of diffusion is the same for both species, we have

$$v_a = -\frac{D^+}{n}\frac{dn}{dx} + K^+ E \qquad (1\text{-}12\text{-}1)$$

and

$$v_a = -\frac{D^-}{n}\frac{dn}{dx} - K^- E \qquad (1\text{-}12\text{-}2)$$

where K^+ and K^- are the mobilities of the ions and electrons, respectively, and D^+ and D^- are their ordinary or "free" diffusion coefficients. All four coefficients are positive numbers. By eliminating E we obtain

$$v_a = -D_a\frac{1}{n}\frac{dn}{dx} \qquad (1\text{-}12\text{-}3)$$

---

* The debye length is a measure of the distance over which deviations from charge neutrality can occur in an ionized gas. It is directly proportional to the square root of the energy and inversely proportional to the square root of the number density of the charged particles in the ionized gas.

where D_a is the coefficient of ambipolar diffusion defined by the equation

$$D_a = \frac{D^+ K^- + D^- K^+}{K^+ + K^-};$$
(1-12-4)

D_a characterizes the diffusive motion of both species.

If we assume that $K^- \gg K^+$ and $T^- \gg T^+$ and use the Einstein relation

$$\frac{D}{K} = \frac{kT}{e},$$
(1-1-5)

we find that

$$D_a \approx D^- \frac{K^+}{K^-} = \frac{kT^-}{e} K^+.$$
(1-12-5)

When $T^+ = T^- = T$, on the other hand,

$$D_a \approx 2D^+ = \frac{2kT}{e} K^+.$$
(1-12-6)

The time-dependent diffusion equation for the ambipolar case is

$$\frac{\partial n}{\partial t} = \mathbf{\nabla} \cdot (D_a \, \mathbf{\nabla} n).$$
(1-12-7)

If D_a is independent of position and the particle number density is assumed to decay as $e^{-t/\tau}$, the time-independent ambipolar diffusion equation is obtained:

$$\nabla^2 n_0 + \frac{n_0}{D_a \tau} = 0.$$
(1-12-8)

This equation is solved for specific problems by the methods in Section 1-11: D_a is given in terms of the decay constant τ, and the appropriate diffusion length Λ, by the equation

$$D_a = \frac{\Lambda^2}{\tau}.$$
(1-12-9)

Hence D_a may be evaluated from a determination of the rate of decay of the charged particle density in a cavity after the ionization source has been turned off. Techniques for measuring D_a are discussed in Chapter 4. If the electron, ion, and gas temperatures are equal, (1-12-6) shows that the ionic diffusion coefficient and mobility may be obtained from the measured value of D_a. The zero-field reduced mobility K_0 is related to the ambipolar diffusion coefficient D_a by the equation

$$K_0 = \frac{D_a p}{T^2} 2.086 \times 10^3. \qquad (1\text{-}12\text{-}10)$$

Here p is the pressure in Torr and T is the gas temperature in degrees Kelvin at which D_a was measured; K_0 is expressed in square centimeters per volt per second and D_a, in square centimeters per second.

REFERENCES

Bederson, B., and L. J. Kieffer (1971), *Rev. Mod. Phys.* **43**, 601.

Crank, J. (1956), *The Mathematics of Diffusion*, Clarendon, Oxford.

Einstein, A. (1905), *Annalen der Physik* **17**, 549. See also Einstein, A. (1908), *Z. f. Elektrochemie* **14**, 235. English translations of these papers appear in *Investigations on the Theory of the Brownian Movement by Albert Einstein* (R. Fürth, Ed.) Dover, New York, 1956.

Elford, M. T. (1971), *Aust. J. Phys.* **24**, 705.

Fick, A. (1855), *Ann. Phys. Lpz.* **170**, 59.

Flannery, M. R. (1972), "Three-body Recombination of Positive and Negative Ions," in *Case Studies in Atomic Collision Physics*, Vol. II (E. W. McDaniel and M. R. C. McDowell, Eds.), North-Holland, Amsterdam.

Gatland, I. R. (1972), *Phys. Rev. Letters* **29**, 9.

Hasted, J. B. (1972), *Physics of Atomic Collisions* (2nd ed.), American Elsevier, New York.

Hirschfelder, J. O., C. F. Curtiss, and R. B. Bird (1964), *Molecular Theory of Gases and Liquids*, Wiley, New York.

Huxley, L. G. H., and R. W. Crompton (1973), *Electron Swarms in Gases*, Wiley, New York.

———, and M. T. Elford (1966), *Bull. Inst. Physics and Physical Society* **17**, 251.

Loeb, L. B. (1960), *Basic Processes of Gaseous Electronics* (2nd ed.) University of California Press, Berkeley, Chapters I and II.

McDaniel, E. W. (1964a), *Collision Phenomena in Ionized Gases*, Wiley, New York, pp. 490–491.

——— (1964b), *op. cit.*, pp. 506–512.

——— (1964c), *op. cit.*, pp. 518–520, 575–582.

——— (1964d), *op. cit.*, chapter 5.

——— (1964e), *op. cit.*, pp. 496–497.

——— (1964f), *op. cit.*, Appendix I.

——— (1972), *Aust. J. Phys.*, **25**, 465.

McLachlan, N. W. (1934), *Bessel Functions for Engineers*, Clarendon, Oxford.

Massey, H. S. W. (1969), *Electronic and Ionic Impact Phenomena* (2nd ed.), Vol. II, Clarendon, Oxford.

——— (1971), *op. cit.*, Vol. III.

———, and E. H. S. Burhop (1969), *op. cit.*, (2nd ed.), Vol. I.

———, and H. B. Gilbody (1973), *op. cit.*, Vol. IV.

Moiseiwitsch, B. L., and S. J. Smith (1968), *Rev. Mod. Phys.* **40**, 238.

Nernst, W. (1888), *Z. phys. Chem.* **2**, 613.

Phelps, A. V. (1968), *Rev. Mod. Phys.* **40**, 399.

Schottky, W. (1924), *Phys. Z.* **25**, 635.

Thomas, E. W. (1972), *Excitation by Heavy Particle Collisions*, Wiley, New York.

Thomson, G. M., J. H. Schummers, D. R. James, E. Graham, I. R. Gatland, M. R. Flannery, and E. W. McDaniel (1973), *J. Chem. Phys., to be published.*

Thomson, J. J., and G. P. Thomson (1969) *Conduction of Electricity Through Gases*, Vol. I (unrevised reprint of 1928 3rd ed.), reprinted by Dover, New York.

Townsend, J. S. (1899), *Phil. Trans.* **193**, 129.

Wannier, G. H. (1953), *Bell System Tech. J.* **32**, 170–254.

2

THE MEASUREMENT

OF DRIFT VELOCITIES AND

LONGITUDINAL DIFFUSION COEFFICIENTS

Direct measurements of drift velocities and longitudinal diffusion coefficients are made with apparatus called drift tubes, descriptions of which are given in Section 2-5. The discussion of drift tubes in this book relates exclusively to their use in the study of ions. Drift tubes are also utilized to investigate the transport properties of electrons in gases and the attachment of electrons to form negative ions, but the techniques in electron studies are somewhat different from those described here. The reader is referred to Huxley and Crompton (1973) for an authoritative discussion of electron drift-tube research.* As we show, the transport of slow ions through gases is closely coupled to the chemical reactions that may occur between the ions and gas molecules, and therefore we must consider in some detail the complicating effects of ion-molecule and charge transfer reactions on drift-tube measurements.† This complication, however, is more than compensated for by the fact that drift tubes may be used for the quantitative study of such reactions,

* It is appropriate to point out that many of the significant advances in electron drift-tube research made during recent years are due to L. G. H. Huxley, R. W. Crompton, and M. T. Elford in Australia and to A. V. Phelps and his colleagues at the Westinghouse Research Laboratories.
† We use the term "ion-molecule reaction" to refer to a heavy particle rearrangement reaction such as $A^+ + BC \rightarrow AB^+ + C$ or $A^- + B + C \rightarrow AB^- + C$. By a "charge transfer reaction" we mean a reaction in which an electron is transferred between the colliding structures, as in $A^+ + B \rightarrow A + B^+$ or $A^- + BC \rightarrow A + BC^-$. Both kinds of reaction are called "chemical."

and drift-tube methods of determining ionic rate coefficients are described in this chapter and in Chapter 3. The use of drift tubes for the measurement of transverse diffusion coefficients is treated in Chapter 3. The techniques described there are quite different from those under consideration here.

Drift tubes were first used at the end of the nineteenth century and have been applied to the study of ionic drift velocities almost continuously since that time (Loeb, 1960; Massey, 1971). The early techniques were understandably crude, and few of the data obtained with drift tubes before 1930 are now of other than historical interest. However, noteworthy advances in instrumentation and gas-purification techniques made by Tyndall and his collaborators at Bristol University in England and by Bradbury and Nielsen in the United States during the 1930's led to a significant amount of good mobility data during that decade. Nonetheless, most of the reliable results on hand now are of much more recent vintage. General techniques for obtaining accurate mobilities of known ionic species in gases in which the ions are involved in reactions with the gas molecules became available only in the 1960's with the development of drift-tube mass spectrometers, examples of which are given in Section 2-5. For reasons discussed in this chapter, accurate ionic diffusion coefficients are much more difficult to measure than mobilities, and the first reliable ionic diffusion data for energies above thermal were reported only in the late 1960's. Drift tubes were used to obtain these results. Few meaningful drift-tube measurements of ionic reaction rates were reported before 1960; the widespread use of drift tubes to obtain ionic rate coefficients began only with the advent of drift-tube mass spectrometers. The rates of several dozen ion-molecule and charge transfer reactions have now been measured with these instruments.

2-1. GENERAL CONSIDERATIONS IN DRIFT-TUBE EXPERIMENTS

A conventional ionic drift tube usually consists of an enclosure containing gas, an ion source positioned on the axis of the enclosure, a set of electrodes that establishes a uniform axial electrostatic field along which the ions drift, and a current-measuring collector in the gas at the end of the ionic drift path. The drift field $\mathbf{E}$ causes the ions of any given molecular composition to "swarm" through the gas with a drift velocity and diffusion rate determined by the nature of the ions and gas molecules, the field strength, and the gas pressure and temperature. For the determination of drift velocities and longitudinal diffusion coefficients the ion source is operated in a repetitive, pulsed mode, and the spectrum of arrival times of the ions at the collector is measured electronically. (Other measurement techniques are required for the determination of transverse diffusion coefficients; they are described in

Chapter 3.) A drift-tube mass spectrometer differs from a conventional drift tube by having the ion collector in the gas replaced by a sampling orifice located in the wall at the end of the drift tube, usually on the axis. Ions arriving at the end plate close to the axis pass through the orifice, out of the drift tube, and into an evacuated region that contains a mass spectrometer and an ion detector (usually a pulse-counting electron multiplier). The mass spectrometer can be set to transmit any one of the various ionic species entering it, the other species of ions being rejected in the mass selection process. Thus the arrival-time spectrum can be mapped separately for each kind of ion arriving at the end of the drift tube.

The pressures at which ionic drift-tube experiments have been performed have ranged from about 2.5×10^{-2} Torr to above 1 atm. On the assumption of a cross section of 50×10^{-16} cm^2 for collisions of the ions with gas molecules, the ionic mean free path corresponding to a pressure of 2.5×10^{-2} Torr is 0.23 cm; the mean free path at 1 atm is 7.4×10^{-6} cm. Most measurements are now made in the pressure range of 7.5×10^{-2} to 10 Torr. At pressures lower than the minimum quoted here the ions would make collisions too few for steady-state conditions to be achieved except in apparatus of uncommonly great length. Ion sampling difficulties may appear in drift-tube mass spectrometer measurements at pressures greater than about 10 Torr.* In any event, measurements made above this pressure would not usually be expected to provide information inaccessible in experiments below 10 Torr. The product of the gas pressure p and the drift distance d should be large enough that the ions will travel a negligible fraction of the total distance before energy equilibration in the drift field is achieved. Drift distances of 0.5 to 44 cm have been employed. Because of the large pd products used, each ion makes many collisions with molecules as it drifts the distance d; the average number of collisions per ion usually lies somewhere between 10^2 and 10^7, depending on the value of pd.

Drift-tube measurements are made at ratios of drift field intensity to gas number density (E/N) as low as about 0.3 Td and as high as about 5000. Below $E/N \approx 6$ Td we would expect to be within the "low-field" region (see Section 1-2), in which most of the applications lie and in which the theory of the transport phenomena is highly developed. In a given experiment the lowest value of E/N to which measurements may be extended will depend on intensity considerations, that is, on the ion current reaching the detector and the sensitivity of the detector. The measured ion signal

---

* By the use of a small exit aperture at the end of the drift tube and fast differential pumping ions may be sampled mass spectrometrically at drift-tube pressures of the order of 1 atm. At high drift-tube pressures, however, molecules may form clusters about the ions in the expanding jet of gas passing through the exit aperture and thereby falsify the mass spectrum. [See Milne and Greene (1967); Hagena and Obert (1972).]

eventually becomes too small for accurate measurements to be made. In many experiments it has proved impossible even to approach the low-field region. The high E/N limit in drift-tube experiments is usually imposed by electrical breakdown within the apparatus. At very high E/N the average ionic energy may attain values of the order of 10 eV* and inelastic collisions may be important. There is little interest in pushing drift-tube measurements to energies of this magnitude, for the data then become difficult to interpret in terms of the transport theory and there appear to be few practical applications in which the results may be used. Indeed, at very high E/N we are more interested in the binary collision properties of the ions than in their transport properties, and ion-molecule collisions in the electron-volt energy range are more accurately studied by beam techniques (Massey et al., 1973; McDaniel et al., 1970) than by the "swarm" methods described here.

At this point, however, it is of interest to note that conventional single beam-static gas or crossed-beam techniques cannot be applied to the study of ionic collisions at laboratory frame ion-beam energies below about 1 eV. This limitation develops from the effects of spurious electric fields produced by space charge in the ion beams, contact potentials, and charge

---

* The equation generally used to calculate the average ionic energy is Wannier's expression

$$\frac{m\overline{v_i^2}}{2} = \frac{mv_d^2}{2} + \frac{Mv_d^2}{2} + \frac{3kT}{2}, \tag{2-1-1}$$

where m and M are the ionic and molecular masses, respectively, $\overline{v_i^2}$ is the mean square of the total ionic velocity, and v_d is the measured drift velocity (Wannier, 1953). This equation was derived for high E/N on the assumption of a constant mean free time scattering model (see Chapter 5). The first term on the right side is the field energy associated with the drift motion of the ion; the second term is the random part of the field energy. The last term represents the thermal energy. This equation illustrates the capacity that light ions in a heavy gas have for storing energy in the form of random motion. For ions traveling in a gas composed of molecules of the same mass as that of the ions the ordered and random field energies are equal. For heavy ions in a light gas the random field energy is negligible.

If m and M are expressed in amu, v_d in units of 10^4 cm/sec, and T in °K, eq. (2-1-1) gives the average ionic energy in eV to be

$$1.036 \times 10^{-4} \frac{m + M}{2} (v_d)^2 + 1.293 \times 10^{-4} T.$$

In those cases in which $m = M$ and resonance charge transfer occurs it is probably better at high field to use the equation

$$\frac{m\overline{v_i^2}}{2} = \frac{\pi}{4} mv_d^2 + \frac{3}{2} kT \tag{2-1-2}$$

(Heimerl et al., 1969) derived from the high-field analysis of Fahr and Müller (1967) for the situation in which resonance charge transfer dominates the picture.

accumulation on insulating surfaces. These fields can disperse and deflect slow ion beams and in the process alter the energy of the ions in the beams by unknown amounts.* Drift tubes, on the other hand, are not subject to this limitation if reasonable precautions are observed because in a well-designed experiment the spurious electric fields are much weaker than the applied electric drift field even in the low-field (thermal energy) region. Hence the spurious fields will not seriously modify the field intentionally employed to produce the drift of the ions, and swarm experiments can be conducted at ionic energies as low as permitted by the thermal contact of the ions with the gas molecules.

It is now appropriate to enumerate some constructional features of drift tubes that should be incorporated into the design if data of the highest quality are to be obtained. First of all it is of great importance that the apparatus be of ultrahigh-vacuum construction and that it be bakeable to a suitably high temperature during pump-down before a series of measurements is commenced. Metal gaskets should be used throughout and only insulators of the best vacuum quality should be employed. The vacuum pumps should be isolated from the drift tube by cooled baffles and sorbent or liquid nitrogen traps. These precautions will minimize the presence of outgassed impurities during measurements and prevent the interior surfaces from becoming coated with insulating films of pump fluid. It is frequently advisable to gold-plate the interior metallic surfaces, particularly in the ion-sampling region of a drift-tube mass spectrometer. This step will provide high conductivity surfaces that can be cleaned up as required by baking and minimize the production of spurious electric fields by charge buildup on insulating films. It is important also to shield all the insulators inside the apparatus so that they cannot be "seen" by the ions.

It may be of interest to discuss the different factors that determine the kinds of ions present in a given experiment. The primary[†] ions produced in the ion source will depend on the gas present in the source and also on the particular ion source used and the conditions under which it is operated; for example, in an electron bombardment ion source containing nitrogen gas

---

* The recently developed merging-beams method may, however, be used to study many kinds of ion-molecule collisions at impact energies well below 1 eV (Neynaber, 1969; Hasted, 1972; Massey et al., 1973). In such studies a fast beam of ions is merged coaxially in a vacuum with a fast beam of molecules whose velocity is slightly different from that of the ions. Each of the two beams is usually given a laboratory frame kinetic energy in the kiloelectron-volt range so that problems with spurious fields will not appear. The center-of-mass impact energy of the ion-molecule collisions which do occur, however, can be made as low as a small fraction of 1 eV. Furthermore, the spread in the impact energy is much smaller than the laboratory frame energy spreads of the merging beams and good energy resolution can be obtained.

† Here the word "primary" is used in the sense of "first in time" not "first in importance."

both N^+ and N_2^+ ions will be formed if the electron beam energy exceeds 24.2 eV, but only N_2^+ ions will be produced at lower energies because the thresholds for production of N^+ and N_2^+ by electron impact are 24.2 and 15.6 eV, respectively. The primary ions may survive over the entire drift distance or they may be converted in charge transfer or ion-molecule reactions to secondary ions of other types. Whether this happens in a given case depends on the value of pd, which determines the number of collisions the primary ions experience and also perhaps on the pressure itself, since some of the reactions may be the three-body variety with reaction frequencies that vary as p^2. Other determining factors are the values of E/N and T, which characterize the energy of the ion-molecule collisions and thus influence the values of the rate coefficients. If secondary ions are formed, they may also undergo further reactions. Whether or not they do so will depend on the values of pd, p, E/N, and T. In most cases there is more than one kind of ion present under given conditions, and it is often difficult to predict their identities and relative abundances.

Still another matter is that of the states of excitation of the ions. Little quantitative information is available concerning their effect on the transport properties of ions, and some of the discrepancies among experimental data may be due to the ions being in different states. The states in which the primary ions are produced will depend on the ion source conditions, and the ability of excited primary ions to retain their excitation energy is governed by their lifetimes against radiative and collisional deexcitation. The excitation states of secondary ions depend on those of the primary ions from which they are produced and the exact nature of the reactions forming them. In most cases the primary ions are probably in the ground electronic state when they enter the drift region from the ion source, but molecular ions may well be vibrationally or rotationally excited. Each case must be examined separately. Of course, what is desired from experiments are mobilities and diffusion coefficients for ions in the ground state or in known states of excitation. Positive ions of the alkalis are particularly interesting in this respect. They may be generated thermionically from coated filaments located within the gas filling the drift tube (McDaniel, 1964), and thermodynamic considerations ensure that they are singly charged and in the ground state. This fact may be partly responsible for the generally good agreement among the mobility data that have been obtained for these ions. Another reason for this good agreement is that the alkali ions do not react to a significant extent with the molecules of many of the gases in which their mobilities have been measured, at least at the temperatures and pressures at which the measurements were made. Hence the arrival-time spectra for these ions usually have a simple shape, and it is an easy matter to infer unambiguous mobilities from them.

2-2. BASIC ASPECTS OF DRIFT VELOCITY MEASUREMENTS

A. EXPERIMENTS ON NONREACTING PRIMARY IONS.   The accurate determination of ionic drift velocities is a rather simple matter if we are dealing with primary ions that do not react with the gas filling the drift tube and consequently retain their identities during their entire transit through the apparatus. We pulse the ion source repetitively to admit bursts of ions into the drift space and measure the times at which the ions arrive at the detector in relation to the times at which they were gated into the drift space. Data are accumulated over many cycles until a statistically satisfactory spectrum of arrival times is obtained. The average arrival time is computed by a procedure appropriate to the apparatus being used, and the drift velocity is calculated from this quantity and the known drift distance. The drift velocity is a unique function of E/N for a given ion-gas combination at a given gas temperature. The width of the pulses admitting ions into the drift space should be short compared with the average transit time in order to obtain high accuracy. These pulses are typically about 1 μsec wide, whereas the transit times are usually of the order of hundreds of microseconds. The pressure of the drift-tube gas should be fairly high (say in the Torr region) to avoid excessive broadening of the arrival-time spectrum due to diffusion and also to permit the pressure to be measured accurately. (As a rule, in drift-tube experiments, the higher the pressure, the more accurately it can be measured in percentages.) Mobilities have been determined with an accuracy of about 1% in the simple cases described here, although the usual accuracy achieved is 4 to 5%.

B. END EFFECTS AND THE SHAPE OF THE ARRIVAL-TIME SPECTRUM. Although the concept of the measurement discussed above is extremely simple, two matters merit mention before we consider more complicated cases. The first is related to the elimination of end effects arising at the ion source or the detector end of the apparatus like those produced by a nonuniformity of the electric field at the entrance to the drift space or the finite time required for the ions to travel from the drift-tube exit aperture to the detector in a drift-tube mass spectrometer. It is desirable to be able to vary the drift distance by moving the ion source or the detector while leaving everything else unchanged. It is then possible to measure the arrival time for various drift distances and accurately determine the drift velocity from the slope of a plot of arrival time versus drift distance if such a plot has a long linear region. (Note that the extension of the linear region will not necessarily pass through the origin of the plot.) An illustration of end effects is provided by Fig. 2-2-1, which shows mean arrival times for $D_3{}^+$ ions in deuterium plotted for various equally spaced ion source positions

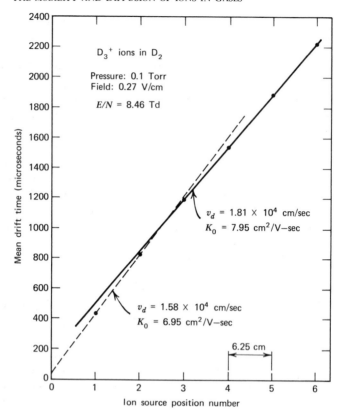

FIG. 2-2-1. An illustration of end effects in drift velocity measurements. The correct drift velocity is given by the solid line. The dashed line yields an erroneous "apparent drift velocity."

in experiments by Miller et al. (1968). Use of the data for the high-numbered, remote source positions gave the correct drift velocity, whereas the data obtained at ion source positions near the end of the drift path yielded a drift velocity that was too low by 12.6%. The utility of a movable ion source in these experiments is apparent. (Another of its advantages, unrelated to the present discussion, is that it facilitates the study of the approach of the various ion populations to equilibrium in reacting systems and permits the use of an attenuation technique in the measurement of transverse diffusion coefficients and reaction rates. This technique is discussed in Chapter 3.)

The other matter that must be covered here is related to the detailed shape of the arrival-time spectrum. A necessary but not sufficient condition for obtaining drift velocities in which we can have confidence is the close matching of the measured arrival-time spectrum with the shape predicted by a solution of the differential equation governing the drift and diffusion of the ions in the particular apparatus used. The differential equation which applies to the Georgia Tech drift-tube mass spectrometer is presented and solved in Section 2-6. Figure 2-2-2a shows experimental arrival-time spectra obtained with this apparatus at low E/N for "nonreacting"* K^+ ions in nitrogen at room temperature, seven spectra corresponding to seven different ion source positions but equal counting times being superimposed on a single drawing (Moseley et al., 1969a). The solid-line curves represent solutions of the drift-diffusion equation which contain the measured value of the drift velocity and the corresponding value of the diffusion coefficient calculated from the Einstein equation (1-1-5). The solid curves are normalized to match the height of the peak calculated for the shortest drift distance to the corresponding experimental peak. The remaining analytical profiles for the other values of the drift distance are not independently normalized. The spectra illustrate the increasing effects of peak broadening by longitudinal diffusion and intensity loss due to transverse diffusion as the drift distance is increased.

Figure 2-2-2b shows a single experimental arrival time spectrum for K^+ ions in N_2 obtained at low pressure and a drift distance of 43.77 cm by Thomson et al. (1973). The smooth curve, in whose calculation the experimentally determined values of the mobility and longitudinal diffusion coefficient were used, gives the prediction of the analysis described in Section 2-6. The agreement between the experimental data and the predictions of the analysis is excellent. Unless this kind of close, quantitative agreement is obtained in a given experiment, there are factors at work that have not been taken into consideration and that might produce faulty results. When good agreement is not achieved, we should probably look first at the question of space charge in the drifting ion swarm. It is widely believed that space charge repulsion will have a negligible effect on the measurement of drift velocities because the effect is usually assumed to be symmetrical about the midplane of the drifting ion cloud perpendicular to the drift-tube axis. Experiments have been performed, however, that indicate that this is not always true. An example of a space charge effect appeared in drift velocity experiments on hydrogen (Miller et al., 1968) in which an ion of low abundance (H^+) was pushed down the drift tube by the dense space charge cloud of a much more abundant ion (H_3^+) of lower mobility and the H^+ arrival-time spectrum was seriously distorted in the process. The drift velocity of H^+ inferred by averaging the arrival times was 10% in error until the space

---

* Actually, K^+ ions can react with N_2 molecules to form $K^+ \cdot N_2$ clusters at room temperature, but the clustering was inappreciable at the relatively low gas pressure used in obtaining the data shown in Fig. 2-2-2a.

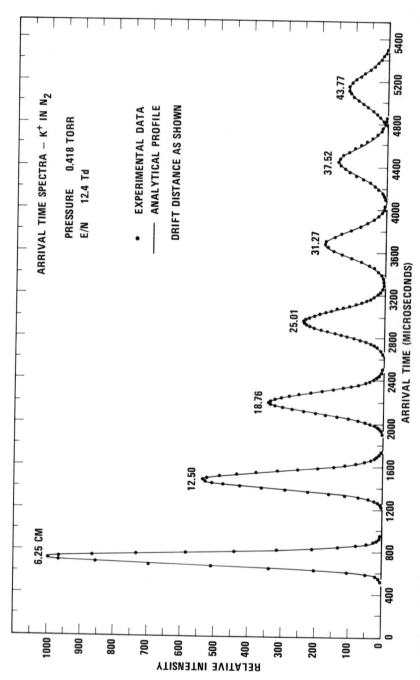

FIG. 2-2-2a. Arrival time spectra of "nonreacting" K$^+$ ions in nitrogen recorded at seven ion source positions (dots) compared with the spectra calculated by the analysis of Section 2-6 (solid curves).

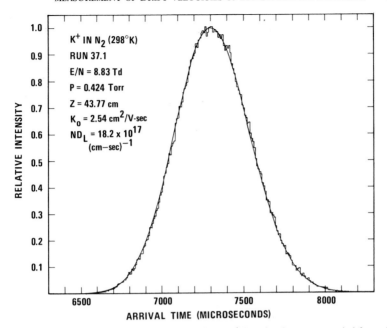

FIG. 2-2-2*b*. Arrival time spectrum of "nonreacting" K^+ ions in nitrogen recorded for a drift distance of 43.77 cm (histogram) compared with the spectrum calculated by the analysis of Section 2-6 (smooth curve). Reactions were insignificant at the gas pressure used here.

charge effect was reduced to negligible proportions by lowering the total ion current.

C. EXPERIMENTS ON REACTING SYSTEMS. The case considered in Section 2-2-A, in which measurements are made on a primary ion whose identity is known and which undergoes no reactions in the gas, was the simplest possible. The next simplest case is that in which a known primary ion can be converted to a known secondary species that does not react and thus retains its identity down the drift space. In this situation it is frequently possible to arrange conditions to obtain accurate mobilities for both the primary and secondary ions in essentially the same straightforward manner described in Section 2-2-A. An example is provided by the experiments on nitrogen ions in nitrogen gas performed by Moseley and his colleagues (Moseley et al., 1969b). In these experiments two primary ions N^+ and N_2^+ were produced by electron bombardment in the ion source; both can be converted into secondary species by reactions with molecules. The conversion proceeds mainly by the three-body reactions

$$N^+ + 2N_2 \longrightarrow N_3^+ + N_2 \qquad (2\text{-}2\text{-}1)$$

and

$$N_2^+ + 2N_2 \longrightarrow N_4^+ + N_2, \qquad (2\text{-}2\text{-}2)$$

and reconversion to the original ions N^+ and N_2^+ is possible by the reverse reactions only at E/N high enough to permit collisional dissociation of the secondary ions. Below the value of E/N at which dissociation of N_3^+ could occur the drift velocity of N^+ can be measured accurately by working at very low pressures, at which conversion of N^+ to N_3^+ is slow, and at such short drift distances that few of the N^+ ions are converted during drift. Similar techniques permit accurate measurements to be made on the other primary ion N_2^+. To get data on N_3^+ we use high pressures and long drift distances, so that the N^+ ions are converted to N_3^+ in a distance that is short compared with the total drift distance. (Most of the N^+ reacts to form N_3^+ in the ion source.) The same procedure is employed for the other secondary ion N_4^+. For all four ions the experimental arrival-time spectra

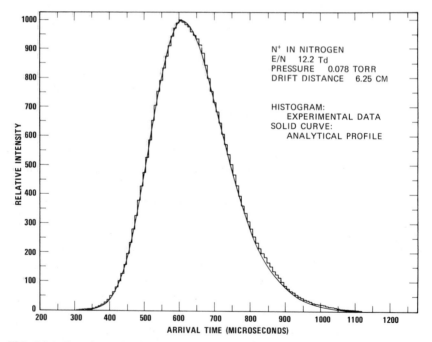

FIG. 2-2-3. Experimental arrival time spectrum for N^+ ions in nitrogen when conversion to N_3^+ is negligible compared with the spectrum calculated by the analysis of Section 2-6. In the calculation of the analytical profile a rate coefficient of 2.0×10^{-29} cm^6/sec was used for the reaction converting N^+ to N_3^+ (2-2-1).

that were used for the determination of drift velocities closely match the spectra predicted by the solution of the drift-diffusion equation for ions not reacting during their drift. Data for N^+ ions taken at such a low pressure and short drift distance that conversion of N^+ to N_3^+ is negligible are shown in Fig. 2-2-3. (When the pressure and the drift distance are increased to the level at which N^+ is converted to N_3^+ at a significant rate but slow enough that the reactions take place all along the drift tube, the arrival-time spectra shown in Fig. 2-2-4 result. Here there is a ramp on the front of the N_3^+ peak produced by ions that left the ion source as N^+ and traveled part of the way in this form but were converted to N_3^+ before they arrived at the detector. Data such as these are not used for drift velocity determinations.)

In the nitrogen experiments of Moseley et al., when end effects were eliminated by the use of multiple source positions, mobilities accurate within 3% were obtained for all four species of ions: N^+, N_2^+, N_3^+, and N_4^+. These measurements illustrate the importance of mass selection, high resolution time analysis, and a movable ion source. In the apparatus used

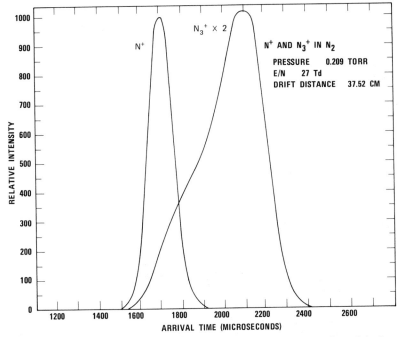

FIG. 2-2-4. Comparison of N^+ and N_3^+ arrival-time spectra showing the effect of the formation of N_3^+ from N^+.

the drift distance can be varied during operation from 1 to 44 cm. An important factor is that both the N^+ and N_2^+ ions are converted to secondary form in three-body reactions for which the reaction frequency varies as p^2. Because of this rapid variation with pressure, it proved possible to go from virtually no reaction at all at low pressure to rapid and essentially complete conversion to secondary form at high pressure. This procedure might not be possible in general if the conversion of primary ions to a secondary species proceeded by a two-body reaction for which the reaction frequency varies more slowly with pressure, only as the first power of p.

Frequently we encounter more complicated situations than those we have described. A common case is one in which primary ions are converted to a secondary species, some members of which are reconverted to the primary form during flight. Figure 2-2-5* is an example that occurs with N_2^+ and N_4^+ ions in nitrogen at room temperature and values of pressure and E/N

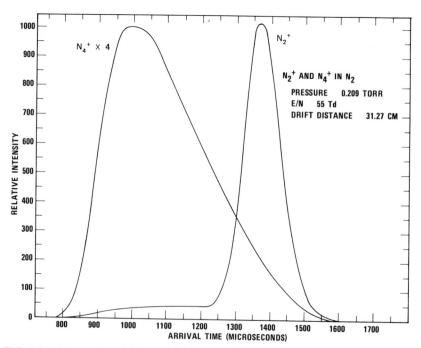

FIG. 2-2-5. Comparison of N_2^+ and N_4^+ arrival-time spectra showing the effects of ion-molecule reactions on each profile.

---

* It may be correctly inferred from Fig. 2-2-5 that N_4^+ ions drift through N_2 gas more rapidly than the lighter N_2^+ ions. The reason for this is that N_2^+ ions are retarded in their drift by resonant charge transfer with N_2 molecules.

in the intermediate range (Moseley et al., 1969b). Here N_2^+ is converted to N_4^+ by the "forward" reaction

$$N_2^+ + 2N_2 \longrightarrow N_4^+ + N_2, \qquad (2\text{-}2\text{-}2)$$

and N_4^+ ions are dissociated in collisions with N_2 molecules by the "backward" reaction

$$N_4^+ + N_2 \longrightarrow N_2^+ + 2N_2. \qquad (2\text{-}2\text{-}3)$$

At some values of E/N and pressure these reactions occur so rapidly that the conversion and reconversion processes take place many times during the passage of the ions down the drift tube and a dynamic equilibrium develops between the two ionic species. The equilibrium concentrations of N_2^+ and N_4^+ will depend on the value of E/N and the gas temperature and pressure.

In another interesting case the primary ions may be converted to secondary ions which can react with the gas molecules to form a tertiary ionic species. This can happen in hydrogen at room temperature and pressures above about 0.4 Torr (Miller et al., 1968) through the sequence of reactions

$$H^+ + 2H_2 \longrightarrow H_3^+ + H_2 \quad \text{and} \quad H_2^+ + H_2 \longrightarrow H_3^+ + H,$$
$$(2\text{-}2\text{-}4)$$

$$H_3^+ + 2H_2 \longrightarrow H_5^+ + H_2. \qquad (2\text{-}2\text{-}5)$$

The H_5^+ ions have a low binding energy and are quickly broken up in collisions with H_2 molecules to form H_3^+, even at low E/N and room temperature.

Despite the complexity of the examples described here, it is possible to obtain reliable drift velocities by solving the drift-diffusion equation with reaction terms included and making a multiparameter fit to the experimental arrival-time spectra. In the process the reaction rate coefficients as well as the drift velocities are evaluated. This procedure has been used by several investigators (see, for example, Beaty and Patterson, 1965; McKnight et al., 1967). Usually the accuracy achieved in the drift velocity determinations is not so high as in nonreacting systems.

Evidently a drift-tube mass spectrometer must be employed in a reacting system of appreciable complexity if true drift velocities characteristic of single known species are to be obtained. If a conventional drift tube is used, drift velocities inferred from a given peak in the composite arrival-time spectrum will usually refer to charge carriers that had spent part of their drift time in one ionic form and the remainder in some other, so that what is measured is some unknown kind of average of the true drift velocities of the various species. However, even if a drift-tube mass spectrometer is used for the measurements, we may obtain drift velocities that are badly in error unless the arrival-time spectra are carefully mapped and properly analyzed. Some investigators have determined drift velocities in reacting systems simply by measuring the times at which the ion intensity peaks. Edelson and his colleagues (Edelson et al., 1967) have demonstrated mathematically how this procedure can lead to large errors.

Before terminating this general discussion we should mention one additional advantage offered by drift-tube mass spectrometers that is unrelated to reactions occurring between the ions and gas molecules. It sometimes happens that two primary ions with very nearly equal mobilities are formed in the ion source. A case in point is that of SF_6^- and SF_5^- in SF_6 gas. Certain drift velocity measurements made on this system before the development of drift-tube mass spectrometers revealed only a single peak in the composite arrival-time spectrum, hence indicated the presence of only a single kind of ion (McDaniel and McDowell, 1959; McAfee and Edelson, 1962). When Patterson studied the negative ions of SF_6 with a drift-tube mass spectrometer and obtained separate arrival-time spectra for each ion, it became apparent that two different ionic species were present with closely similar mobilities (Patterson, 1970).

2-3. THE DETERMINATION OF LONGITUDINAL DIFFUSION COEFFICIENTS

Let us consider again the simple situation in which some particular ionic species is formed entirely within the ion source and travels the length of the drift tube without undergoing reactions with the gas molecules. We would expect that the shape and width of the arrival-time spectrum would depend strongly on the value of the longitudinal diffusion coefficient D_L and that it should prove possible to determine D_L from the shape of the observed spectrum. This is indeed the case, although its determination is much more difficult than that of the drift velocity and its published values date back only to 1968 (Moseley et al., 1968).

To determine D_L accurately we must have a solution of the transport equation describing the drift and diffusion of the ions in the apparatus used for the experiment. Such a solution will provide the expected functional form of the arrival-time spectrum in terms of D_L, the drift velocity v_d, and the transverse diffusion coefficient D_T. The dependence on v_d will be strong, since it determines the length of time the ions spend diffusing in the gas, but, as we have seen in Section 2-2-A, the drift velocity can be measured accurately in the simple case considered here and is assumed to be known. The shape of the arrival-time spectrum depends only weakly on the value of D_T, and if D_T is not known from independent measurements we may use its value obtained from the low-field mobility by application of the Einstein equation (1-1-5). Although this value of D_T will apply strictly only in the low-field region, it may be used in determinations of D_L even in the high-field region without introducing a significant error. To obtain slightly greater accuracy we may use the value of D_T calculated for the appropriate E/N from the modified Wannier equations (5-2-30) or (7-4-4)* (see McDaniel and Moseley, 1971; Thomson et al., 1973; Whealton and Mason, 1973). Then all the quantities appearing in the analytical expression for the arrival-time spectrum may be considered known except for D_L. We may insert a trial value of D_L into the analytical time profile and adjust this value until the analytical profile and the experimental arrival-time spectrum best correspond when normalized to agree at their points of maximum intensity. In the usual procedure a computer performs a least-squares or least-cubes fit to the experimental data by varying D_L. (It is necessary to shift the analytical profile along the time axis to match up the positions of the peaks if end effects are of significant size in the experiment.)

It is most important in obtaining the experimental arrival-time spectrum that the charge density in the ion swarm be low enough that no detectable space charge expansion of the swarm will occur. A small amount of space charge expansion will not necessarily affect the measurement of v_d to an appreciable extent, but it will materially increase the value of D_L obtained by the fitting procedure described here. As the ion current is increased from an initially low value the peak in the arrival-time spectrum will be noticeably broadened by mutual repulsion long before the measured drift velocity has begun to be seriously affected. A good way to determine whether the ion current is high enough to produce space charge effects is to compare the measured values of the mobility and the diffusion coefficient at low E/N to see whether the two results are related by the Einstein equation. If the ratio D/K is significantly greater than predicted by this equation, then space charge effects are probably operative and the ion current should be reduced.

The technique we have described for determining D_L may also be applied to the case in which the primary ions formed in the ion source undergo depleting reactions as they drift through the gas, provided that the rate coefficient for the reaction is known and the effect of the reaction is included in the solution of the transport equation (Moseley et al., 1968). If the primary ions are capable of reacting during their drift, the arrival-time spectrum should be mapped at low pressure so that the conversion of the primary ions to secondary form will occur only slowly. Operation at low pressure

---

* These equations, however, do not allow for the effect of resonant charge transfer, hence should not be used for those ion-molecule combinations for which this effect occurs.

also offers the advantage of accentuating the effect of diffusion and producing a broad spectrum that can be fitted more accurately with an analytical
profile than a narrow spectrum can.

2-4. THE DETERMINATION OF REACTION RATE COEFFICIENTS FROM ARRIVAL-TIME SPECTRA

It has already been stated that the drift tube offers a useful method of determining the rate coefficients of ion-molecule and charge transfer reactions.
The accuracy that can be achieved in such determinations is probably as high
as that accessible in any of the alternative methods in use (McDaniel et
al., 1970; Franklin, 1972). In favorable cases accuracies of about 5%* can be
realized with drift tubes if proper precautions are observed (McDaniel, 1970).
The drift tube offers the advantage that data can be obtained not only for
thermal energy ions but also for ions of average energy ranging up to several
electron volts. The capability of measuring ionic rate coefficients in the
suprathermal energy range has excited a great deal of interest, since
conventional beam experiments cannot be performed at these energies
because of the difficulties present in space charge and stray fields. Merging
beams techniques (see Section 2-1) can be applied at suprathermal energies
for the study of two-body reactions, but they cannot be utilized for the investigation of reactions of the three-body kind. Hence the drift tube can
bridge an important energy gap in the study of many important reactions.

Up to this point in this chapter we have been discussing the methods of
determining drift velocities and longitudinal diffusion coefficients from
measured arrival-time spectra. It may be appropriate, then, to extend the
discussion slightly and show how reaction rates may also be obtained from
such spectra. Other drift-tube methods of determining rate coefficients which
require the measurement of total ion currents are described in Section 2-5-C
and Chapter 3.

Let us consider the case in which a primary ion may be converted to a
nonreacting secondary species by reactions with the gas molecules during its
drift. The shape of the arrival-time spectrum for the primary ion will not
depend strongly on the value of the rate coefficient, although the area under
the primary ion spectrum will. The effect of the reaction, however, will be
clearly manifested in the shape of the arrival-time spectrum for the secondary
ion (provided its drift velocity is not close to that of the primary ion), and
often the value of the rate coefficient can be accurately determined from an
analysis of this shape. The technique requires the use of a solution of the
transport equation for the secondary ion containing a source term for the

* For a discussion of experiments in which this degree of accuracy was achieved, see
Graham, E. (1974), Ph.D. thesis, Georgia Institute of Technology, Atlanta.

creation of these ions from the primaries. The rate coefficient for the reaction appears as a parameter in the solution, and a value for this quantity may be assumed and then varied until the analytical profile matches the experimental arrival-time spectrum for the secondary ion.

Figure 2-4-1 presents the data used to determine the rate coefficient for the two-body reaction $O^- + O_2 \rightarrow O_2^- + O$ at $E/N = 124$ Td and a gas

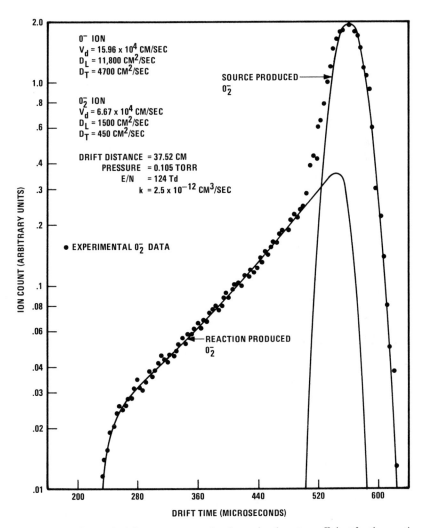

FIG. 2-4-1. An O_2^- arrival-time spectrum used to determine the rate coefficient for the reaction $O^- + O_2 \rightarrow O_2^- + O$.

temperature of 301°K (Snuggs et al., 1971a). The dots represent the experimental arrival-time spectrum for the O_2^- ion. The sharp peak at the right corresponds to O_2^- ions that were produced in the ion source and that traveled all the way down the drift tube in this form. The points near the top of the ramp leading to the sharp peak are for O_2^- ions that were formed from O^- early in their drift, whereas those at the bottom of the ramp correspond to ions that underwent a "death-bed conversion" from O^- to O_2^- at the end of the drift, just in time to be detected as O_2^- by the mass spectrometer. The smooth curve is the analytical profile predicted from the solution of the transport equation for the O_2^- ion. The values of the drift velocities and diffusion coefficients used in the analysis are shown on the drawing. Fitting the analytical profile to the experimental data yielded a reaction rate coefficient of 2.5×10^{-12} cm^3/sec.

The determination of the rate coefficient for a different reaction is now discussed in greater detail to illustrate the strong interplay between the transport properties and reactions of ions in certain drift-tube experiments. The reaction discussed is that of the conversion of CO^+ ions (which we call species A) into the dimer ions $CO^+ \cdot CO$ (species B) in three-body collisions with CO molecules:

$$CO^+ + 2CO \longrightarrow CO^+ \cdot CO + CO. \qquad (2\text{-}4\text{-}1)$$

The measurements were made at 300°K and over the pressure range 0.078–0.159 Torr. The range of E/N extended from 75 to 150 Td. The drift-tube mass spectrometer described in Section 2-5-A was employed in the measurements (Schummers, 1972; Schummers et al., 1973b).

An analytic solution of the transport equation for the product ion B, developed by Gatland (1972), was matched with the experimental arrival-time spectrum for this ion. For the present case in which dissociation of species B is negligible this solution has the form

$$\Phi_B(z, t) = sa \int_0^t du [f_B \, \delta(t - u) + f_A \alpha_{BA}](\pi r_L^2)^{-1/2}$$
$$X \left[\frac{2D_{LB}}{r_L^2} (z - r_d) + v_{dB} \right] \exp \left[-\alpha_{BA}(t - u) - \frac{(z - r_d)^2}{r_L^2} \right] \left[1 - \exp \left(-\frac{r_0^2}{r_T^2} \right) \right].$$

$$(2\text{-}4\text{-}2)$$

Here $\Phi_B(z, t)$ is the flux of ions of species B arriving at the detector at time t for a drift distance z. The area of the drift-tube exit aperture is a. Ions of planar density s enter the drift region at time $t = 0$ from the ion source through an aperture of radius r_0. Among these ions a fraction f_A are of species A and a fraction f_B belong to species B. The time of drift spent by an ion in form B is denoted by u. The reaction frequency is $\alpha_{BA} = kN^2$, where

k is the rate coefficient; $r_L^2 = 4D_{LA}(t - u) + 4D_{LB}u$ and $r_T^2 = 4D_{TA}(t - u) + 4D_{TB}u$, where D_L and D_T represent longitudinal and transverse diffusion coefficients, respectively. The drift velocity is v_d, and $r_d = v_{dA}(t - u) + v_{dB}u$.

Figure 2-4-2a shows a typical experimental arrival-time spectrum (solid curve) for the product ion $CO^+ \cdot CO$ (species B) obtained at a drift distance of 43.77 cm, a pressure of 0.133 Torr, and an E/N of 75 Td. [Since the product ion here has a higher mobility than the parent ion, this spectrum differs markedly in shape from that shown in Fig. 2-4-1, for which the product ion drifts more slowly than the parent.] Also shown as a dashed curve in Fig. 2-4-2a is a plot of the analytic expression (2-4-2) for the same experimental conditions. The drift velocities v_{dA} and v_{dB}, which were used in plotting (2-4-2), were determined in separate measurements, as was D_{LA}. The modified Wannier equations (7-4-3) and (7-4-4) (see Section 7-4) were used to compute D_{LB} and D_{TB} from v_{dB}. These equations, however, cannot be used to calculate D_{LA} and D_{TA} because the diffusion of CO^+ ions in CO is dominated by resonant charge transfer. The best estimate of D_{TA} (given by $ND_{TA} = 17.1 \times 10^{17}$/cm-sec) was obtained by an iterative curve-fitting procedure that utilized arrival-time spectra for the product ion mapped at low pressures, where the effects of transverse diffusion are accentuated and D_{TA} can be determined most accurately. With this value of D_{TA} [see (2-4-2)] a value of 1.35×10^{-28} cm^6/sec for the rate coefficient k gives the best fit to the experimental arrival-time spectrum mapped at the higher pressure of 0.133 Torr, where the effects of the reaction dominate those of transverse diffusion. In the curve-fitting procedure the analytic expression is normalized to agree with the experimental spectrum at its maximum, since the source density s is not accurately known. Two other assumed values for D_{TA} and the corresponding best-fit values of k are indicated on the graph to demonstrate the sensitivity of the evaluation of k to the value of D_{TA} which is used in plotting (2-4-2). These values of D_{TA} are given by $ND_{TA} = 12.8 \times 10^{17}$/cm-sec and $ND_{TA} = 23.5 \times 10^{17}$/cm-sec. They produce analytic spectra that do not agree nearly so well with the experimental spectra obtained at low pressures as the value of D_{TA} given by $ND_{TA} = 17.1 \times 10^{17}$/cm-sec (see Fig. 2-4-2b).

The measured values of the rate coefficient k for the reaction (2-4-1) range from 1.35×10^{-28} cm^6/sec at $E/N = 75$ Td to 1.1×10^{-28} cm^6/sec at E/N between 110 and 150 Td. The accuracy of these results is estimated as $\pm 14\%$. No theoretical equation appears to be available for an accurate calculation of the average energy of the reacting CO^+ ions in the intermediate range of E/N explored here. As pointed out in Section 2-1, however, Fahr and Müller (1967) have developed a theory applicable to the high-E/N regime for the case in which resonant charge transfer collisions dominate the transport behavior of the ions, as they do here. Their analysis gives the

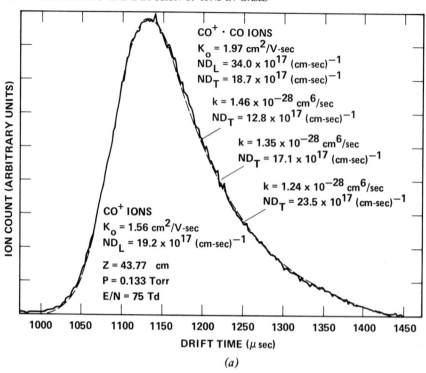

FIG. 2-4-2. Arrival-time spectra of $CO^+ \cdot CO$ ions used to determine the rate coefficient for the reaction $CO^+ + 2\,CO \rightarrow CO^+ \cdot CO + CO$.

average energy of the ions derived from the electric field in excess of the thermal value as $\pi m v_d^2/4$, where m is the mass of the ions [see (2-1-2)]. In the absence of a more appropriate expression this result has been used to estimate the average "field energy" of the reacting CO^+ ions above the thermal value in the intermediate range of E/N studied here. The result is that the average "field energy" is estimated as 60% of the thermal value at $E/N = 75$ Td and 184% of the thermal value at $E/N = 150$ Td.

2-5. DESCRIPTIONS OF DRIFT TUBES

Scores of drift tubes, differing greatly in principles of operation and in details of design and construction, have been used for studies of slow ions in gases, and many of them are described in books on atomic collisions and gaseous electronics [see, in particular, Loeb (1960) and Massey (1971)]. Here we concentrate our attention on drift-tube mass spectrometers in current use.

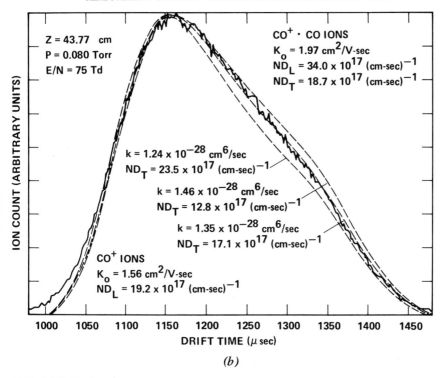

FIG. 2-4-2. *Continued.*

We also describe briefly several conventional drift tubes that possess particularly interesting features and have been used in experiments of special importance.

Our reason for emphasizing drift tubes that incorporate mass analysis is probably apparent at this point in the discussion, but it can be summarized briefly. Even though the identities of the primary ions produced by an ion source are known, reactions of these ions with molecules of the gas filling the drift tube can produce an unexpected assortment of ions, and the inter-relationships among the drifting ions can be complicated. Usually it is impossible to disentangle the behavior of the individual species without the use of mass selection and careful analysis of the shapes of the separate arrival-time spectra, and in most cases the use of a drift-tube mass spectrometer is essential if we are to obtain unambiguous results.

Since the early days of mobility research the importance of excluding impurities from drift tubes has been realized because it has long been known that the primary ions produced in the gas intended for study can be efficiently converted to ions of other types in reactions with impurity molecules. Special attention has been paid to polar and highly polarizable impurities (which can produce strong clustering of molecules about the ions) and to impurities of low ionization potential (which can become ionized by charge transfer at the expense of the primary ions). Strangely enough, however, many years elapsed before the complicating effects of reactions with molecules of the *parent* gas were fully appreciated, and it is probably this fact that was responsible for the late appearance of apparatus employing mass analysis. The first drift-tube mass spectrometers were developed in the early 1960's at Georgia Tech (Barnes et al., 1961; McDaniel et al., 1962; Martin et al., 1963) and the Bell Telephone Laboratories (McAfee and Edelson, 1963a, 1963b; Edelson and McAfee, 1964). Both instruments have been replaced by improved apparatus, described in this section along with instruments developed elsewhere.

A. THE GEORGIA TECH DRIFT-TUBE MASS SPECTROMETER.*   The experimental facility now in use at the Georgia Institute of Technology for the measurement of drift velocities, diffusion coefficients, and reaction rates was constructed during 1965 and 1966 and is described in a series of papers (Albritton et al., 1968; Miller et al., 1968; Moseley et al., 1968, 1969a, 1969b; Snuggs et al., 1971a, 1971b; Volz et al., 1971; Thomson et al., 1973; Schummers et al., 1973a, 1973b). It consists of a large ultra-high-vacuum enclosure containing a drift tube and ion-sampling apparatus, plus associated circuitry (see Figs. 2-5-1 and 2-5-2). The gas to be studied is admitted to the drift tube through a servo-controlled leak, and the sample gas flows continuously from the tube through an exit aperture on the axis at the bottom of the tube. The pressure in the drift tube is held constant during operation at some desired value in the range 0.025 to several Torr. A pulsed electron-impact ion source is used to create repetitive, short bursts of primary ions at a selected source position on the drift-tube axis. Each burst of ions moves downward out of the source and migrates along the axis of the drift tube under the influence of a uniform electric field produced by electrodes (the drift-field guard rings) inside the tube. When the ions reach the bottom of the drift tube, those close to the axis are swept out through the exit aperture and the core of the emerging jet of ions and gas molecules is cut out by a conical skimmer and allowed to pass into an rf quadrupole mass spectrometer. Ions of a selected charge-to-mass ratio traverse the length of the spectrometer, all other ions being rejected in the mass selection process. The selected ions are then detected individually by a nude electron multiplier operated as a pulse counter, and the resulting pulses are electronically

* The detailed engineering and construction of this apparatus was carried out by D. L. Albritton and T. M. Miller.

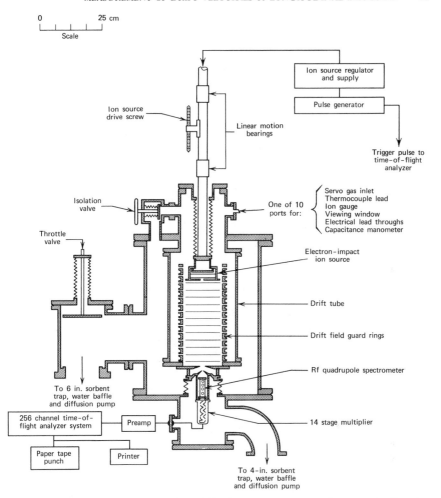

0   25 cm
Scale

Ion source regulator
and supply

Pulse generator

Ion source
drive screw

Linear motion
bearings

Trigger pulse to
time-of-flight
analyzer

Isolation
valve

One of 10
ports for:
{ Servo gas inlet
Thermocouple lead
Ion gauge
Viewing window
Electrical lead throughs
Capacitance manometer

Throttle
valve

Electron-impact
ion source

Drift tube

Drift field guard rings

Rf quadrupole spectrometer

To 6 in. sorbent
trap, water baffle
and diffusion pump

256 channel time-of-
flight analyzer system

Preamp

14 stage multiplier

Paper tape
punch

Printer

To 4-in. sorbent
trap, water baffle
and diffusion pump

FIG. 2-5-1. Overall schematic view of the drift-tube mass spectrometer in use at the Georgia Institute of Technology. The objects connecting the drift-tube exit aperture plate to the housing of the conical skimmer above the spectrometer are posts which do not significantly impede the action of the pump shown at the left in disposing of the gas flowing from the drift tube.

sorted according to their arrival times by a 256-channel time-of-flight analyzer. Because of transverse diffusion and geometrical losses, only an extremely small fraction of the ions originally present in each burst reach the detector. A statistically significant histogram of arrival times can, however, be built up by accumulating data from 10^5 to 10^6 ion bursts for a given source position (see Fig. 2-2-3, for example). This procedure is repeated for

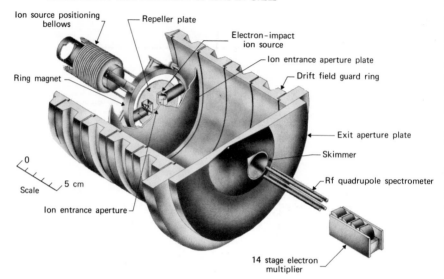

FIG. 2-5-2. Isometric view of the drift tube, ion source, and ion sampling apparatus shown in Fig. 2-5-1.

various other positions of the source along the axis of the drift tube, following which, the mass spectrometer is tuned successively to other ionic masses and arrival-time spectra are acquired for each type of ion present in the drift tube. Finally, the sequence of measurements is repeated for other values of gas pressure and drift field intensity.

If a given type of primary ion travels all the way from the ion source to the detector without undergoing chemical reactions with the gas molecules, its arrival-time spectrum should consist of a "Gaussian" peak slightly skewed toward later arrival times (see Section 2-6). Typical experimental arrival-time spectra with this expected shape have been shown superimposed in Fig. 2-2-2a for seven different positions of the ion source. Only slight deviations from the peak shape shown in Fig. 2-2-2 are to be expected if the primary species undergoes reactions at a moderate rate as it drifts through the gas, provided that this ion is produced only in the ion source and not, in addition, by reactions along the drift path. Situations occur, however, in which reactions can produce arrival-time spectra of quite different shape for primary ions as well as secondaries. Examples have been displayed in Figs. 2-2-5 and 2-4-1 and are described in the text accompanying them.

The main vacuum chamber shown in Fig. 2-5-1 is constructed of stainless steel and is evacuated by oil diffusion pumps which are separated from the chamber by water-cooled baffles and molecular sieve traps. Metal gaskets provide the vacuum seals, and the chamber is baked at a temperature of 200°C during pumpdown before measurements. The stainless-steel drift tube inside this chamber is heated to 300°C during pumpdown and base vacua below 10^{-9} Torr are achieved within it. The background drift-tube pressure after the isolation valve is closed never exceeds 2×10^{-8} Torr. Important to the achievement of low base pressures are (a) the use of organic materials is scrupulously avoided, (b) all insulators are made either of alumina (Al_2O_3) or steatite $(MgO \cdot SiO_2)$, and (c) all welds are heliarc welds made from the inside. After the drift tube has been evacuated and filled with the gas to be studied the pressure in the drift tube is measured with a capacitance manometer which has been calibrated by a trapped McLeod gauge. Thermocouples attached to the exterior of three of the drift-field guard rings indicate the temperature inside the drift tube.

Details of the ion source construction are shown in Fig. 2-5-2. The source contains two nonmagnetic stainless-steel boxes, one mounted on each side of a ring magnet which produces a field of about 100 gauss in the magnet gap. Electrons are evaporated from a filament* in the box on the left side and are periodically admitted into the ionization region through a slit in a control plate which is used to gate the passage of the electrons. Because of the magnetic field, the electrons are constrained to move in tight helices. The electron beam has the shape of a narrow ribbon perpendicular to the drift-tube axis, and thus the primary ionization is restricted to a narrow, well-defined region in the gas. After traversing the ionization gap the electrons are collected in the box on the right side of the source. The source frame is maintained at the local equipotential in the drift field, and a suitable potential is applied to the repeller plate at the top of the source to cause the ions formed to move toward the ion entrance aperture plate. A Tyndall gate, not shown, is mounted in the $\frac{3}{4}$-in. hole in this plate. This gate consists of two closely spaced wire meshes mounted perpendicular to the drift-tube axis. Normally a potential is applied between these meshes to prevent the ions from passing through the hole. Periodically, however, this potential is removed to allow the ions to flow through the grid and enter the drift space. The repetition rate of the pulses applied to the control plate and the Tyndall gate is 10^2 to 10^4/sec, and the width of the pulses is usually less than 1 μsec, a negligible time compared with typical drift times. Short pulse widths and an electron beam current of less than 1 μA during each pulse are utilized to restrict the number of ions in each burst to a value that does not produce appreciable space charge effects in the drift space.

---

* Thoriated iridium is normally used as the filament material, but in experiments on oxygen a platinum-rhodium filament coated with a $BaZrO_3$-$BaCO_3$-$SrCO_3$ mixture was employed because of its longer lifetime in this reactive gas. This filament, however, has the disadvantage of requiring a great deal of power for its operation and can cause an undesirable temperature gradient along the drift tube.

The electron beam in the ion source has a fairly small energy spread (several electron volts), and its energy can be set at any desired value within wide limits. This feature provides closer control on the production of excited and multiply charged ions than is possible with many other kinds of ion sources.

Both positive and negative ions can be produced by the ion source. Positive ions are created by the ejection of bound electrons into the continuum. In an electronegative gas negative ions can be produced directly by beam electrons in various processes or as the result of capture by gas molecules of electrons ejected from other gas molecules by the ion source beam.

A coated filament is also located inside the ion source on the axis, directly in front of the repeller plate. It can be used to produce positive ions of the alkalis by thermionic emission (see Section 2-1).

The ion source is mounted at the bottom of a long stainless-steel bellows whose length can be varied from outside the apparatus during operation. By changing the length of the bellows we may place the ion source within a few thousandths of an inch at any of 16 positions along the drift-tube axis to provide drift distances that range from 1 to 44 cm.

The drift space which the ions enter after leaving the ion source is bounded by a set of 14 guard rings with 17.5 cm id. The rings are similar to those used by Crompton, Elford, and Gascoigne (1965) and maintain an axial electric field which is uniform to a fraction of a percent in the region traversed by the ion swarm.* Concealed alumina spacers and dowel pins electrically separate the guard rings and provide alignments to a few thousandths of an inch. All surfaces exposed to the ions here and in other parts of the apparatus are gold-plated to reduce surface potentials. Guard rings of unusually large diameter are used so that the drifting ions will not be able to reach them by transverse diffusion. Thus uncertainties concerning the fate of ions when they strike a surface are avoided, and the mathematical solution of the transport equation describing the ionic motion is considerably simplied because the ions can be assumed to drift and diffuse in a space of unbounded radial extent (see Section 2-6).

Ions leave the drift space through a knife-edged hole, 0.035 cm in diameter, in the exit aperture plate at the bottom of the drift tube. This aperture is similar to a simple molecular beam effusion orifice, and its design minimizes mass discrimination effects in the ionic mass sampling (Parkes, 1971). Strong differential pumping is applied between the drift-tube

---

* The calculation of the electric field pattern inside the drift tube appears in the thesis of Albritton (1967), which also contains much detailed information concerning the construction of the apparatus described here.

exit aperture and the skimmer and between the skimmer and the mass spectrometer so that few collisions occur after the ions have passed out of the drift tube. The apparatus may be operated with no guiding or focusing field applied between the drift-tube exit aperture plate and the skimmer so that the ions will not be accelerated until they have passed through the skimmer and entered the analysis region where the pressure is below 10^{-6} Torr. This procedure guards against the possibility that weakly bound molecular ions may be dissociated in energetic collisions with gas molecules.* When tests on a given ion in a given gas indicate that it is safe to do so, however, a drawout potential of 35 V is usually applied to the skimmer to increase the detected ion current and reduce the spread of transit times between the drift tube and the mass spectrometer. The hole in the tip of the skimmer has a diameter of 0.079 cm.

After the ions have passed through the skimmer they are brought to an energy of about 4 eV for analysis in a quadrupole mass filter. Electron multipliers of several different types have been used as pulse counters to detect the ions transmitted through the mass spectrometer. Pulse-counting techniques permit the collection of data even though the signal may be extremely weak. This capability allows the apparatus to be operated at low E/N and with ion currents small enough to prevent space charge effects from appearing. A few counts per second is the practical lower limit on the detector sensitivity. The maximum counting rate that may be used with a dynode-type multiplier (see Fig. 2-5-2) is about 10^5/sec. With a capillary-type multiplier ("channeltron") the maximum counting rate is only about 10^3/sec, but this kind of detector offers the advantage of requiring less voltage for its operation than the dynode-type. Furthermore, its performance does not deteriorate when it is exposed to air and it is much more convenient to use.

The time interval between the injection of an ion swarm into the drift region and the detection of one of its members is measured and stored by a 256-channel time-of-flight analyzer (channel widths: 0.25 to 64 μsec). The sweep of the analyzer is triggered by the pulse applied to the Tyndall gate at the upper boundary of the drift space.

The drift tube may be operated over a wide pressure range (about 0.025 to several Torr). The lower limit is imposed by the requirement that the product of the gas pressure and drift distance be great enough to permit the ions to make many collisions during their drift. The upper limit arises from the poor performance of the electron bombardment ion source at pressures much above 1 Torr. With an ion source of the electrical

---

* This dissociation effect has been observed in the sampling of N_4^+ ions from a drift tube containing nitrogen gas (McKnight et al., 1967).

discharge type measurements with this apparatus could be extended up to drift-tube pressures of about 10 Torr before the gas-handling capacity of the pumps would be exceeded.* It is important to be able to operate over a wide range of pressure to determine accurately the pressure dependence of ion-molecule reactions occurring in the drift tube. Furthermore, the reaction pattern in a given gas may make it mandatory to operate at either a high or a low gas pressure to obtain easily interpretable results for a given ionic species in that gas (see Section 2-2-C).

The foregoing description of the Georgia Tech drift-tube mass spectrometer applied up to the end of 1971, but in 1972 several important changes were made (Schummers, 1972; Schummers et al., 1973a, 1973b; Thomson et al., 1973). The first change was the removal of the magnet from the ion source to eliminate the fringe field it produced in the top of the ionic drift region. This field had distorted certain arrival-time spectra for secondary ions formed early during the drift of the parent species, although it had not affected any of the published data to a significant extent. With the magnet removed there is sometimes a slight leakage of ions into the drift space when the Tyndall gate is closed, but this leakage has not been a problem.

Next, the conventional cold trap in the gas feed line was replaced by a "refrigerating vapor bath" developed by Puckett et al. (1971). The gas to be admitted to the drift tube can be stored in this device at any temperature between 0 and −196°C, a technique that results in some cases in much more efficient trapping of impurities than was possible with the conventional cold trap.

Finally, a "multipulsing" scheme was developed which permits the sampling of one ion cloud arriving at the bottom of the drift tube while several other ion swarms are in transit down the tube. This technique is feasible whenever the drift time is much greater than the spread in arrival times of the ions in a given cloud. A major advantage of multipulsing is that with it data can be accumulated at a much faster rate than is possible if only one ion swarm is in transit at a time.

B. THE CANBERRA DRIFT-TUBE MASS SPECTROMETER: THE BRADBURY-NIELSEN METHOD.   This apparatus was constructed by Creaser and Elford at the Australian National University in Canberra during 1966–1968 (Creaser, 1969). In some of its constructional features it is similar to the apparatus described in Section 2-5-A; therefore its description here is

---

* In measurements made with this apparatus before 1970 the exit aperture at the end of the drift tube had a diameter of 0.079 cm, and it was possible to pressurize the drift tube to only 2 or 3 Torr before the pumps were overloaded.

brief. The Canberra apparatus offers the important advantage of allowing measurements to be made over a wide range of gas temperatures. The entire drift section is surrounded by a dewar (see Fig. 2-5-3), which may be filled with liquid nitrogen at 78°K, with boiling water at 373°K, or with other substances to achieve intermediate temperatures.

The ion source shown in Figs. 2-5-3 and 2-5-4 is the coated filament type (Section 2-1) and is designed to produce positive ions of the alkalis. Ions from the filament are drawn into the drift region through a knife-edged hole, of 0.318 cm in diameter and located in a metal plate at the top of this region. Guard rings of 4.95 cm id provide a drift field that is

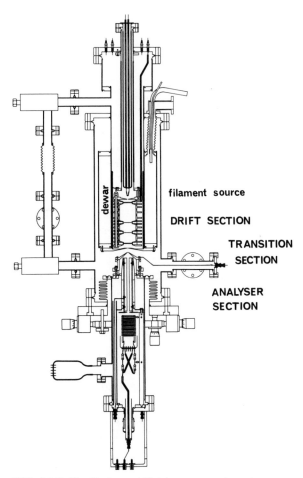

FIG. 2-5-3. The Canberra drift-tube mass spectrometer.

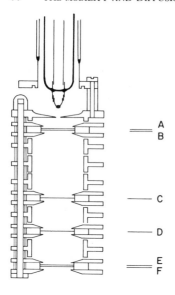

FIG. 2-5-4. The ion source and drift region of the Canberra drift-tube mass spectrometer. The grids described in the text are labeled A-F here.

nearly uniform in the region near the axis of the drift space. No provision is made for moving the ion source along the axis of the apparatus, but the effective drift distance may be changed by using various combinations of the electric grid shutters A to F shown in the drift space to time the flight of the ions in a manner described below. These shutters are made of wires 0.008 cm in diameter and spaced 0.04 cm apart. The distance between the top and bottom shutters is 12.0 cm. All surfaces exposed to the ion swarm are gold-plated. The exit aperture at the bottom of the drift space is a knife-edged hole 0.05 cm in diameter; the hole in the tip of the conical skimmer in the transition section has a diameter of 0.1 cm. An rf quadrupole mass filter is used for mass analysis, and an electron multiplier is employed as the detector of the ions passing through the mass filter.

In the experiments of Creaser and Elford (Creaser, 1969) drift velocities of K^+ ions were measured in room-temperature nitrogen, hydrogen, helium, neon, and argon over a wide range of E/N at pressures up to several Torr. The majority of the measurements were made by the Bradbury-Nielsen technique (Bradbury and Nielsen, 1936) with only the drift section of the apparatus, the mass spectrometer playing no role at all. The grids labeled B and C in Fig. 2-5-4 were used as Bradbury-Nielsen shutters and the grid D was employed as the collector in these measurements. The distance between B and C was 5.99 cm. A sine wave potential whose mean value at each grid was the value of the uniform drift-field potential at that location was applied between alternate wires of the grids B and C. The frequency of the alternating potential could be varied between wide limits. During most of each alternating cycle ions arriving at grid B from the ion source were swept out laterally to the grid wires and prevented from passing through the grid. Ions that reached the grid at times when the instantaneous value of the alternating voltage was nearly zero were transmitted, however, so that regularly spaced bursts of ions were allowed to pass down the drift space to grid C. The alternating voltage applied to grid C had the same frequency, amplitude, and phase as that applied to B. If a given burst of ions reached C at any time other than when the voltage was passing through zero, it was swept out to the grid wires, and no current would be passed to the collector grid D, which was held electrically closed by a dc potential of 15 V applied across the alternate wires of this grid. If, however, the drift velocity was such that the ions reached grid C in exactly one-half cycle, or an integral multiple thereof, the sweeping voltage was zero and the burst was transmitted to the collector D. Sharp maxima were therefore observed in the collector current corresponding to those frequencies at which the ion transit time through the drift space corresponded to an integral number of half-cycles. Since the drift distance was known, the drift velocity could be obtained at various values of the drift-field intensity and gas pressure. The current arriving at grid D was typically of the order of 10^{-12} A.

To utilize the mass spectrometer grids D, E, and F were kept electrically open so that the ions arriving at D could pass on to the exit aperture plate. A small fraction of these ions was then transmitted through the transition section and mass spectrometer (provided it was set to pass K^+ ions) and arrived at the electron multiplier, where they were detected. Pulses from the multiplier were counted by a scaler, and the scaler counting rate was measured as a function of the alternating frequency applied to the grids B and C. In this method of operation the collection of data was very slow, and because of the small currents arriving at the multiplier the measurement of the drift velocities was not so accurate as when grid D was used as the ion collector. For this reason only a few measurements were made with the mass spectrometer, but since K^+ was the only ion ever observed there was no reason at the time to suspect the data obtained by using only the drift section of the apparatus. Subsequent experiments (Beyer and Keller, 1971; McDaniel, 1972; Thomson et al., 1973), however, showed that a significant amount of clustering of K^+ ions takes place in room-temperature nitrogen and argon at the pressures used by Creaser and Elford; $K^+ \cdot N_2$ clusters were observed in the Georgia Tech drift-tube mass spectrometer even at nitrogen pressures as low as 0.1 Torr. In addition, small traces of $K^+ \cdot Ne$, $K^+ \cdot He$, and $K^+ \cdot D_2$ clusters were observed when the drift tube was filled with neon, helium, and deuterium, respectively. It is likely that the weakly bound clusters of the latter types that were observed were formed only as the K^+ ions emerged from the drift tube and that they experienced no collisions; otherwise they probably would have been dissociated.

In the Creaser-Elford experiments it was planned to utilize various combinations of the grids shown in Fig. 2-5-4 to obtain different drift distances. Measurements showed that if a drift distance such as $B–D$ was used the resulting value of the measured drift velocity would be slightly higher than that obtained by using distances such as $B–C$ or $C–D$. When the drift distance $B–D$ was employed, grid C was held at the dc potential appropriate to its position in the electrode structure, and the use of a grid in this fashion is presumably the source of the difficulty. To avoid this source of error Creaser and Elford made all their measurements over the drift distance $B–C$.

C. THE REVERSIBLE FIELD APPARATUS OF BIONDI AND HIS COWORKERS.* Biondi and his colleagues at the University of Pittsburgh have constructed a drift-tube mass spectrometer that possesses two unusual and especially interesting features. This apparatus was designed primarily for the study of ion-molecule and charge transfer reactions, but it has also been used for some important measurements of mobilities (Heimerl et al., 1969; Johnsen et al., 1970).

The apparatus is shown schematically in Fig. 2-5-5. One of its unusual features relates to the construction of the ion source, which is provided with a separate gas inlet and which communicates with the drift region only through a number of very small holes. This arrangement facilitates the study of ions in a foreign gas in a manner described below. The other unusual feature of the apparatus is that the electric field in the drift region may be reduced to zero or reversed in direction for a variable length of time during the passage of a burst of ions through the tube. This feature is not utilized in the measurement of drift velocities but has proved to be useful in the study of ionic reactions by a new technique which Biondi and his associates have called the "additional residence time method." This method is based on the measurement of the loss of parent (i.e., primary) ion current caused by reactions and is discussed at the end of this section.

The ion source in Biondi's apparatus employs an electrical discharge between a plane anode and a plane cathode, both of which are made of stainless steel. The cathode is a disk in which a single hole was drilled at its center and six other holes were spaced 60° apart on a 1.3-cm diameter circle. These seven holes, each of which has a diameter of 0.045 and a thickness of 0.013 cm, permit ions generated in the active discharge to

---

* An improved version of this apparatus is described by Brown (1972).

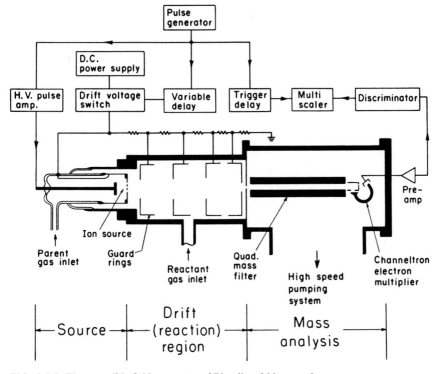

FIG. 2-5-5. The reversible field apparatus of Biondi and his coworkers.

enter the drift region. The disk, 1.9 cm in diameter, which forms the anode is mounted on a threaded assembly that permits variation of the anode-cathode distance from 0 to 1.3 cm by a magnetic coupling to four soft iron pieces. A high-voltage pulse (1 to 3 kV, 5 to 10 μsec long, repeated every 1.3 msec) is applied to the ion source and generates a discharge plasma in the gap between the anode and cathode. The separation distance between the two electrodes is set to optimize the discharge conditions for the particular gas filling used at the time. A negative bias voltage of 50 to 100 V applied to the anode prevents ions from leaving the source during the afterglow period following the active discharge and only a narrow pulse of ions may enter the drift region.

The drift tube is made of stainless steel and is 15.5 cm long. The inside diameter of the guard rings defining the drift field is 2.5 cm. Ions at the end of the drift space are sampled through a hole 0.045 cm in diameter, mass analyzed in an rf quadrupole mass filter, and detected by a pulse-counting Channeltron electron multiplier. The data-handling technique is similar to that described in Section 2-5-A. A multichannel analyzer is used to accumulate data over many cycles of operation. A time-of-flight logic unit feeds each ion count into the appropriate channel of a 256-channel computer memory so that the channel address provides a measure of the elapsed time between the discharge pulse in the ion source and registration of an ion. After a sufficient number of cycles have been completed the content of the memory is printed out in digital form to provide the basis of the data analysis.

Ultra-high-vacuum techniques were employed in the construction of the apparatus and gas-handling system. The entire vacuum system can be baked at 300°C, and a base pressure lower than 10^{-8} Torr is usually achieved after bakeout.

In the experiments of Johnsen, Brown, and Biondi with this apparatus (Johnsen et al., 1970) measurements were made of the mobilities of N^+, N_2^+, O^+, and O_2^+ ions in helium. For these measurements a mixture of helium and nitrogen or oxygen was continuously admitted to the ion source through the parent gas inlet at a total pressure of 1 Torr, the nitrogen or oxygen being present only to about 1 part in 10^4. Pure helium was admitted to the drift region through its gas inlet to maintain a pressure of 1 Torr in this region. Various processes occurring in the ion source led to the production of nitrogen or oxygen ions: direct ionization of the parent molecules by electron impact, ion-molecule and charge transfer reactions between helium ions and the nitrogen or oxygen molecules, and Penning ionization of these molecules by metastable helium atoms produced in the discharge. Arrival-time spectra were mapped for the minority nitrogen or oxygen ions, and in each case the spectrum exhibited an easily identified "source" peak along with a very small tail produced by reactions occurring in the drift region. The source peaks were used to determine the ions' time of flight. Because the gas in the drift region contained a trace of the molecular gas admitted to the ion source an insignificant error in the measured mobilities was produced. In arriving at their final mobility data Johnsen and his associates took into account the time required for an ion to traverse the distance between the sampling plane and the electron multiplier, the difference between the time of arrival of maximum current and the true time of flight, and the effect of spatial nonuniformities in the applied drift field and made appropriate small corrections for these effects.

Johnsen et al. measured the rates of reaction of O^+ ions with CO_2, O_2^+ with NO, N^+ with O_2, and N_2^+ with O_2. The parent ions were generated in the manner described above by admitting a mixture of about 10^{-4} Torr of oxygen or nitrogen in about 1 Torr of helium to the ion source. A small amount of the reactant gas in each case was admitted only into the drift region in a mixture with helium which served as an inert buffer gas to inhibit diffusion and quickly establish energy equilibrium. The pressure in the drift region was maintained slightly lower than that in the ion source so that molecules of the reactant gas could not enter the source. (Diffusion from the drift region into the source is greatly inhibited by the small size of the orifices connecting these regions and the consequent high flow velocity in the opposite direction.) The additional residence-time technique was used for these measurements. In this technique the drift field is turned off or reversed for a selected time interval while the parent ion swarm is near the center of the drift space, thus causing these ions to remain in the drift (reaction) region for an additional residence time Δt. By measuring the decrease in the number of parent ions reaching the mass spectrometer as a function of Δt, first with and then without reactant gas added to the helium in the drift region, the reaction rate could be determined (Heimerl et al., 1969, describe the analysis used.) The net loss of the parent ions due to reaction with the reactant gas during Δt is thus separated from the loss due to transverse diffusion and possible reactions with molecules of the parent gas. The range of E/p covered in these measurements extended from 0 to about 20 V/cm-Torr.

When a gas mixture is used in a drift-tube mass spectrometer, as in the experiments described above, a problem may develop in the determination of the composition of the mixture. If the gases have appreciably different molecular weights, they will be pumped out of the drift tube through the exit aperture at significantly different speeds, and the actual composition in the drift tube will differ from that of the mixture admitted to the tube. If this fact is ignored, large errors can result. In the experiments described in this section this factor was taken into account, and the correction that had to be applied typically amounted to about 50%. A detailed analysis of this problem appears in Heimerl's thesis (Heimerl, 1968).

D. MCKNIGHT'S DRIFT-TUBE MASS SPECTROMETER. An apparatus constructed by McKnight at the Bell Telephone Laboratories for the study of mobilities and reactions of negative ions in oxygen is shown schematically in Fig. 2-5-6 (McKnight, 1970). The apparatus is divided into three separately pumped regions. The first is the drift cell that contains the electron source, the electron attachment region in which the negative ions are formed, and the ion drift space. An exit aperture at the end of the drift space leads to the second region (the ion-focusing region) that contains electrodes for accelerating and focusing the emerging ions. From this part of the apparatus the ions enter the mass-analysis region, which contains the mass analyzer and ion detector.

The electrodes in the drift cell are made of platinum and supported on high-purity alumina rods. Electrons are produced from a platinum filament

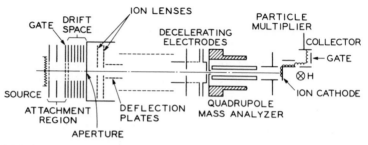

FIG. 2-5-6. McKnight's drift-tube mass spectrometer, used for the study of negative ions in oxygen

coated with a $BaZrO_3$-$BaCO_3$-$SrCO_3$ mixture. Under the influence of an electric field electrons from the filament pass through a platinum grid into the attachment region, which is 3 cm long and within which the electrons react with oxygen molecules to form negative ions. At the end of the attachment region either the remaining unattached electrons or the negative ions or both species of charge carrier are admitted to the drift space by pulsing a double-grid Tyndall gate. (This method of selecting the type of particles admitted to the drift space depends on the application of a pulse to the grid nearest the filament and is based on the difference in the drift velocities of the electrons and negative ions.) The Tyndall gate is made of two fine platinum-mesh grids separated by 1 mm. Gate-open times range from 1 to 20 μsec and are set at about 1% of the ion drift time. The drift space is 3 cm long and 2.54 cm in diameter. A series of guard rings along the drift space ensures that a uniform field is achieved in this space.

Drift-tube pressures between 0.5 and 4 Torr were used in the experiments on oxygen, and data were obtained at E/N between 2 and 400 Td. Temperatures in the drift space were measured with thermocouples at the beginning and the end of this region. The use of the filament electron source increased the temperature above ambient and caused a temperature gradient down the drift cell. Measurements were made with drift space temperatures between 40 and 100°C and gradients of 30 to 40° along the drift space.

The exit aperture plate is made of platinum foil 0.007 cm thick and pierced by 25 holes of 0.0025 cm in diameter that open up at an angle of about 60° on the side away from the drift space to facilitate extraction of the ions. The holes are concentrated in an area about 0.5 cm in diameter around the axis of the drift cell. Ions emerging through the aperture are extracted into the ion-focusing region in which the pressure is less than 10^{-4} Torr. Here the ions are accelerated and focused so that they cross this region at about 400 eV and are then decelerated into the mass-analysis region in which the pressure is below 10^{-5} Torr. There the ions enter a quadrupole mass filter that transmits ions of only a single selected charge-to-mass ratio. Ions emerging from this device are accelerated to about 1.5 keV and directed onto the cathode of a resistive strip particle multiplier which multiplies electrons produced by ion collisions on the cathode. The electron gain is between 10^6 and 10^7. Electron current in the multiplier is controlled by gate electrodes that direct current to the anode collector when a pulse is applied. Time resolution of the ion current is accomplished by changing the time between the pulse that opens the Tyndall gate at the entrance to the drift space and the pulse that directs current into the anode of the particle multiplier. The output current from the anode is measured on an electrometer whose output is chopped and capacitively coupled to ground, where it is rectified and applied to the Y axis of an X-Y recorder.

The delay time between the pulse that opens the Tyndall gate and the pulse that diverts current into the multiplier anode is measured with a time-interval counter accurate to 0.2 μsec. This delay time is converted to analog form and displayed on the X axis of the X-Y recorder. The delay time between the pulses is slowly changed and the ion current is measured as a function of the delay time. Correcting the delay time for the time spent by the ions in the ion optical elements and the mass filter (16 to 28 μsec) makes it possible to determine the current from the drift cell as a function of time for each ionic species present. Typical plots of ion current from the drift space versus delay time are shown in Fig. 2-5-7. These data were taken at a drift-cell pressure of 1.7 Torr and $E/N = 5.6$ Td when only ions were admitted to the drift space. The peaks for O^- and O_2^-

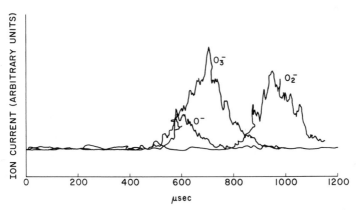

FIG. 2-5-7. Arrival-time spectra for O^-, O_2^-, and O_3^- ions in McKnight's experiment.

are due to ions that entered the drift space at the opening of the Tyndall gate and crossed the drift space without reacting. The drift velocities of these ions can be derived from the arrival times of the peak positions of the ion current. For O_3^- the shape of the ion current transient and the variation in current intensity with drift-space conditions indicate that virtually all of the O_3^- in Fig. 2-5-7 were formed in a three-body reaction of O^- with oxygen molecules within the drift space. Under these conditions the peak current is not indicative of the true drift velocity but is a function of the drift-cell pressure and length as well. To obtain the true drift velocity of O_3^- the numerical method described by Edelson et al. (1967) was used to generate ion current transients that would be expected from the drift space for a given set of parameters. Comparisons of the observed with the computed currents then give true drift velocities of the O_3^- ion.

Briefly, the numerical method of generating the ion currents consists of simulating the drift cell on a digital computer by dividing the drift space along the direction of drift into a large number of regions (usually between 50 and 200) and calculating the changes in ion density for each ion in each of the regions as a function of the time after opening the Tyndall gate. Initially it is assumed that the ions are concentrated in a few regions at the beginning of the drift space as they would be after being admitted by the Tyndall gate. Thereafter the ions progress from region to region in a manner depending on the drift velocity of the ion in question and the iteration time of the calculation, which is determined by the number of iterations required for the last of the slowest ions to cross the drift space. Reactions of the ions are treated by converting a fraction of the reacting species in each region of the drift cell into the species to which the reaction takes place, subtracting this fraction from the reacting ion population, and adding it to the other ion population. Diffusion is handled on a region-by-region basis, assuming that the ions in each region diffuse independently of those around this region. The number of ions in a given region after each iteration time is the sum of the number of ions that has diffused into it from neighboring regions minus the sum of those that have diffused out. For each calculation the drift velocities, reaction rates, diffusion coefficients, and initial concentrations of each ion must be specified. The output flux density is then the ion density at the end of the drift space multiplied by the drift velocity of the ion or, alternatively, the integral of the ion density that moves past the end of the drift space during each iteration. The calculation is continued until the ion density remaining in the drift space is a small fraction ($< 10^{-8}$) of that admitted through the gate.

E. THE APPARATUS OF EDELSON AND HIS COLLEAGUES.   Edelson and his coworkers at the Bell Telephone Laboratories have constructed a drift-tube mass spectrometer that utilizes a time-of-flight mass analyzer for the detection and identification of the ions (Young et al., 1970). This apparatus, which has been used for the measurement of mobilities and ion-molecule reaction rates, is shown schematically in Fig. 2-5-8. Primary ions are produced in the source region by bombardment of gas molecules with thermionic electrons. Ions are extracted from this region into a "cooling" region in which they are allowed to equilibrate with the gas at a specified E/p before being injected into the drift region proper through a double-grid Tyndall gate. The drift region is defined by a series of guard rings to which appropriate potentials are applied and an end wall kept at ground potential. A sample of the ions arriving at this wall is withdrawn for analysis through a knife-edged exit orifice 0.1 mm in diameter. These ions are accelerated and focused by the lens and skimmer electrodes into the source region of a Bendix Model 14 time-of-flight spectrometer whose

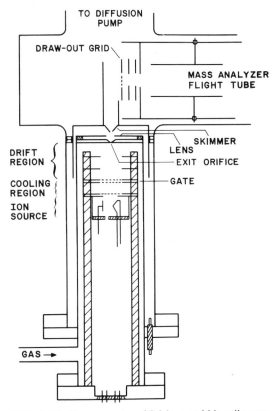

FIG. 2-5-8. The apparatus of Edelson and his colleagues.

axis is oriented perpendicularly to the drift direction. Since this mass analyzer is inherently a time-resolved device, no terminal gate for the ions is required. Appropriate potentials applied to the mass-analyzer source region electrodes confine the ions to a trajectory along the drift-tube axis until a sampling pulse is applied to the draw-out grid which causes the ions immediately behind the grid aperture to be injected into the mass analyzer. The transit time from the exit orifice to the mass-analyzer source is separately measured and applied as a correction to the drift time. Potentials applied to deflection plates in the mass analyzer compensate for the lateral component of velocity with respect to the analyzer-tube axis; these potentials also provide a degree of discrimination against those ions that have suffered collisions after leaving the drift region.

Several stages of differential pumping are used to permit pressures of several Torr to be maintained in the drift tube. The first stage of pumping is in the region between the exit orifice and the skimmer cone, which has a 2-mm orifice. The mass-analyzer source region and the flight tube are separately pumped.

The drift-tube electrode assembly is constructed by stacking the grids and guard rings on posts and separating them with accurately ground ceramic spacers. A range of drift distances of 1 to 10 cm can be achieved. The grids are fabricated by gold-fusing electroformed copper mesh of 85% transparency across a hole in a kovar support plate. The differential coefficient of expansion results in the mesh being stretched tightly across the aperture and allows close control of cell geometry. All metal parts are gold-plated.

The time-of-flight mass analyzer provides a complete spectrum of the ions present in the analyzer source region on every cycle of the experiment. With the small currents obtained it is necessary to use a counting technique to record the data. The instrumentation, shown in Fig. 2-5-9, is designed for automatic simultaneous recording of the sampled currents of as many as four ionic species as a function of drift time. This system employs a multi-channel scaler with a memory divided into four quadrants of 100 channels each. Address advance occurs simultaneously in all four quadrants, and an analog voltage proportional to the channel address provides a means of controlling the time t_1 between the Tyndall gate pulse and the spectrometer ion grid pulse. This time is accurately measured by a digital time-interval meter to 0.1 μsec. A repetition rate generator initiates the pulse sequence by simultaneously triggering the gate pulser and the swept delay unit which generates a ramp voltage proportional to time. The spectrometer ion grid is pulsed at the time of coincidence between the ramp voltage and the level determined by the current memory address. Proper baseline and span of the address analog voltage is achieved with an operational amplifier, thus producing a scan through the desired range of drift times t_1 as the

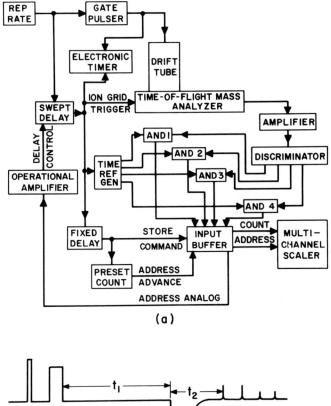

(a)

(b)

FIG. 2-5-9. The electronics used by Edelson and his colleagues.

memory address advances. Relative to the sharp leading edge of the ion grid sampling pulse, four marker pulses at times t_2 appropriate to four ions of interest are fed to the coincidence (AND) units for the corresponding memory quadrants. A pulse in the mass spectrum coincident with a marker pulse causes a count to be stored in the appropriate channel of the input buffer. Sufficient time (20 μsec) is allowed for arrival of the highest mass of interest, after which a store command empties the buffer contents into the corresponding memory quadrants. The sequence starting with the reference pulse is repeated; the channel address is advanced after a given number of times, controlled by a preset counter. In practice dwell intervals of a few thousand counts, corresponding to times of the order of 1 sec, are used. Thus the entire range of drift times is swept through many times during a typical run of 30 to 60 minutes and provides a good average over small fluctuations in drift-cell currents. Relative count rates of different species are not affected by variations in total current. Since the input buffer has single-count capacity, sampled ion currents must be low enough to make the probability of detecting a given ion per sampling cycle much less than unity.

F. BEATY'S DRIFT TUBE.   An extensive and important series of mobility measurements was made by Tyndall and his collaborators at Bristol University during the 1930's with the four-gauze, or four-grid, electrical shutter method developed by Tyndall, Starr, and Powell (Tyndall, 1938; Loeb, 1960). This method has also been used more recently by Beaty (1962), who took advantage of the subsequent advances in electronics and vacuum techniques to improve on the original version of the apparatus. Neither Tyndall nor Beaty employed a mass spectrometer in his apparatus, but both worked with systems that were simple enough to produce results that for the most part could be interpreted in a straightforward manner. The four-gauze method is discussed here in terms of Beaty's apparatus (Fig. 2-5-10): (a) is a schematic representation of the electrode geometry. A typical potential distribution applicable to a portion of the measurement cycle is shown in (b) and the timing sequence, in (c). The two pairs of closely spaced grids form two electrical shutters (Tyndall gates), a distance of about 1 mm separating the grids composing each shutter. The space between the two shutters is the drift space; it has guard rings with holes 1.0 in. in diameter placed 1 cm apart to maintain a uniform electric field. The potential difference across the drift space is labeled V_1. Between the discharge-type ion source and the first shutter is a space in which the ions from the discharge are degraded in energy before their drift velocity is measured. This thermalizing space is divided into two regions by a grid, and, as indicated, the potentials across these two regions are designated V_2 and V_3; V_1, V_2, and V_3 can be varied independently.

The gas in the ion source is ionized periodically for intervals of 7 μsec by a gated 20-Mc oscillator. The ions diffuse to the walls under the influence of the space charge fields, and some of the ions pass through the holes at the front of the source and into the thermalizing region. After a time T_2 a negative voltage pulse, which is applied to the entire drift space, opens the first shutter and allows ions to enter the drift space. After another delay T_1 a positive pulse opens the second shutter to admit ions to the collector. The advantage of this arrangement of pulsing is that the electric field is not affected by the pulses anywhere except in the shutters.

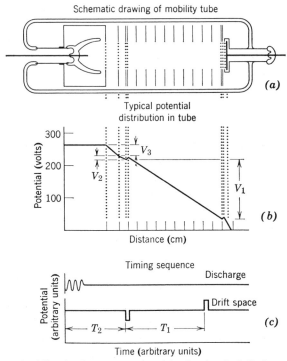

Schematic drawing of mobility tube

(a)

Typical potential
distribution in tube

(b)

Timing sequence

(c)

FIG. 2-5-10. Beaty's drift tube. Ions are produced in the electrical discharge ion source at the left and are collected by the plane electrode at the right following their drift across the tube. No mass analysis is employed.

The output of an electrometer attached to the collector is connected to an X-Y recorder. The electrometer current can be plotted as a function of V_1, V_2, V_3, T_1, or T_2, and if the reaction pattern in the gas is simple the drift velocity can be determined as a function of E/N by a straightforward procedure.

The gas-handling apparatus consists of a pyrex vacuum system which can be evacuated to a pressure of 10^{-9} Torr, a mercury cutoff to disconnect the pump, a bakeable metal valve to admit gas from a cataphoresis purification tube, a mercury manometer, a McLeod gage for reading pressures less than 6 Torr, and a liquid nitrogen trap for removing mercury vapor. After sealing the cutoff, and with the ionization gage operating at low emission, it takes several days for the pressure to reach 10^{-6} Torr.

Beaty used this apparatus to measure the mobility of positive ions of argon in the parent gas over an E/p_0 range extending from about 1 to 80 V/cm-Torr (Beaty, 1962). Pressures of 0.4 to 17 Torr were used. Beaty and

Patterson, who have also made studies of helium and neon with this apparatus, obtained mobilities and reaction rates for conversion of the atomic to molecular ions (Beaty and Patterson, 1965; 1968). Patterson has constructed a drift-tube mass spectrometer that contains a drift tube similar to the one described here (Patterson, 1970).

G. HORNBECK'S PULSED TOWNSEND DISCHARGE APPARATUS. Another drift tube which makes no provision for mass analysis but which nevertheless has been used in a number of significant experiments on the noble gases is that of Hornbeck (Hornbeck, 1951a, 1951b; Varney, 1952). This apparatus is shown in Fig. 2-5-11. A 0.1-μsec burst of photo-electrons is released from a cathode C by ultraviolet light from a spark source operated at a repetition rate of 60 cps. These electrons are accelerated through the gas at high field and produce a Townsend avalanche. The primary and avalanche electrons are collected at the anode A in a time of the order of a few tenths of a microsecond, but the exponential distribution of positive ions formed in the avalanche is swept across to the cathode much more slowly. A voltage transient is developed across the resistor R and displayed on an oscilloscope. The transient consists of a sharp spike, attributable to the photoelectrons and their progeny electrons, and a smaller component produced by the positive ions. The ionic drift velocity is determined by measuring the arrival time of ions formed "at" the anode in the discharge; this time is signaled by a break in the voltage trace on

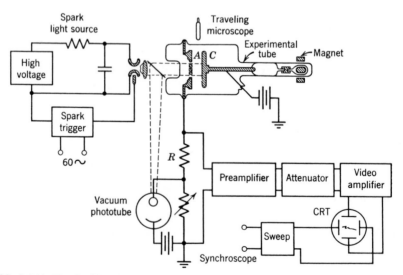

FIG. 2-5-11. Hornbeck's pulsed Townsend discharge apparatus. No provision is made for mass analysis of the ions.

the oscilloscope. The separation between the cathode and anode may be varied by means of an external magnet and measured by a traveling microscope. The gap spacing is typically about 1 cm. Pressures in the range of 0.1 to 30 Torr have been used, and ionic drift times of about 2 to 20 μsec result. The tube current is of the order of 0.1 μA. Measurements have been made over a range of E/p_0 extending from about 10 to 2000 V/cm-Torr.

The Hornbeck technique is particularly useful for obtaining data at high E/p; however, it is intrinsically not well suited to low-field measurements and indeed with this technique it is difficult to reach low values of E/p. Another disadvantage associated with this method is that it does not permit much control over the initial ionization of the gas, since ions are produced by electrons with a wide energy spread and the ionization is not closely confined spatially. The analysis of the ion production and transport is rather involved because of the complicated nature of the physical processes; however, the mathematics is not intractable because the drift-tube geometry and method of generating ions lead to an essentially one-dimensional problem.

The Hornbeck technique has been utilized with mass analysis of the ions arriving at the cathode by McAfee and Edelson in the first drift-tube mass spectrometer built at the Bell Telephone Laboratories (McAfee and Edelson, 1963a). To obtain data at lower E/p than would have otherwise been possible, McAfee and his colleagues subsequently modified their drift tube. They added a grid to confine the Townsend discharge to a smaller volume and thus permit the use of a weaker electric field in the drift region than that required in the discharge region (McAfee et al., 1967).

2-6. THE MATHEMATICAL ANALYSIS OF IONIC MOTION IN DRIFT TUBES

A. INTRODUCTION.   In Sections 2-2 and 2-3 we discussed in general terms the experimental techniques involved in the determination of ionic drift velocities and longitudinal diffusion coefficients. We saw that a necessary condition for obtaining a drift velocity in which we can have confidence is that the experimental arrival-time spectrum from which the drift velocity is derived must closely resemble the spectrum calculated by solving the transport equation for the ion in question in the apparatus used for the experiment. Furthermore, we saw that a solution of the transport equation is required for the accurate determination of the longitudinal diffusion coefficient. Therefore it is now appropriate to show how the transport equation may be set up and solved for a typical experimental situation existing in the measurement of v_d and D_L with a particular apparatus. The solution we obtain can also be used for the determination of the transverse diffusion coefficient and the rate of chemical reactions between the ions and the molecules filling the drift tube, as we show in Chapter 3.

The apparatus for which the analysis applies is the Georgia Tech drift-tube mass spectrometer described in Section 2-5-A. We analyze the space-time behavior of a swarm of primary ions which are diffusing and reacting with molecules as they drift through the apparatus. The case in which no reactions occur may be described by putting the reaction frequency equal to zero in the final solution. We assume that all the primary ions are produced in the ion source and none is derived from other ionic species by reactions along the drift path. This is the usual situation prevailing in the measurement of v_d and D_L.

B. SOLUTION OF THE TRANSPORT EQUATION FOR PRIMARY IONS IN THE GEORGIA TECH DRIFT-TUBE MASS SPECTROMETER (Moseley, 1968; Moseley et al., 1969a). Consider a population of ions of a single type created at one end of a cylindrically symmetric drift space filled with gas of uniform pressure p in which there exists an externally applied uniform electric field $\mathbf{E}$ along the axis. Assume that the number density of the ion swarm $n(\mathbf{r}, t)$ is low enough to make the space charge field negligible. In the general case in which E/p is not assumed to be small the ionic flux density is given by the equation

$$\mathbf{J}(\mathbf{r}, t) = \mathbf{v_d}\, n(\mathbf{r}, t) - \mathbf{D} \cdot \nabla n(\mathbf{r}, t) \qquad (2\text{-}6\text{-}1)$$

(cf. Section 1-1). Here $\mathbf{v_d}$ is the drift velocity and $\mathbf{D}$, the diffusion tensor characterizing the motion of the ions through the gas. If we do not allow for the possibility of gain of the ionic species under consideration during the movement down the drift tube but do consider the loss of these ions by chemical reactions with the frequency α, the ion swarm is subject to a continuity equation of the form

$$\frac{\partial n}{\partial t} + \nabla \cdot \mathbf{J} + \alpha n = 0 \qquad (2\text{-}6\text{-}2)$$

[cf. (1-9-1)] or, in light of (2-6-1),

$$\frac{\partial n}{\partial t} - \nabla \cdot \mathbf{D} \cdot \nabla n + \mathbf{v_d} \cdot \nabla n + \alpha n = 0. \qquad (2\text{-}6\text{-}3)$$

(cf. Gatland and McDaniel, 1970). With the explicit form for $\mathbf{D}$ suggested in (1-1-7) and the addition of a source term $\beta(\mathbf{r}, t)$ to represent an input of ions at the beginning of the drift space (2-6-3) becomes, in rectangular coordinates,

$$\frac{\partial n(x, y, z, t)}{\partial t} = D_T\left(\frac{\partial^2 n}{\partial x^2} + \frac{\partial^2 n}{\partial y^2}\right) + D_L \frac{\partial^2 n}{\partial z^2} - v_d \frac{\partial n}{\partial z} - \alpha n + \beta(x, y, z, t).$$

$$(2\text{-}6\text{-}4)$$

In the apparatus under consideration the ions enter the drift space through a circular aperture lying in a plane normal to the drift-tube axis and centered on the axis. If the coordinate system has its origin at the center of this aperture and the electric field along z is in a direction that causes the ions to drift in the positive z direction, the solution to (2-6-4) in unbounded space* is

$$n(x, y, z, t) = \int_{-\infty}^{t} dt' \int_{-\infty}^{\infty} dx' \int_{-\infty}^{\infty} dy' \int_{-\infty}^{\infty} dz' \frac{\beta(x', y', z', t')}{[4\pi(t - t')]^{3/2} D_T D_L^{1/2}}$$

$$\times \exp\left\{-\alpha(t - t') - \frac{(x - x')^2 + (y - y')^2}{4D_T(t - t')} - \frac{[z - z' - v_d(t - t')]^2}{4D_L(t - t')}\right\}.$$

$$(2\text{-}6\text{-}5)$$

That (2-6-5) is indeed a solution to (2-6-4) has been demonstrated by Moseley (1968). To express (2-6-5) in cylindrical coordinates let

$$x' = r' \cos \theta'; \qquad y' = r' \sin \theta'; \qquad x = r \cos \theta; \qquad y = r \sin \theta. \quad (2\text{-}6\text{-}6)$$

Then

$$n(r, \theta, z, t) = \int_{0}^{\infty} r' dr' \int_{0}^{2\pi} d\theta' \int_{-\infty}^{\infty} dz' \int_{-\infty}^{t} dt' \frac{\beta(r', \theta', z', t')}{[4\pi(t - t')]^{3/2} D_T D_L^{1/2}}$$

$$\times \exp\left\{-\alpha(t - t') - \frac{r^2 + r'^2 - 2rr' \cos(\theta - \theta')}{4D_T(t - t')} - \frac{[z - z' - v_d(t - t')]^2}{4D_L(t - t')}\right\}.$$

$$(2\text{-}6\text{-}7)$$

Clearly, if the input β is cylindrically symmetric, the ion number density n will be cylindrically symmetric. Suppose

$$\beta(r', z', t') = \frac{b}{\pi r_0^2} S(r_0 - r') \delta(z') \delta(t'), \qquad (2\text{-}6\text{-}8)$$

where $S(\phi) = 0$ if $\phi < 0$, $S(\phi) = 1$ otherwise. This function describes an axially thin disk source of b ions with uniform surface density and radius r_0, created instantaneously at $t' = 0$ in the plane $z' = 0$. Let $b/\pi r_0^2 = s$, the planar source density. Then

$$n(r, z, t) = \frac{se^{-\alpha t} \exp\left[-(z - v_d t)^2/4D_L t\right]}{(4\pi t)^{3/2} D_T D_L^{1/2}}$$

$$\times \int_{0}^{r_0} r' dr' \int_{0}^{2\pi} d\theta' \exp\left(-\frac{r^2 + r'^2 - 2rr' \cos \theta'}{4D_T t}\right). \quad (2\text{-}6\text{-}9)$$

* In the Georgia Tech apparatus the drift-field guard rings have such a large inside diameter that few ions ever reach these rings by diffusion. Hence the drift tube may be considered to have an infinite radial extent (Schummers, 1972).

Now (Gray et al., 1952)

$$I_0(x) = \frac{1}{\pi} \int_0^\pi d\theta e^{\pm x \cos \theta} = \sum_{m=0}^{\infty} \frac{(x/2)^{2m}}{(m!)^2}. \tag{2-6-10}$$

Hence

$$n(r, z, t) = \frac{2\pi s e^{-\alpha t} \exp\left[-(z - v_d t)^2/4D_L t\right]}{(4\pi t)^{3/2} D_T D_L^{1/2}}$$

$$\times \int_0^{r_0} r' \, dr' \exp\left(-\frac{r^2 + r'^2}{4D_T t}\right) \sum_{m=0}^{\infty} \frac{1}{(m!)^2} \left(\frac{rr'}{4D_T t}\right)^{2m}. \tag{2-6-11}$$

If we let $x = r'^2/4D_T t$, then

$$n(r, z, t) = \frac{s}{(4\pi D_L t)^{1/2}} \exp\left[-\alpha t - \frac{(z - v_d t)^2}{4D_L t} - \frac{r^2}{4D_T t}\right]$$

$$\times \sum_{m=0}^{\infty} \frac{(r^2/4D_T t)^m}{(m!)^2} \int_0^{(r_0^2/4D_T t)} x^m e^{-x} \, dx. \tag{2-6-12}$$

The remaining integral can be done by integrating by parts m times. The result is

$$\int_0^a x^m e^{-x} \, dx = m!\left(1 - e^{-a} \sum_{i=0}^{m} \frac{a^i}{i!}\right). \tag{2-6-13}$$

Substituting in (2-6-12), we obtain

$$n(r, z, t) = \frac{s}{(4\pi D_L t)^{1/2}} \exp\left[-\alpha t - \frac{(z - v_d t)^2}{4D_L t} - \frac{r^2}{4D_T t}\right]$$

$$\times \sum_{m=0}^{\infty} \frac{(r^2/4D_T t)^m}{m!} \left[1 - e^{-r_0^2/4D_T t} \sum_{i=0}^{m} \frac{(r_0^2/4D_T t)^i}{i!}\right] \tag{2-6-14}$$

or

$$n(r, z, t) = \frac{s}{(4\pi D_L t)^{1/2}} \exp\left[-\alpha t - \frac{(z - v_d t)^2}{4D_L t}\right]$$

$$\times \left[1 - \sum_{m=0}^{\infty} \sum_{i=0}^{m} \frac{1}{m! \, i!} \left(\frac{r^2}{4D_T t}\right)^m \left(\frac{r_0^2}{4D_T t}\right)^i \exp\left(-\frac{r_0^2 + r^2}{4D_T t}\right)\right]. \tag{2-6-15}$$

Equation 2-6-15 is an expression for the ion number density at any given time at any point in space for an ion swarm which (a) was instantaneously created with uniform density across an axially thin disk, (b) drifts in unbounded space under the influence of a constant electric field, and (c) possibly undergoes a depleting reaction with the neutral gas molecules.

Since the ions detected in this apparatus are those that exit the drift tube on the axis, the result of interest is the *axial ionic number density*

$$n(0, z, t) = \frac{se^{-\alpha t}}{(4\pi D_L t)^{1/2}} \left[1 - \exp\left(-\frac{r_0^2}{4D_T t} \right) \right] \exp\left[-\frac{(z - v_d t)^2}{4D_L t} \right], \qquad (2\text{-}6\text{-}16)$$

where s is the initial ion surface density of the delta-function input of ions and r_0 is the radius of the ion entrance aperture.

Two limiting cases are also of interest. For a "point source" in which $r_0 \to 0$ but the total number of ions $b = s(\pi r_0^2)$ in the disk remains finite application of L'Hospital's rule to the transverse factor in (2-6-15) yields

$$\lim_{r_0 \to 0} \frac{1 - \sum_{m=0}^{\infty} \sum_{i=0}^{m} (1/m!i!) \exp\left[-(r^2 + r_0^2)/4D_T t \right](r^2/4D_T t)^m (r_0^2/4D_T t)^i}{r_0^2}$$

$$= \frac{e^{-r^2/4D_T t}}{4D_T t}. \qquad (2\text{-}6\text{-}17)$$

Then the *point source* solution is

$$n(r, z, t) = \frac{b}{(4\pi t)^{3/2} D_T D_L^{1/2}} \exp\left[-\alpha t - \frac{r^2}{4D_T t} - \frac{(z - v_d t)^2}{4D_L t} \right]. \qquad (2\text{-}6\text{-}18)$$

For an *infinite plane source* $r_0 \to \infty$, and now it is the source density s that remains finite. Then from (2-6-5)

$$n(r, z, t) = \frac{s}{(4\pi D_L t)^{1/2}} \exp\left[-\alpha t - \frac{(z - v_d t)^2}{4D_L t} \right]. \qquad (2\text{-}6\text{-}19)$$

Note, as expected, that (2-6-19) is independent of r and the transverse diffusion coefficient does not appear.

The quantity measured experimentally is the flux Φ of ions leaving the drift tube through the exit aperture of area a at a fixed distance z from the source plane:

$$\Phi(0, z, t) = aJ(0, z, t), \qquad (2\text{-}6\text{-}20)$$

where $J(0, z, t)$ is the z component of the ionic flux density in the drift tube, on the axis, at the end of the drift distance z.

The ionic flux density is related to the ionic number density by (2-6-1), and

$$J(0, z, t) = -D_L \left(\frac{\partial n}{\partial z} \right) + v_d n, \qquad (2\text{-}6\text{-}21)$$

where $n(0, z, t)$ is given by (2-6-16). Differentiation and substitution into (2-6-20) gives

$$\Phi(0, z, t) = \left(\frac{a}{2} \right)\left(v_d + \frac{z}{t} \right)n(0, z, t) \qquad (2\text{-}6\text{-}22)$$

or, in full,

$$\Phi(0, z, t) = \frac{sae^{-\alpha t}}{4(\pi D_L t)^{1/2}} \left(v_d + \frac{z}{t} \right)$$

$$\times \left[1 - \exp\left(-\frac{r_0^2}{4D_T t} \right) \right] \exp\left[-\frac{(z - v_d t)^2}{4D_L t} \right]. \qquad (2\text{-}6\text{-}23)$$

In summary, (2-6-23) gives the flux of ions of the single species considered passing through the exit aperture of the drift tube as a function of the time t and drift distance z. All the ions of the single species under consideration are assumed to be introduced from the ion source in periodic delta-function bursts and none is produced by reactions in the drift space. Loss of the ions in reactions producing other species is allowed, however, the rate of loss being described by the frequency α.

Equation 2-6-23 was used to calculate the expected shapes of the arrival-time spectra used for comparison with experimental data in Figs. 2-2-2 and 2-2-3. The foregoing analysis, and the other analyses of the ionic transport equations utilized by the Georgia Tech group, are the work of I. R. Gatland.

C. OTHER SOLUTIONS OF TRANSPORT EQUATIONS FOR IONS IN DRIFT TUBES. In addition to the analysis given in Section 2-6-B, analyses that apply to other apparatus or to different reaction patterns have been published. They are listed below with brief comments concerning the scope of the analysis:

Burch and Geballe (1957). This is a one-dimensional analysis that applies to transient ion currents in a Townsend discharge. The effects of diffusion are not included, but the forward reaction of primary ions to form two kinds of secondary ion is considered.

Edelson and McAfee (1964). This paper discusses the determination of Townsend ionization and electron attachment coefficients and ionic mobilities in pulsed discharge experiments. The analysis is one-dimensional; the effects of diffusion and reactions are considered.

Edelson et al. (1964). In this one-dimensional analysis of a pulsed Townsend discharge ionization, electron attachment, and secondary-electron production at the cathode are considered. The transport of positive and negative ions is treated with and without diffusion. The results are useful in the determination of mobilities.

Frommhold (1964). In this paper electron detachment from negative ions is discussed as a process occurring in avalanches in various gases. Frommhold also considered two interconverting, drifting species in the absence of diffusion and obtained results later derived independently by Edelson, et al. (1967).

Beaty and Patterson (1965). The basic analysis is for a cylindrical geometry in which the ion source and collector form the ends of the drift chamber. Drift, diffusion (with $D_L = D_T$), and a depleting reaction to form a secondary species are considered. Both the primary and secondary ions are assumed to have the same temperature. The analysis was applied to experiments on helium to determine the mobilities of the primary and secondary ions and the forward reaction rate.

Barnes (1967). This analysis, basically one-dimensional, deals with drift, diffusion, and depleting reactions. Statistical estimates are developed for the mobility, diffusion coefficient, and reaction rate.

Edelson et al. (1967). This paper contains a one-dimensional analytic solution for a system composed of two kinds of drifting, interconverting ions. Ions of one species are introduced initially by a delta-function source. Diffusion is not included in the model. Numerical solutions, also discussed, contain the effects of diffusion and can be used to determine mobilities and reaction rates.

Edelson (1968). In this analysis the effects of radial diffusion on the collected ion current due to different residence times are determined. The effect of a reduced collector area is also discussed.

Whealton and Woo (1968). This three-dimensional analysis pertains to the determination of reaction rates. The effects produced by the primary and secondary ions with unequal temperatures are discussed.

Woo and Whealton (1969). This analysis, also three-dimensional, considers drift, transverse and longitudinal diffusion, and a forward reaction to a secondary species. The source term corresponds to a disk with arbitrary radial ion density. Expressions are obtained for both the primary and secondary ion densities.

Keller et al. (1970). This paper suggests mathematical modifications to the solution of the Woo-Whealton model which facilitate calculations.

Snuggs (1970). In this thesis a comprehensive analysis is made of an interreacting system of two ion swarms. The geometry is three-dimensional, and drift, transverse and longitudinal diffusion, and reactions are considered. The reaction from each ionic species to the other is taken into account, as well as other depleting reactions. Comparison of experimental data with analytic arrival time spectra permits the determination of mobilities, diffusion coefficients, and reaction rates. Many simplifying assumptions are also considered, and appropriate formulas to aid in data reduction are obtained.

Snuggs et al. (1971b). An analysis is presented for a cylindrical drift tube in which drift, transverse and longitudinal diffusion, and depleting reactions are occurring. The ion source is an axially thin disk and the ion current is measured on the axis of the drift tube. For the primary ion species a closed-form solution for the time-integrated ion current is obtained and used to determine transverse diffusion coefficients and reaction rates by an attenuation method (see Section 3-1). An analytic solution is presented for the arrival-time spectrum of the secondary ions.

Woo and Whealton (1971). This paper deals mainly with the error introduced by neglecting transverse diffusion when calculating reaction rates from arrival-time spectra. Three-dimensional geometry was used and various ion source configurations were considered. Longitudinal diffusion was neglected and only a forward reaction was included.

Schummers (1972). This thesis describes an analysis of the drift, longitudinal and transverse diffusion, and interreaction of two ion species in cylindrical geometry. The ion source is an axially thin disk and the ion current is evaluated on the axis. This analysis can be used for both ionic species to determine mobilities, diffusion coefficients, and forward-backward reaction rates from arrival-time spectra.

Gatland (1972). This paper describes a Green's function solution of the ion transport equations. This type of solution is particularly suitable when the drift-tube boundaries have only a slight effect on the observed ions. Solutions are worked out for a variety of reaction patterns. Ionic drift, longitudinal and transverse diffusion, and reactions are treated exactly. Boundary effects can be included as perturbations. Some of the analysis summarized here is reproduced in the thesis by Schummers (1972).

In closing this section it is important to point out that even when no analytic solution is available we may always resort to numerical methods to solve the relevant transport equations.

REFERENCES

Albritton, D. L. (1967), Ph.D. thesis, Georgia Institute of Technology, Atlanta.
———, T. M. Miller, D. W. Martin, and E. W. McDaniel (1968), *Phys. Rev.* **171**, 94.
Barnes, W. S. (1967), *Phys. Fluids* **10**, 1941.
———, D. W. Martin, and E. W. McDaniel (1961), *Phys. Rev. Letters* **6**, 110.
Beaty, E. C. (1962), *Proc. 5th Intern. Conf. Ionization Phenomena Gases* (Munich, 1961), Vol. 1, p. 183, North-Holland, Amsterdam.
———, and P. L. Patterson (1965), *Phys. Rev.* **137**, 346.
——— (1968), *Phys. Rev.* **170**, 116.
Beyer, R. A., and G. E. Keller (1971), *Trans. Am. Geophys. Union* **52**, 303.
Bradbury, N. E., and R. A. Nielsen (1936), *Phys. Rev.* **49**, 388.
Brown, H. L. (1972), Ph.D. thesis, University of Pittsburgh, Pittsburgh.
Burch, D. S., and R. Geballe (1957), *Phys. Rev.* **106**, 188.
Creaser, R. P. (1969), Ph.D. thesis, Australian National University, Canberra.
Crompton, R. W., M. T. Elford, and J. Gascoigne (1965), *Australian J. Phys.* **18**, 409.

Edelson, D. (1968), *J. Appl. Phys.* **39**, 3497.

———, and K. B. McAfee (1964), *Rev. Sci. Instr.* **35**, 187.

Edelson, D., J. A. Morrison, and K. B. McAfee (1964), *J. Appl. Phys.* **35**, 1682.

Edelson, D., J. A. Morrison, L. G. McKnight, and D. P. Sipler (1967), *Phys. Rev.* **164**, 71.

Fahr, H., and K. G. Müller (1967), *Z. Physik* **200**, 343.

Franklin, J. L., Ed. (1972), *Ion-Molecule Reactions*, Plenum, New York.

Frommhold, L. (1964), *Fortschr. Physik* **12**, 597.

Gatland, I. R. (1972), U. S. Air Force Office of Scientific Research Technical Report, Georgia Institute of Technology, Atlanta. An expanded version will be published in the journal *Case Studies in Atomic Physics* (1974).

———, and E. W. McDaniel (1970), *Phys. Rev. Letters* **25**, 1603.

Gray, A., G. B. Mathews, and T. M. MacRoberts (1952), *Bessel Functions* (2nd ed.), Macmillan, London, pp. 20, 46

Hagena, O. F., and W. Obert (1972), *J. Chem. Phys.* **56**, 1793.

Hasted, J. B. (1972), *Physics of Atomic Collisions* (2nd ed.), Americal Elsevier, New York.

Heimerl, J. M. (1968), Ph.D. thesis, University of Pittsburgh, Pittsburgh.

———, R. Johnsen, and M. A. Biondi (1969), *J. Chem. Phys.* **51**, 5041.

Hornbeck, J. A. (1951a), *Phys. Rev.* **83**, 374.

——— (1951b), *Phys. Rev.* **84**, 615.

Huxley, L. G. H., and R. W. Crompton (1973), *Electron Swarms in Gases*, Wiley, New York.

Johnsen, R., H. L. Brown, and M. A. Biondi (1970), *J. Chem. Phys.* **52**, 5080.

Keller, G. E., M. R. Sullivan, and M. D. Kregel (1970), *Phys. Rev. A* **1**, 1556.

Loeb, L. B. (1960), *Basic Processes of Gaseous Electronics* (2nd ed.), University of California Press, Berkeley, Chapter I.

McAfee, K. B., and D. Edelson (1962), *Bull. Am. Phys. Soc.* **7**, 135.

———, (1963a) *Proc. 6th Intern. Conf. Ionization Phenomena Gases* (Paris), Vol. **1**, p. 299, SERMA, Paris.

———, (1963b), *Proc. Phys. Soc. (London)* **81**, 382.

———, D. Sipler, and D. Edelson (1967), *Phys. Rev.* **160**, 130.

McDaniel, E. W. (1964) *Collision Phenomena in Ionized Gases*, Wiley, New York, pp. 683–684.

———, (1970) *J. Chem. Phys.* **52**, 3931.

———, (1972) *Australian J. Phys.* **25**, 465.

———, V. Čermák, A. Dalgarno, E. E. Ferguson, and L. Friedman (1970), *Ion-Molecule Reactions*, Wiley, New York.

McDaniel, E. W., and M. R. C. McDowell (1959), *Phys. Rev.* **114**, 1028.

McDaniel, E. W., D. W. Martin, and W. S. Barnes, (1962) *Rev. Sci. Instr.* **33**, 2.

McDaniel, E. W., and J. T. Moseley (1971), *Phys. Rev. A* **3**, 1040.

McKnight, L. G. (1970) *Phys. Rev. A* **2**, 762. See also L. G. McKnight, and J. M. Sawina, (1971), *Phys. Rev. A* **4**, 1043.

———, K. B. McAfee, and D. P. Sipler (1967), *Phys. Rev.* **164**, 62.

Martin, D. W., W. S. Barnes, G. E. Keller, D. S. Harmer, and E. W. McDaniel (1963), *Proc. 6th Intern. Conf. Ionization Phenomena Gases* (Paris), Vol. **1**, p. 295, SERMA, Paris.

Massey, H. S. W. (1971), *Electronic and Ionic Impact Phenomena* (2nd ed.), Vol. III, Clarendon, Oxford, Chapter 19.

———, E. H. S. Burhop, and H. B. Gilbody (1973), *ibid.*, Vol. IV.

Miller, T. M., J. T. Moseley, D. W. Martin, and E. W. McDaniel (1968), *Phys. Rev.* **173**, 115.

Milne, T. A., and F. T. Greene (1967), *J. Chem. Phys.* **47**, 4095.

Moseley, J. T. (1968), Ph.D. thesis, Georgia Institute of Technology, Atlanta.

———, I. R. Gatland, D. W. Martin, and E. W. McDaniel (1969a), *Phys. Rev.* **178**, 234.

Moseley, J. T., R. M. Snuggs, D. W. Martin, and E. W. McDaniel (1968), *Phys. Rev. Letters* **21**, 873.

———, (1969b), *Phys. Rev.* **178**, 240.

Neynaber, R. H. (1969), "Experiments with Merging Beams," in *Advances in Atomic and Molecular Physics*, Vol. 5 (D. R. Bates and I. Estermann Eds.), Academic, New York.

Parkes, D. A. (1971), *Trans. Faraday Soc.* **67**, 711.

Patterson, P. L. (1970), *J. Chem. Phys.* **53**, 696.

Puckett, L. J., M. W. Teague, and D. G. McCoy (1971), *Rev. Sci. Instr.* **42**, 580.

Schummers, J. H. (1972), Ph.D. thesis, Georgia Institute of Technology, Atlanta.

———, G. M. Thomson, D. R. James, I. R. Gatland, and E. W. McDaniel (1973a), *Phys. Rev. A* **7**, 683.

Schummers, J. H., G. M. Thomson, D. R. James, E. Graham, I. R. Gatland, and E. W. McDaniel, (1973b), *Phys. Rev. A* **7**, 689.

Snuggs, R. M. (1970), Ph. D. thesis, Georgia Institute of Technology, Atlanta.

Snuggs, R. M., D. J. Volz, J. H. Schummers, D. W. Martin, and E. W. McDaniel (1971a), *Phys. Rev. A* **3**, 477.

———, D. J. Volz, I. R. Gatland, J. H. Schummers, D. W. Martin, and E. W. McDaniel (1971b), *Phys. Rev. A* **3**, 487.

Thomson, G. M., J. H. Schummers, D. R. James, E. Graham, I. R. Gatland, M. R. Flannery, and E. W. McDaniel (1973), *J. Chem. Phys. To be published.*

Tyndall, A. M. (1938), *The Mobility of Positive Ions in Gases*, Cambridge University Press, Cambridge.

Varney, R. N. (1952), *Phys. Rev.* **88**, 362.

Volz, D. J., J. H. Schummers, R. D. Laser, D. W. Martin, and E. W. McDaniel (1971), *Phys. Rev. A* **4**, 1106.

Wannier, G. H. (1953), *Bell System Tech. J.* **32**, 170.

Whealton, J. H. and E. A. Mason (1973), *Annals of Physics, To be published.*

Whealton, J. H., and S. B. Woo (1968), *Phys. Rev. Letters* **20**, 1137.

Woo, S. B., and J. H. Whealton (1969), *Phys. Rev.* **180**, 314. Errata published (1970) *Phys. Rev. A* **1**, 1558.

———, (1971), *Phys. Rev. A* **4**, 1046.

Young, C. E., D. Edelson, and W. E. Falconer (1970), *J. Chem. Phys.* **53**, 4295.

3

THE MEASUREMENT OF

TRANSVERSE DIFFUSION COEFFICIENTS

Measurements of D_T are performed in drift tubes but by techniques different from those used to determine v_d and D_L and discussed in Chapter 2. Two experimental approaches have been utilized in measurements of D_T to date—the attenuation method and the Townsend method. Both are described in this chapter. The remarks made in Section 2-1 concerning desirable features of drift-tube construction apply with equal force here but are not repeated. Likewise it is hardly necessary to repeat the earlier arguments (Sections 2-2-B and 2-3) concerning the requirement of low current densities and negligible space charge if accurate data are to be obtained.

3-1. THE ATTENUATION METHOD

This method was introduced by the Georgia Tech group (Miller et al., 1968; Moseley et al., 1969a, 1969b) and is described in terms of their drift-tube mass spectrometer (see Section 2-5-A). It is convenient to divide the discussion into two parts: the first applies to cases in which the ions being investigated do not react chemically with the gas filling the drift tube; the second to cases in which reactions occur.

A. NO REACTIONS. Let us suppose that we are dealing with an ionic species produced only in the ion source in bursts that are negligibly short compared with the drift time. Then the flux $\Phi(0, z, t)$ of the ions passing through the exit aperture of the drift tube is given as a function of the time t and the drift distance z by (2-6-23). Integrating this expression over time gives

$$I = \int_0^\infty \Phi(0, z, t) \, dt, \qquad (3\text{-}1\text{-}1)$$

the total number of ions flowing through the exit aperture following each pulse of the ion source. Analytically, I may be expressed as

$$I(z) = \frac{sa \exp\left(zv_d/2D_L\right)}{4D_L^{1/2}} \left\{ \left[2D_L^{1/2} + \frac{v_d}{(\alpha + v_d^2/4D_L)^{1/2}} \right] \exp\left[\frac{-z}{D_L^{1/2}} \left(\frac{v_d^2}{4D_L} + \alpha \right)^{1/2} \right] \right.$$

$$- \left[\frac{z}{(z^2/4D_L + r_0^2/4D_T)^{1/2}} + \frac{v_d}{(\alpha + v_d^2/4D_L)^{1/2}} \right]$$

$$\left. \exp\left[-2\left(\frac{z^2}{4D_L} + \frac{r_0^2}{4D_T} \right)^{1/2} \left(\frac{v_d^2}{4D_L} + \alpha \right)^{1/2} \right] \right\} \qquad (3\text{-}1\text{-}2)$$

(Snuggs, 1970). In the case under discussion here the ions are assumed to be incapable of reacting with the drift-tube gas; therefore the reaction frequency α equals zero. The drift velocity v_d and the longitudinal diffusion coefficient D_L are assumed to be known from the measurements described in Chapter 2.* The area of the drift-tube exit aperture a and the radius of the ion entrance aperture r_0 are also known quantities. The planar source density s depends on the operating conditions of the ion source. It now becomes apparent that if all the operating parameters of the apparatus are maintained constant except the ion source position I is a function of the drift distance z alone and (3-1-2) can be used to determine the value of the transverse diffusion coefficient. The technique is to measure the ionic current reaching the detector for various source positions, assume a value of D_T in (3-1-2), and adjust the assumed value of D_T until the best agreement between the analytical and experimental variations of I with z is achieved. During the measurements the ion source may be operated in either the pulsed or the dc mode, but in either one the source output must be maintained constant at all source positions.

---

* Actually the value of D_T which is obtained by the analysis described here is insensitive to the value of D_L which is used. Likewise, as stated in Section 2-3, the transverse diffusion only very weakly affects the determination of D_L, and the two diffusion coefficients may be obtained essentially independently of one another. In experimental determinations of D_T, if D_L has not been measured as a function of E/N, the low-field value D obtained from the mobility by the Einstein equation may be used for D_L without significant error. A better procedure would be to use Wannier's modified equation (5-2-31) or (7-4-3) to calculate D_L as a function of E/N (see Sections 5-2-A and 7-4), provided that the ions under consideration do not undergo resonant charge transfer with the gas molecules.

Figure 3-1-1 shows typical data used in the determination of D_T for K^+ ions in nitrogen under conditions in which reactions are negligible. The symbols indicate the experimental decrease in count rate as the drift distance is increased at five different drift-tube pressures. (As the drift distance is increased, the enhanced effect of transverse diffusion causes a smaller fraction of the ions to leave the drift tube through the exit aperture and reach the detector.) The smooth curves in each case are plots of I versus z obtained from (3-1-2), with the value of D_T which gives the best fit with the

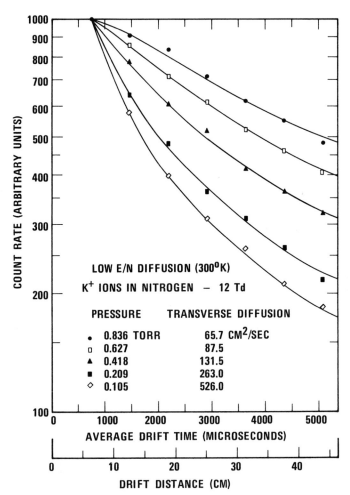

FIG. 3-1-1. The experimental decrease in count rate as the drift distance is increased, compared with the predictions of (3-1-2) for K^+ ions in N_2 (Moseley et al., 1969a).

experiment being used at each pressure. Since the absolute intensity of the ion swarm entering the drift tube and the ion sampling efficiency at the end of the drift tube are not known, the curves are normalized to agree with the experimental points obtained at the shortest drift distance. It will be noticed that the measured values of D_T vary as $1/N$, where N is the drift-tube gas number density, as should be the case. The experimental results obtained for D_T at low E/N are in agreement with the value predicted by the Einstein equation from the measured value of K (see Fig. 7-4-B-2).

Even for the simple situation described here the accuracy achieved (typically 10 to 20%) is not so high as would be hoped for, due in large measure to the difficulty in meeting the requirement that the ion source output be held constant as the source is moved along the axis of the drift tube.

B. DEPLETING REACTIONS.   Let us continue to restrict our attention to a species of ions created only in the ion source but now allow for the possibility of the destruction of these ions by chemical reactions with the gas molecules. Then (3-1-2) applies if some appropriate nonzero value is given to α, the frequency of the reactions depleting the primary ion population. At a given gas temperature α is a function of E/N and the gas pressure. The evaluation of D_T is considerably complicated by the presence of reactions because now both the reactions and the transverse diffusion contribute to the decrease in counting rate for the primary ion as the drift distance is increased. The problem is to separate the two effects, and we find that in the process of doing so we can evaluate the reaction frequency α (and rate coefficient k) as well as the transverse diffusion coefficient. Let us assume that the reaction of the ions with the gas molecules proceeds through a single channel, as is frequently the case, and that α refers to a single type of reaction. We take advantage of the fact that D_T varies as $1/N$, whereas α varies as kN in a two-body reaction or as kN^2 in a three-body reaction; this enables us to emphasize the effect of diffusion by operating at low pressures and the effect of reactions by working at high pressures. For small N the effect of D_T on $I(z)$ is much greater than the effect of α; for large N the situation is reversed.

We begin by measuring $I(z)$ at low E/N, where D_T can be calculated from the measured value of the mobility by the Einstein equation (1-1-5). Then α, the only unknown, may be evaluated directly at low E/N (thermal energy) by assuming a value for this quantity and varying it in (3-1-2) until the best fit is obtained with the experimental attenuation data. After this is done E/N is increased slightly and a first approximation for D_T at this higher E/N is made by assuming that it has its low E/N value. This first approximation for D_T is used with high-pressure experimental data to obtain a

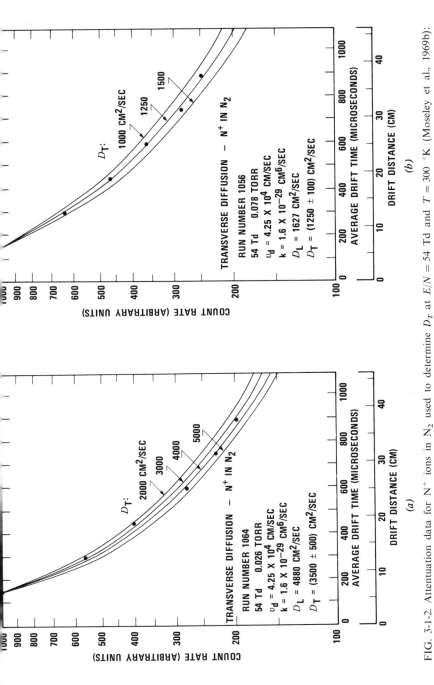

FIG. 3-1-2. Attenuation data for N^+ ions in N_2 used to determine D_T at $E/N = 54$ Td and $T = 300$ °K (Moseley et al., 1969b): (a) Pressure of 0.026 Torr; (b) pressure of 0.078 Torr.

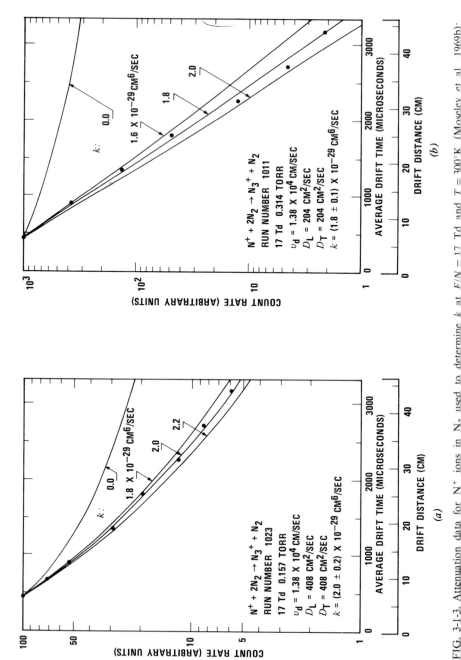

FIG. 3-1-3. Attenuation data for N^+ ions in N_2 used to determine k at $F/N = 17$ Td and $T = 300°K$ (Moseley et al. 1969b).

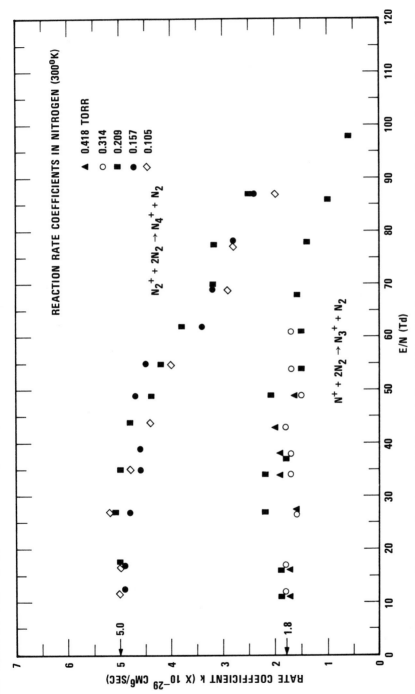

FIG. 3-1-4. The rate coefficients as functions of E/N for the indicated N^+ and N_2^+ reactions in N_2 at 300 K (Moseley et al., 1969b).

first approximation for α at the higher value of E/N. The first approximation for α is then used with low-pressure experimental data to obtain a second approximation for D_T, which in turn can be used to determine a second approximation for α from the high-pressure data, and so on until stationary values of D_T and α are obtained for this higher value of E/N. E/N is increased to successively higher values, and the procedure described above is used to evaluate D_T and α as functions of E/N up to the highest value of this parameter that can be reached in the experiment. From the pressure dependence of the measured values of α the order of the reaction can be determined and the reaction rate coefficient evaluated.

Figure 3-1-2 shows typical attenuation data for N^+ ions in N_2 which were used to determine D_T at $300°K$ and $E/N = 54$ Td. The dots represent experimental points; the curves are plots of (3-1-2) for various assumed values of D_T. The final values of D_T derived from these data are (3500 ± 500) cm^2/sec at 0.026 Torr and (1250 ± 100) cm^2/sec at 0.078 Torr.

N^+ ions react with nitrogen molecules according to the reaction $N^+ + 2N_2 \rightarrow N_3^+ + N_2$. Attenuation data used to determine the rate coefficient for this reaction at $E/N = 17$ Td and $T = 300°K$ are shown in Fig. 3-1-3, in which the experimental points are indicated by dots and plots of (3-1-2) by curves. The rate coefficients for the N^+ reaction and for the $N_2^+ + 2N_2 \rightarrow N_4^+ + N_2$ reaction as well are shown in Fig. 3-1-4. The attenuation method was utilized to obtain these data. The accuracy of the low-field determinations of k was estimated to be about 11%.

3-2. THE TOWNSEND METHOD

This method represents the application to ions of techniques developed many years ago by Townsend for measurements of the transport properties of slow electrons in gases (McDaniel, 1964; Huxley and Crompton, 1973). Judging by the sparsity of the published data on ions obtained by the Townsend method and the generally poor agreement among them, we conclude that reliable ionic measurements based on the Townsend technique are considerably more difficult to make than the corresponding electron measurements, which have produced a large volume of accurate results on a variety of gases. This situation is due mainly to the greater difficulty of producing satisfactory inputs of ions to the diffusion cell and to the continually vexing problems associated with chemical reactions between the ions and molecules of the gas through which they are moving. The Townsend method does not provide values of D_T directly but rather the ratio of the transverse diffusion coefficient to the mobility D_T/K; D_T can be obtained, however, at a given E/N if the mobility is separately determined at the same E/N for the same ionic species. The earliest published data on

D_T/K for ions obtained by the Townsend method are those of Llewellyn-Jones (1935) for unselected argon ions in argon; subsequently, little work of this type was done until the 1960's. In this section we describe three variants of the Townsend method used recently for measurements of D_T/K for ions.

A. UNSELECTED IONS: MOVABLE ION COLLECTOR. The apparatus of Fleming et al. (1969a. 1969b) at the University of Liverpool is described here to illustrate this method. It consists of a cylindrical chamber containing a dc ion source on the axis, a drift region in which the ions are allowed to come into energy equilibrium at the chosen value of E/N, an entrance aperture plate through which the ions then enter the diffusion region, and a collector assembly mounted perpendicular to the axis at the far end of the diffusion region. It is assumed that only a single, nonreacting species of ion is created by the ion source. A uniform axial electric field is provided in both the drift and diffusion regions by sets of thick cylindrical electrodes of the type shown in Fig. 2-5-2. The ion collector assembly (Fig. 3-2-1) consists of a pair of semicircular electrodes surrounded by a grounded electrode and mounted on a sheet of flat plate glass. The semicircular electrodes are separated from one another by a 0.0025-cm gap and from the grounded electrode by a 0.01-cm gap. The entire collector assembly is movable as a unit in a direction perpendicular to the axis of the electrode system, its position being adjusted by an external screw drive. The entrance aperture which allows the ions to pass from the drift region into the diffusion region is a slit 1 cm long and 0.01 cm wide, aligned parallel with the

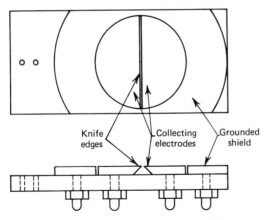

FIG. 3-2-1. The ion collector assembly in the transverse diffusion apparatus of Fleming, Tunnicliffe, and Rees (Fleming et al., 1969a, 1969b).

gap between the semicircular collecting electrodes. The distance between the entrance aperture plate and the collector assembly is 10 cm.

The ion current to each of the two collector electrodes is measured as a function of the off-axis displacement of the collector assembly with a pair of electrometer amplifiers. A curve of the ratio of the two currents versus the displacement is then plotted, and values of D_T/K are obtained from the analysis of this curve. Skullerud (1966a, 1966b) has also constructed an apparatus similar to that of Fleming and his associates and applied it to measurements of D_T/K.

Evidently the success of this method hinges on the assumption that only a single known ionic species is present in the apparatus, and the technique cannot be employed with confidence when the ion source may produce more than one kind of ion or when reactions may occur between the ions and the gas molecules. By comparison the attenuation method described in Section 3-1 is not subject to this limitation.

B. DRIFT-SELECTED IONS: STATIONARY SOURCE AND COLLECTOR. This method was developed in an attempt to overcome the problems associated with the production of a multiplicity of ionic species in the ion source and can be described in terms of the apparatus at the University College of Swansea (Dutton et al., 1966; Dutton and Howells, 1968). The apparatus (Fig. 3-2-2) consists of two sections. The first is a four-gauze electrical shutter drift tube similar to that described in Section 2-5-F and is shown schematically at the left of the plate J. Ions are created in the glow discharge ion source S, and their drift velocities are determined by measuring the ion currents through the system as functions of the frequency of the pulses applied to the shutters B-C and F-G.* The drift section is also used to select the desired species of ion for admission through a slit in J to the second section, which consists of a Townsend-type transverse diffusion cell used for the determination of D_T/K for the selected species. The collector assembly at the end of this cell consists of two D-shaped electrodes, P_1 and P_2, and a circular electrode, P_3. The ratio D_T/K at a given value of E/N is determined from the measured value of the ratio R of ion current received at P_3 to the total ion current collected by P_1, P_2, and P_3. Obviously reliable measurements for the selected species can be made only if its mobility is significantly different from those of the other species and if the selected ions do not react with the gas during their passage through the diffusion cell. Even then the utility of the value of D_T/K obtained depends on the unambiguous identification of the selected species through its measured mobility.

---

* This method of measuring the drift velocity is similar to that described in Section 2-5-B.

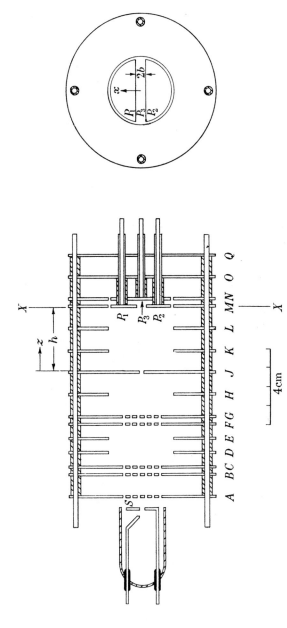

FIG. 3-2-2. The Swansea apparatus used for determinations of the drift velocity and transverse diffusion coefficient (Dutton et al., 1966).

In the work of Dutton and his associates diffusion in the field direction was ignored in the analysis of the data, but consideration of a more accurate expression for R containing an explicit dependence on D_L indicated that this procedure would probably lead to an error of the order of only 1%. A more serious difficulty came to light when the measured values of D_T/K at low E/N were found to be in significant disagreement with the predictions of the Einstein equation (1-1-5). This discrepancy was attributed to the distortion of the electric field in the vicinity of the collecting electrodes due to the finite thickness (0.06 cm) of P_1 and P_2. A rather large empirical correction was applied to the data at all values of E/N to correct for this effect.

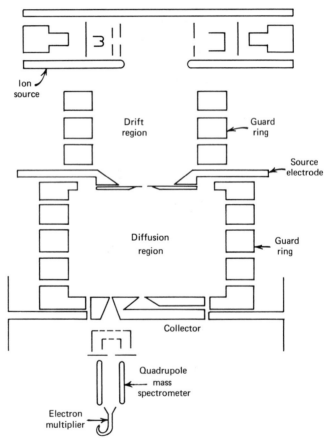

FIG. 3-2-3. The Townsend-type apparatus of Gray and Rees (1972).

C. THE APPARATUS OF GRAY AND REES. Features of both instruments described in Sections 3-2-A and 3-2-B have been combined by Gray and Rees (1972) at the University of Liverpool into the apparatus shown in Figs. 3-2-3 and 3-2-4. Ions are produced in an electron bombardment ion source at the top of the apparatus and pushed downward into the drift region, which has a length of 6 cm. This region can be operated as a Bradbury-Nielsen drift tube (see Section 2-5-B) when electrical shutters are inserted at the top and bottom of the drift space. Hence the apparatus can provide either unselected ions or drift-selected ions for admission through the source electrode into the diffusion region. The orifice in the source electrode is a 2.5 cm × 0·02 cm slit perpendicular to the plane of drawing in Fig. 3-2-3. The length of the diffusion region is 10 cm. The collector assembly at the bottom of this region consists of two semicircular disk electrodes shown in Fig. 3-2-4, and in one of these electrodes there is a

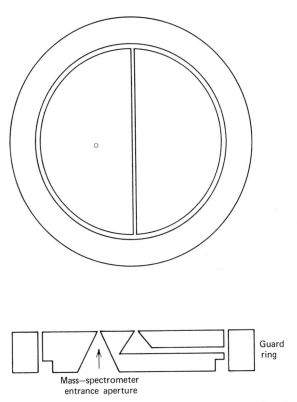

Mass—spectrometer
entrance aperture

Guard
ring

FIG. 3-2-4. The collector assembly in the apparatus of Gray and Rees (1972).

small hole leading to a differentially pumped quadrupole mass spectrometer. The source electrode is mounted on ruby balls 2 mm in diameter and can be moved perpendicular to the axis of the apparatus along a line in the plane of Fig. 3-2-3. All components above the source electrode are rigidly connected to this electrode and move with it. The ion currents to each of the collector electrodes are measured as a function of the lateral displacement of the slit in the source electrode. In addition, the mass spectrometer scans the diffusing stream of ions as it is translated by moving the source electrode. Analysis of the collector and mass spectrometer currents, when combined with drift velocity data, will yield reliable values of D_T, provided that reactions do not occur to an appreciable degree within the diffusion region. When only one species of ion is present, the apparatus of Gray and Rees should be able to provide data of high quality.

REFERENCES

Dutton, J., and P. Howells (1968), *J. Phys. B* **1**, 1160.

——, F. Llewellyn Jones, W. D. Rees, and E. M. Williams (1966), *Phil. Trans. Roy. Soc. London* **A-259**, 339.

Fleming, I. A., R. J. Tunnicliffe, and J. A. Rees (1969a), *Brit. J. Appl. Phys. (J. Phys. D)* **2**, 551.

—— (1969b), *J. Phys. B* **2**, 780.

Gray, D. R., and J. A. Rees, (1972), *J. Phys. B* **5**, 1048.

Huxley, L. G. H., and R. W. Crompton (1973), *Electron Swarms in Gases*, Wiley, New York.

Llewellyn Jones, F. (1935), *Proc. Phys. Soc.* **47**, 74.

McDaniel, E. W. (1964), *Collision Phenomena in Ionized Gases*, Wiley, New York, pp. 524–528.

Miller, T. M., J. T. Moseley, D. W. Martin, and E. W. McDaniel (1968), *Phys. Rev.* **173**, 115.

Moseley, J. T., I. R. Gatland, D. W. Martin, and E. W. McDaniel (1969a) *Phys. Rev.* **178**, 234.

——, R. M. Snuggs, D. W. Martin, and E. W. McDaniel (1969b), *Phys. Rev.* **178**, 240.

Skullerud, H. R. (1966a), Technical Report GDL 66-1, Gassutladningslaboratoriet, Institutt for Teknisk Fysikk, Norges Tekniske Högskole, Trondheim, Norway.

—— (1966b), *Proc. 7th Intern. Conf. Phenomena in Ionized Gases* (Belgrade, 1965), Vol. **1**, p. 50, Gradevinska Knjiga, Belgrade.

Snuggs, R. M. (1970), Ph.D. thesis, Georgia Institute of Technology, Atlanta.

4

AFTERGLOW TECHNIQUES

The main theme in this chapter is the determination of zero-field ionic diffusion coefficients from observations on a decaying plasma* during the "afterglow" period, although the evaluation of other quantities from such studies is briefly discussed. The afterglow is the time regime following the removal of the ionizing source that produces the plasma during which the gas remains ionized to an appreciable extent. Except for times late in the afterglow, the ionization densities are much higher than those in the drift-tube measurements described in Chapters 2 and 3 and high enough to make the attractive forces between the electrons and positive ions important. Under these conditions the diffusion is ambipolar (see Section 1-12) and the diffusion coefficient measured is the coefficient of electron-ion ambipolar diffusion D_a.† The afterglow experiments discussed here are made with no applied electric field and usually under conditions in which the ion and

---

* By a "plasma" we mean an ionized gas whose dimensions are large compared with its debye shielding length. The differences between a plasma and an ordinary ionized gas are discussed in detail by McDaniel (1964a).

† In an electronegative gas such as nitric oxide the electrons produced in the ionization process may be captured to form negative ions before they diffuse to the chamber walls. Hence late in the afterglow the ambipolar diffusion may actually involve negative and positive ions rather than electrons and positive ions. This type of ion-ion ambipolar diffusion has been observed and is discussed in Section 4-1-C.

electron temperatures equal the gas temperature. Under these conditions the ordinary, or free, diffusion coefficient for the positive ions D^+ equals one-half of D_a [see (1-12-6)], hence may be determined from the measured value of the ambipolar diffusion coefficient.

Afterglow experiments designed to provide information on atomic collisions and transport phenomena are of two basically different types— stationary and flowing. In a stationary afterglow experiment, the only kind treated in detail here, we observe the temporal variation of the charged-particle number densities following an ionizing pulse in a gas that is either at rest or flowing with a negligible speed. Measurements on stationary afterglows have provided momentum transfer cross sections for electrons, electron attachment coefficients, recombination coefficients, and rate coefficients for chemical reactions between ions and molecules as well as ambipolar diffusion coefficients (McDaniel, 1964b; Biondi, 1968; Oskam, 1969; and McDaniel et al., 1970). Flowing afterglow experiments yield mainly information on chemical reactions (Ferguson et al., 1969; McDaniel et al., 1970; Franklin, 1972) and nothing on diffusion. In such experiments we sample the reaction products in a rapidly flowing field-free gas at some location downstream from the point at which the reactant was introduced into the flow and substitute a spatial measurement for a temporal measurement. This kind of measurement has been extraordinarily fruitful in studies of ion-molecule and charge transfer reactions. To date, however, it has the limitation of having been applied only to the study of thermal energy processes. The future may see a combination of flowing afterglow and drift-tube techniques in which the ions are heated by a drift field during their flow and suprathermal data are obtained.*

At present stationary afterglow techniques can provide zero-field ionic diffusion coefficients with an accuracy of about 8 to 20%. Much better accuracy (1/2 to 4%) can usually be obtained in the evaluation of these coefficients if we resort to the indirect method of calculating them from experimental zero-field mobilities by use of the Einstein equation (1-1-5). Nonetheless, stationary afterglow methods are of considerable interest and warrant a rather detailed treatment in this book. We discuss three quite different techniques, all of which have produced important results.

4-1. THE TECHNIQUE OF LINEBERGER AND PUCKETT

Lineberger and Puckett have constructed a stationary afterglow apparatus at the Aberdeen Proving Ground and applied it to the study of diffusion and reactions in nitric oxide (Lineberger and Puckett, 1969a). They

---

* An apparatus to accomplish this is being constructed by D. L. Albritton, A. L. Schmeltekopf, and M. McFarland (private communication, 1973).

subsequently used the apparatus for some interesting experiments on reactions of positive and negative ions in NO-H$_2$O mixtures (Lineberger and Puckett, 1969b; Puckett and Lineberger, 1970), but we shall confine our attention to the work reported in their first paper.

A. APPARATUS. The apparatus of Lineberger and Puckett is shown schematically in Fig. 4-1-1. The afterglow cavity is a bakeable, gold-plated, stainless-steel cylinder, 18 in. in diameter and 36 in. long, which is sealed by metal gaskets. When the cavity is filled with nitric oxide, NO$^+$ ions are produced in periodic bursts by the krypton discharge lamp. Ions of this type and other species formed in secondary processes diffuse to the wall and are mass spectrometrically sampled as a function of time through the orifice shown halfway along the wall of the cavity. The cavity and the differential pumping regions shown at the left are pumped by oil diffusion pumps equipped with water-cooled baffles and sorbent traps. The cavity can be isolated from its pumping station by means of an all-metal, bakeable valve. Typical base pressures within the cavity following a 48-hr, 200°C bake are about 10^{-9} Torr, with a rate of rise less than 10^{-10} Torr/sec. Gas is admitted into the cavity through a servo-driven variable leak valve coupled to a feedback control system that regulates the cavity pressure to a value set by the capacitance manometer used to measure the pressure.

The discharge lamp, which is microwave-powered, produces the primary NO$^+$ ions by photoionization of the nitric oxide molecules with krypton resonance radiation (123.6 and 116.5 nm). The photons have insufficient energy to form any other ionic species in NO. The lamp body is quartz with a MgF$_2$ window 1 mm thick which transmits approximately 50% of the resonance radiation. The intensity of the light inside the afterglow cavity is about 10^{15} photons/sec. With this level of radiation the initial charged-particle number densities are low enough that electron-positive ion recombination makes a negligible contribution to the decay of the plasma, and this fact greatly simplifies the analysis of the data. In the midplane of the cavity, where the ions diffusing to the wall are sampled, the radial distribution of the light is approximately a truncated cosine distribution that fills about 80% of the chamber. This distribution is similar to a zeroth-order Bessel function J_0 (see Section 1-11-D), and the ions quickly stabilize into a fundamental mode spatial distribution.

The sampling orifice, which has a 0.60-mm diameter, is contained in a disk 6 cm in diameter, machined to match the contour of the cavity wall. The orifice disk is insulated from the wall so that it can be electrically biased and the transient current to the disk is measured with a computer of average transients. This feature permits tests to be made to determine the optimum potential that should be applied to the sampling disk to produce a count rate of mass-analyzed ions directly proportional to their number density in the cavity throughout the plasma decay. Lineberger and Puckett consider this matter in great detail and present data to show the pronounced effect of orifice potentials on the measured time constants of the afterglow decay. Most of their measurements are made with attractive potentials of several tens of millivolts applied to the orifice plate.

The relatively high pressures used in the afterglow cavity (up to about 300 mTorr) make advisable the use of differential pumping in the ion sampling apparatus. As Fig. 4-1-1 shows, two stages of differential pumping are provided, each affording a pressure reduction by a factor of about 10^3. These two stages are separated by an insulated conical skimmer with a hole 1 mm in diameter in its tip. The skimmer is normally operated at an attractive potential of 25 to 300 V to enhance the ion sampling efficiency and to reduce mass discrimination in the sampling. Checks are made to ensure that molecular ions are not dissociated in energetic collisions during the sampling process. The mass spectrometer is a General Electric monopole spectrometer, modified to permit operation with ions formed at

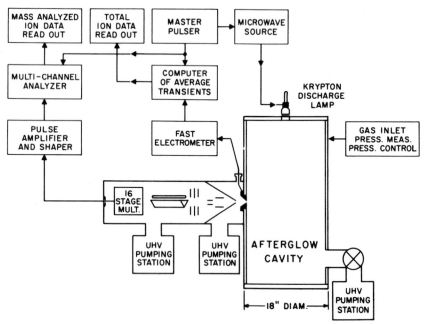

FIG. 4-1-1. Schematic diagram of the stationary afterglow apparatus of Lineberger and Puckett.

ground potential. Those ions that pass through the spectrometer are counted with a 16-stage Ag-Mg electron multiplier whose output pulses are sorted by a 1024 channel time analyzer. A master pulser simultaneously triggers the discharge lamp, the time analyzer, and the computer of average transients. Repetitive pulsing of the apparatus permits the accumulation of a sufficient number of counts in each time channel to provide a statistically significant history of the plasma decay.

B. DATA ANALYSIS.   In the analysis of their data Lineberger and Puckett assume fundamental mode diffusion and negligible recombination losses and demonstrate the validity of these assumptions except for times early in the afterglow. Here we reproduce their analysis. Let us suppose that electron-ion ambipolar diffusion of NO^+ and a reaction with NO molecules having a frequency v_l are the only significant loss processes for NO^+ ions. Then, if there are no sources of NO^+ during the afterglow, the continuity equation for the NO^+ ions is

$$\frac{\partial}{\partial t}[NO^+(r, t)] = D_a \nabla^2 [NO^+(r, t)] - v_l[NO^+(r, t)], \qquad (4\text{-}1\text{-}1)$$

where the brackets [ ] denote number densities and D_a is the NO^+ electron-ion ambipolar diffusion coefficient. For a long cylindrical cavity of radius R and an initial NO^+ distribution given by

$$[NO^+(r, 0)] = [NO^+(0, 0)]J_0\left(2.405\ \frac{r}{R}\right) \qquad (4\text{-}1\text{-}2)$$

the solution of (4-1-1) is

$$[NO^+(r, t)] = [NO^+(0, 0)]J_0\left(2.405\ \frac{r}{R}\right) \exp\left[-\left(\frac{D_a p}{\Lambda^2 p} + v_l\right)t\right], \qquad (4\text{-}1\text{-}3)$$

where Λ is the characteristic diffusion length of the cavity, p is the gas pressure, and $[NO^+(0, 0)]$ is the initial axial number density. (The reader may find it helpful to review Section 1-11-D at this point.)

In the Lineberger-Puckett apparatus the direct physical observable is the count rate at the mass spectrometer, not the ionic volume number density. If we assume that the NO^+ count rate $CR(NO^+)$ is proportional to the NO^+ wall-current density and that the wall-current density is simply a diffusion current, driven by the ion density gradient, then

$$CR(NO^+) \propto D_a[NO^+(0, 0)] \exp\left[-\left(\frac{D_a p}{\Lambda^2 p} + v_l\right)t\right]$$

$$\propto D_a[NO^+(0, 0)]e^{-v t}, \qquad (4\text{-}1\text{-}4)$$

where v is the total NO^+ loss frequency. Thus, under the assumptions made above, the NO^+ count rate is directly proportional to the NO^+ number density in the cavity. This simple proportionality does not necessarily hold, however, in the event of a time-dependent spatial distribution, surface charging effects, or the use of a drawout voltage on the orifice plate. Lineberger and Puckett discuss experimental tests made to check the validity of each of their assumptions.

It is now apparent that both D_a and v_l may be determined if v is measured as a function of the pressure. The data presented in the next section indicate that NO^+ is lost through the reaction

$$NO^+ + 2NO \longrightarrow NO^+ \cdot NO + NO, \qquad (4\text{-}1\text{-}5)$$

where v_l and the reaction rate coefficient k are related by the equation

$$v_l = k[NO]^2. \qquad (4\text{-}1\text{-}6)$$

Hence the p^{-1} (diffusion) and p^2 (reaction) contributions to v can be separated and D_a and k may be evaluated individually.

C. EXPERIMENTAL RESULTS.   At room temperature and over the pressure range of 10 to 200 mTorr the dominant positive ions present in pure NO are NO^+ and the dimer $NO^+ \cdot NO$, whereas NO_2^- dominates the negative ion spectrum by a factor of about 100. (At pressures above 200 mTorr hydrated impurity ions begin to appear in significant amounts.) Figure 4-1-2 shows the NO^+ count rate as a function of time at 20 mTorr pressure. The data were accumulated over 10^4 light pulses and extend over more

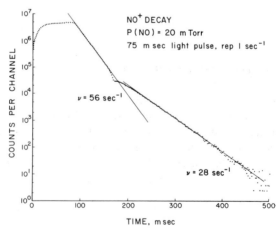

FIG. 4-1-2. Typical NO^+ decay curve at 20 mTorr pressure in the experiments of Lineberger and Puckett.

than six decades of afterglow decay. The maximum ionic number density at the center of the cavity, while not directly measured, is estimated to be about $10^7/cm^3$. The decay between about 90 and 160 msec is associated with electron-positive ion ambipolar diffusion and is exponential, as expected. The loss frequency during this period, $v = 56/sec$, is the reciprocal time constant for this period. The break in the NO^+ count rate near 170 msec is due to the sudden transition from electron-positive ion ambipolar diffusion domination (characterized by a coefficient that we label $D_{+,e}$) to domination by negative ion-positive ion ambipolar diffusion (characterized by $D_{+,-}$). These data represent the first reported observation of such a transition. The structure in the count rate in the vicinity of the transition is the result of the decay of the electron and ion densities to a level too small to sustain the space-charge field that produces the electron-positive ion ambipolar diffusion and of the release of negative ions whose diffusion (unlike that of the positive ions) is opposed by the electron-positive ion space charge field. The diffusion of negative ions influences the positive ion wall current by the establishment of a weak positive ion-negative ion space-charge field through which positive and negative ions are coupled in ambipolar diffusion.

The experimental data for the pressure variation of the loss frequency of NO^+ ions are displayed in Fig. 4-1-3 and were obtained with orifice disk potentials varied between zero and 100 mV negative with respect to the cavity wall. Over the pressure range of 5 to 200 mTorr the NO^+ loss frequency may be expressed as

$$v = \frac{a}{p} + bp^2, \tag{4-1-7}$$

which implies that the dominant NO^+ loss processes are diffusion and a three-body reaction between NO^+ ions and NO molecules. Since the only significant secondary ion observed in this pressure range was the dimer, the reaction occurring is that described by (4-1-5). The rate coefficient for this reaction was determined from b and found to be $(5 \pm 1) \times 10^{-30}$ cm^6/sec.

Since it was determined that electron-ion ambipolar diffusion was the process operative during the early afterglow between 90 and 160 msec, the observed diffusion loss frequency, together with the diffusion length of the cavity,* provides enough information for the determination of the electron-

---

* The diffusion length used in the data analysis is that for the lowest order radial and axial modes of the cavity,

$$\Lambda^2 = \left[\left(\frac{2.405}{R} \right)^2 + \left(\frac{\pi}{L} \right)^2 \right]^{-1} = 78.5 \text{ cm}^2, \tag{4-1-8}$$

where R and L are, respectively, the cavity radius and length in centimeters.

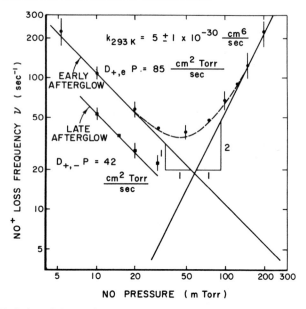

FIG. 4-1-3. Variation of the NO^+ loss frequency in pure NO as a function of NO pressure. The closed circles represent early (electron-ion) afterglow data, the closed squares, late (ion-ion) afterglow data. The dashed line is the sum of the asymptotic p^{-1} and p^2 lines.

ion ambipolar diffusion coefficient D_a. The result obtained is $D_a p = 85 \, cm^2 \, Torr/sec$. Since the electron, ion, and gas temperatures are the same in this experiment, the free diffusion coefficient for NO^+ ions D^+ is given by $D_a/2$.

Decay rates for NO^+ ions were also measured late in the afterglow (after the transition at 170 msec) to obtain information on the ion-ion ambipolar diffusion. The analysis presented above is applicable here if the ion-ion ambipolar diffusion coefficient $D_{+,-}$ is substituted for D_a. Lineberger and Puckett have shown that if the ion and gas temperatures are assumed to be equal

$$D_{+,-} = 2 \frac{D^+ D^-}{D^+ + D^-}, \tag{4-1-9}$$

where D^+ and D^- are the free diffusion coefficients for the positive and negative ions, respectively. It is apparent that $D_{+,-}$ must have a value intermediate between D^+ and D^- and that if $D^+ = D^-$ the ambipolar diffusion process is effectively one of free diffusion. Three species of diffusing ions—NO^+, $NO^+ \cdot NO$, and NO_2^-, the last two being dominant—appear late in the afterglow. All three species have approximately the same free diffusion coefficients,* in contrast with the early afterglow in which the dominant charge carriers (NO^+ and electrons) have vastly different free diffusion coefficients. Hence in the late afterglow the ion-ion ambipolar diffusion coefficient will be nearly equal to the free diffusion coefficient for each of the three ionic species. Then at low pressures, where the only significant loss process is through diffusion, the late afterglow loss frequency (which is the same for all three types of ion) should be one-half the early afterglow loss frequency for NO^+ if in the early afterglow the electron and ion temperatures equal the gas temperature. That this is indeed the case is verified in Fig. 4-1-3, which shows that in the late afterglow $D_{+, -} p = 42$ cm^2 Torr/sec. (Here the plus sign refers equally well to NO^+ and to $NO^+ \cdot NO$ and the minus sign refers to NO_2^-.) This is an important observation, for it shows that the electrons are not preferentially heated in the afterglow.

Lineberger and Puckett also sampled the $NO^+ \cdot NO$ and NO_2^- ions as a function of time following the light pulse. Early afterglow measurements of the $NO^+ \cdot NO$ decay frequency gave a $D_a p$ product of 84 cm^2-Torr/sec to describe the electron-ion ambipolar diffusion for this species; the estimated uncertainty was $\pm 15\%$. The uncertainty in the determination of the electron-ion ambipolar diffusion coefficient quoted above for NO^+ was estimated to be slightly lower—$\pm 12\%$. About one-quarter of this total uncertainty was associated with the fact that variations in the potential applied to the sampling orifice affected the results, and it was impossible to determine precisely the proper potential to use.

4-2. THE TECHNIQUE OF SMITH AND HIS COLLEAGUES

David Smith and his colleagues at the University of Birmingham have developed some useful stationary afterglow techniques and applied them to an extensive study of ambipolar diffusion, chemical reactions, and electron-ion recombination in the noble gases, oxygen, and nitrogen. Their apparatus differs from that of Lineberger and Puckett, described in Section 4-1-A, in several important respects. Their afterglow cavity is much smaller and is made of glass. The initial ionization is produced by a pulsed rf discharge, and Langmuir probes (Chen, 1965) are used, as well as mass spectrometric sampling of the ions, to obtain information on the decaying plasma. Furthermore, the apparatus may be operated over a wide range of temperature.

---

* The NO^+ ion has a free diffusion coefficient as low as that of the other much heavier ions because of the retarding effect of resonance charge transfer it experiences.

A. APPARATUS. A typical experimental tube used by Smith and his associates is shown in Figs. 4-2-1 and 4-2-2. It is a pyrex cylinder which has an internal diameter of 11.0 cm and an active length of 18.0 cm. Energy to ionize the gas in the tube is supplied as repetitive 10 μsec pulses of 10 MHz power coupled capacitively to the gas by two external sleeve electrodes, the power in each pulse normally being about 20 kVa. This amount of power produces an initial electron density of the order of

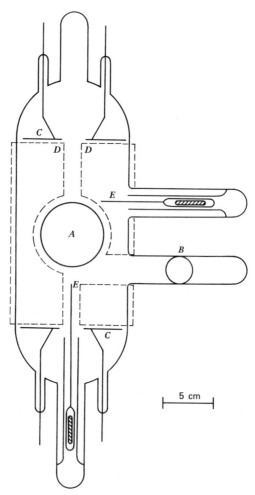

FIG. 4-2-1. Top view of afterglow tube used by Smith et al. *A* is the mass spectrometer side-arm; *B* the pumping and gas admittance side-arm; *C*, the nickel internal electrodes; *D*, the external sleeve electrodes; and *E*, the Langmuir probes.

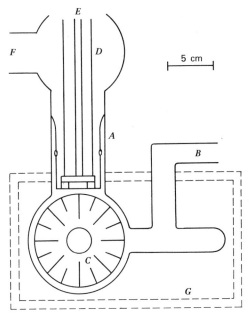

FIG. 4-2-2. End view of afterglow tube used by Smith et al. *A* is the mass spectrometer side-arm; *B*, the pumping and gas admittance side-arm; *C*, the nickel internal electrode with slits for uniform rf induction heating; *D*, the quadrupole mass spectrometer; *E*, to electron multiplier; *F*, to differential pumping system; and *G*, oven or cooling bath.

$10^{10}/cm^3$, as indicated by Langmuir probe measurements (Smith et al., 1968). Electrical contact with the plasma is provided by two plane circular nickel electrodes mounted inside the tube. These electrodes, which also serve in conjunction with the glass wall to define the plasma boundary, are positioned normal to the cylinder axis with a separation of 18.0 cm so that the characteristic diffusion length for the fundamental mode is $\Lambda = 2.11$ cm ($\Lambda^2 = 4.44$ cm^2). The ion sampling orifice is a hole 0.10 mm in diameter in a Nilo-K alloy disk 0.05 mm thick which is sealed to the cylindrical wall midway between the internal disk electrodes. The sampling disk is maintained at the same potential as the internal electrodes. Ions effusing from the plasma are mass-selected by a quadrupole mass spectrometer and detected by an electron multiplier. The variation of the ion current with time in the afterglow is observed by displaying the amplified signal on an oscilloscope (Fig. 4-2-3).

The entire vacuum system with the exception of the pumping lines is baked for several hours at 350°C before experiments are performed and the metal electrodes are outgassed by induction heating. The concentration of

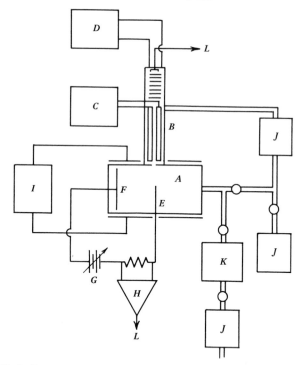

FIG. 4-2-3. Block diagram of apparatus of Smith et al. A is the afterglow tube; B, the mass spectrometer side-arm; C, the quadrupole mass spectrometer power supply; D, the electron multiplier power supply; E, the Langmuir probe; F, the internal electrode; G, the probe bias; H, the probe amplifier; I, the pulsed rf supply; J, the vacuum pumps; K, the gas-handling system; and L, to oscilloscope input.

impurities is further reduced by flushing the discharge tube with a sample of the gas to be used and running a discharge in this gas for several minutes to displace residual impurity atoms absorbed on the interior surfaces. This procedure results in residual gas pressures of the order of 5×10^{-7} Torr, the residual gas consisting mainly of atoms of the gas used to flush the tube. When appropriate, the gas used in the experiment is purified by cataphoresis before being admitted to the tube.

B. STUDIES OF AMBIPOLAR DIFFUSION AND CHEMICAL REACTIONS. For measurements of ambipolar diffusion and reaction rates the ion decay rates in the afterglow are measured with the mass spectrometer as a function of the gas pressure under the assumption that the sampled ion current is directly proportional to the ion number density in the body of the plasma throughout the observation (Smith and Cromey, 1968; Smith and Copsey, 1968; Smith and Fouracre, 1968; Smith et al., 1970). Such measurements are described here in terms of recent experiments on argon (Smith et al., 1972) in which the atomic ion is converted to the molecular ion in the three-body reaction

$$Ar^+ + 2Ar \longrightarrow Ar_2^+ + Ar. \qquad (4\text{-}2\text{-}1)$$

Studies of the atomic ion are facilitated because the loss of this ion by recombination with electrons occurs at a negligibly slow rate compared with that due to diffusion and reaction. Hence the rate of loss of the Ar^+ ions is governed by the equation

$$\frac{\partial n}{\partial t} = D_a \nabla^2 n - vn, \qquad (4\text{-}2\text{-}2)$$

where n is the atomic ion number density, D_a is the electron-ion ambipolar diffusion coefficient for Ar^+, and v is the frequency of the reaction (4-2-1). Observations are made late enough in the afterglow that diffusion occurs almost entirely in the fundamental mode. The solution of (4-2-2) for fundamental mode diffusion indicates an exponential decrease in n with time; the decay constant λ is given by

$$\lambda = \frac{D_a}{\Lambda^2} + \beta p_0^2, \qquad (4\text{-}2\text{-}3)$$

where Λ is the fundamental-mode characteristic diffusion length, β is the rate coefficient for (4-2-1), and p_0 is the gas pressure reduced to the standard temperature of 273°K; β and v are related by the equation

$$v = \beta p_0^2. \qquad (4\text{-}2\text{-}4)$$

Since D_a is inversely proportional to p_0, Smith and his colleagues find it convenient to express (4-2-3) as

$$\lambda p_0 = \frac{D_a p_0}{\Lambda^2} + \beta p_0^3. \qquad (4\text{-}2\text{-}5)$$

Their method of data analysis is somewhat different from that of Lineberger and Puckett (Section 4-1-B).

Experimental data on the variation of λp_0 with reduced pressure p_0 are shown in Fig. 4-2-4. The curve drawn through the experimental points is flat in region B; this region corresponds to domination of the Ar^+ ion decay rate by ambipolar diffusion. The ambipolar diffusion coefficient is obtained by solving (4-2-5) for D_a by using the value of λp_0 obtained in

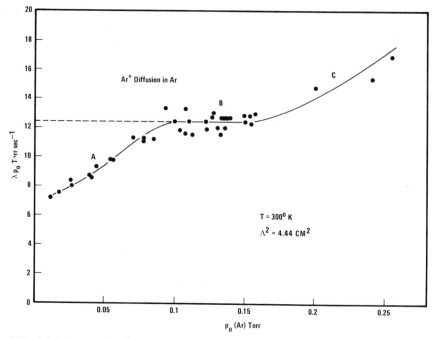

FIG. 4-2-4. λp_0 as a function of p_0 for pure argon. A is the diffusion cooled region; B, normal ambipolar diffusion; and C, the onset of reaction.

region B and neglecting the term βp_0^3. At higher pressures in region C diffusion is inhibited, and the reaction (4-2-1) begins to dominate the decay; β is determined from the slope of a plot of λp_0 versus p_0^3 by using the data of region C and its extension to still higher pressures. In region A the phenomenon of "diffusion cooling," first observed and discussed by Biondi (1954), is in evidence. This term describes the situation in which the average energy of the electrons is reduced by the diffusion of the faster electrons to the walls. In a low-pressure argon afterglow the thermal contact between the electrons and the gas atoms is poor, and the rapid diffusion loss of the faster component of the electrons causes a large reduction of the electron temperature with respect to the gas temperature.

The ambipolar diffusion coefficient of the molecular ions Ar_2^+ is somewhat more difficult to obtain than that of the atomic ions because Ar_2^+ undergoes recombination with electrons at a rate orders of magnitude more rapid than Ar^+ (McDaniel, 1964c); D_a for Ar_2^+, however, may be determined from observations of the Ar_2^+ decay rate late in the afterglow when the electron density and the recombination rate are small.

In their work on the noble gases Smith and his colleagues consider several effects not yet mentioned here and show that it is unlikely that they affected their experiments significantly. In particular, they discuss the production of free electrons in collisions between pairs of metastable atoms, the heating of electrons in superelastic collisions with metastables, and the heating of the gas atoms by the discharge.

C. STUDIES OF ELECTRON-ION RECOMBINATION.   The Langmuir probes shown in Figs. 4-2-1 and 4-2-2 may be used to measure electron number densities and electron temperatures and have been employed to determine coefficients of dissociative electron-ion recombination for molecular ions (Smith and Goodall, 1968; Smith et al., 1970). If the measurements are made under conditions in which some nonreacting molecular ion is the dominant species, the rate of change of the electron number density n_e may be described by the equation

$$\frac{\partial n_e}{\partial t} = D_a \nabla^2 n_e - \alpha n_e^2, \qquad (4\text{-}2\text{-}6)$$

where D_a is the ambipolar diffusion coefficient of the molecular ions and α is the two-body coefficient that describes dissociative recombination between these ions and electrons. (The latter coefficient is defined such that the number of recombination events per unit volume and unit time is $\alpha n_e n_+$ but here $n_+ \approx n_e$.) If the gas pressure is high enough that diffusion has a small effect and if the ionization density is also sufficiently high, the afterglow may be said to be recombination-controlled.* Under these conditions the approximate solution to (4-2-6) is

$$\frac{1}{n_e(t)} = \frac{1}{n_e(0)} + \alpha t, \qquad (4\text{-}2\text{-}7)$$

and observations of the electron density as a function of time in the recombination-controlled afterglow can yield the recombination coefficient.

* In a very important paper, which has been summarized by McDaniel (1964c), Gray and Kerr (1962) considered in detail the effects of diffusion and recombination in spheres and infinitely long cylinders and derived explicit criteria for domination of the plasma decay by each process. One important conclusion gained from these studies is that linearity of the plot of reciprocal electron number density versus time does not in itself imply that a meaningful value for α can be deduced from the slope of the plot; linearity can be observed even when the decay is controlled by diffusion. The Gray-Kerr criteria should be carefully heeded in any stationary afterglow experiment. The reader is also referred to a paper by Oskam (1958) for an analysis of plasma decay in infinite parallel plate geometry.

4-3. MICROWAVE TECHNIQUES

Microwave techniques for studying atomic collisions and transport phenomena were introduced by M. A. Biondi and S. C. Brown at the Massachusetts Institute of Technology during the late 1940's. These techniques have undergone continuous development since that time, particularly by Biondi and his coworkers at the Westinghouse Research Laboratories and the University of Pittsburgh, and have yielded a great deal of extremely important information. They have been especially valuable in the investigation of electron-ion recombination and ambipolar diffusion, and more has been learned about the former phenomenon by microwave techniques than by any other method. Because of the vast literature on microwave studies of plasmas, we cite here only a few of the recent papers by Biondi and his colleagues:

Frommhold et al., 1968
Weller and Biondi, 1968
Kasner and Biondi, 1968
Mehr and Biondi, 1968
Mehr and Biondi, 1969
Philbrick et al., 1969
Frommhold and Biondi, 1969

References to Biondi's earlier work and the work of other investigators are listed in recent reviews (Biondi, 1968; Oskam, 1969).

The microwave studies of interest here involve the use of small cavities whose dimensions are typically a few centimeters. The gas or gas mixture to be investigated is contained in the cavity at a pressure that might be as low as a fraction of a Torr for diffusion studies or as high as tens of Torr in recombination measurements. Frequently provision is made for varying the cavity temperature over a wide range. The gas is ionized periodically, usually by short pulses of microwave radiation, and the decrease in electron number density n_e is measured as a function of time in the decaying afterglow by a microwave probing signal. This measurement, at any given time, is accomplished by determining the shift in the resonant frequency of the cavity produced by the presence of the free electrons (McDaniel, 1964c). Since the sensitivity of the probing technique is not great enough to allow determination of electron number densities below about $10^7/cm^3$, the quantity measured in diffusion studies is the coefficient of ambipolar diffusion D_a, as in the other stationary afterglow experiments discussed in this chapter. In some studies the ions diffusing to the walls are sampled mass spectrometrically as a function of time in the afterglow to ascertain the identities and relative abundances of the ions present. In addition, the spectral distribution and temporal variation of the light emitted by the plasma are sometimes measured. Such measurements have proved useful in studies of dissociative recombination (Frommhold and Biondi, 1969). In another useful technique that has been employed in recombination experiments a separate microwave input is applied to heat the electrons during the afterglow so that the recombination coefficient can be obtained as a function of the electron temperature.

In the determination of ambipolar diffusion coefficients by microwave techniques it is important, of course, to ensure that the plasma decay is dominated by diffusion and not affected appreciably by electron-ion recombination. If the decay is diffusion-dominated and fundamental mode diffusion has been established, a plot of n_e versus time on a semilogarithmic scale should give a straight line and D_a may be obtained from its slope. Meaningful results will be obtained in general only if a single type of positive ion is dominant and the electrons are not subject to appreciable attachment to form negative ions. Techniques for obtaining both D_a and the electron attachment coefficient when attachment occurs in a three-body collision are discussed by Weller and Biondi (1968).

Microwave techniques have been used to study a feature of diffusion which is of general interest in this book and should be mentioned at this point. The discussion in Section 1-11 indicates that the higher

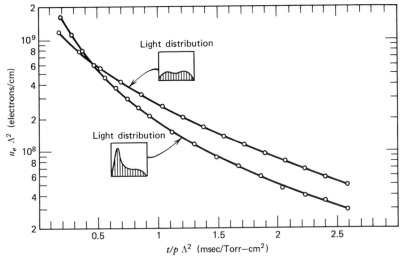

FIG. 4-3-1. The influence of the initial spatial distribution on the decay of the electron number density n_e (Persson and Brown, 1955).

diffusion modes decay much faster than the fundamental mode and that even if higher modes are strongly excited by breaking down a gas in an asymmetric fashion the fundamental mode will eventually dominate the diffusion. This fact is illustrated in Fig. 4-3-1, which shows data obtained by Persson and Brown (1955) in a microwave afterglow experiment. The two plots shown here correspond to different discharge conditions and different initial combinations of diffusion modes. Note that the curves become straight and parallel to one another at large t, each having the slope corresponding to the fundamental mode. The "light distributions" illustrated are qualitative measures of the initial spatial electron density at the start of the decay period obtained by scanning the discharge with a photomultiplier and slit system and displaying the signal on an oscilloscope.

REFERENCES

Biondi, M. A. (1954), *Phys. Rev.* **93**, 1136.

——— (1968), "Afterglow Experiments: Atomic Collisions of Electrons, Ions, and Excited Atoms" in *Methods of Experimental Physics*, Vol. **7-B** (edited by B. Bederson and W. L. Fite), Academic, New York.

Chen, F. F. (1965), "Electric Probes" in *Plasma Diagnostic Techniques* (R. H. Huddlestone and S. L. Leonard, Eds.), Academic, New York. See also Smith, D., and I. C. Plumb (1972), *J. Phys. D* **5**, 1226.

Ferguson, E. E., F. C. Fehsenfeld, and A. L. Schmeltekopf, (1969), "Flowing Afterglow Measurements of Ion-Neutral Reactions" in *Advances in Atomic and Molecular Physics*, Vol. **5** (D. R. Bates and I. Estermann, Eds.), Academic, New York.

Franklin, J. L. (Ed.) (1972), *Ion-Molecule Reactions*, Plenum, New York.

Frommhold, L., and M. A. Biondi (1969), *Phys. Rev.* **185**, 244.

———, and F. J. Mehr (1968), *Phys. Rev.* **165**, 44.

Gray, E. P., and D. E. Kerr (1962), *Ann. Phys.* (New York), **17**, 276.

Kasner, W. H., and M. A. Biondi (1968), *Phys. Rev.* **174**, 139.

Lineberger, W. C., and Puckett, L. J. (1969a) *Phys. Rev.* **186**, 116.

——— (1969b), *Phys. Rev.* **187**, 286.

McDaniel, E. W. (1964a), *Collision Phenomena in Ionized Gases*, Wiley, New York, Appendix I.

——— (1964b), *ibid.*, pp. 120–122, 514–518, 605–607.

——— (1964c), *ibid.*, Chapter 12.

———, Čermák, V., A. Dalgarno, E. E. Ferguson, and L. Friedman (1970), *Ion-Molecule Reactions*, Wiley, New York.

Mehr, F. J., and M. A. Biondi (1968), *Phys. Rev.* **176**, 322.

——— (1969), *Phys. Rev.* **181**, 264.

Oskam, H. J. (1958), *Philips Res. Rept.* **13**, 335, 401.

——— (1969), "Recombination of Rare Gas Ions with Electrons" in *Case Studies in Atomic Collision Physics*, Vol. I (E. W. McDaniel and M. R. C. McDowell, Eds.), North-Holland, Amsterdam.

Persson, K. B., and S. C. Brown (1955), *Phys. Rev.* **100**, 729.

Philbrick, J., F. J. Mehr, and M. A. Biondi (1969), *Phys. Rev.* **181**, 271.

Puckett, L. J., and W. C. Lineberger (1970), *Phys. Rev. A* **1**, 1635.

Smith, D., and M. J. Copsey (1968), *J. Phys. B* **1**, 650.

Smith, D., and P. R. Cromey (1968), *J. Phys. B* **1**, 638.

Smith, D., A. G. Dean, and N. G. Adams (1972), *Z. Phys.* **253**, 191.

Smith, D., and R. A. Fouracre (1968), *Planet. Space Sci.* **16**, 243.

Smith, D., and C. V. Goodall (1968), *Planet. Space Sci.* **16**, 1177.

———, N. G. Adams, and A. G. Dean (1970), *J. Phys. B* **3**, 34.

Smith, D., C. V. Goodall, and M. J. Copsey (1968), *J. Phys. B* **1**, 660.

Weller, C. S., and M. A. Biondi (1968), *Phys. Rev.* **172**, 198.

5

KINETIC THEORY

OF DIFFUSION AND MOBILITY

5-1. DEFINITIONS AND GENERAL RESULTS

The phenomenological definitions of D and K have been given in Section 1-1. The only additional remark needed here is that the ions being present only in trace quantities avoids the trouble involving the impossibility of simultaneously having zero net flux and zero total pressure gradient, which arise in describing diffusion in a general mixture (Mason and Marrero, 1970).

A. EINSTEIN RELATION. The Einstein relation between D and the zero-field value of K is quite general and depends essentially only on the fact that the phenomenological equations relating current, concentration gradient, and electric field are linear and that departures from equilibrium are small. Thus it holds not only for the dilute gases of concern here but also for dense gases, liquids, and solids. Probably the most elegant derivation of the Einstein relation would use the technique of time-correlation functions. A readable survey of this method has been given by Zwanzig (1965). The following argument is somewhat less elegant but involves the same essential concepts. Because of the linearity, the flux density in the presence of a concentration gradient and an electric field is given by

$$\mathbf{J} = -D \, \nabla n + n K \mathbf{E}. \qquad (5\text{-}1\text{-}1)$$

This equation is general, and to find a relation between D and K we can pick any special case that is convenient. We choose equilibrium, for which $\mathbf{J} = 0$ and for which $\mathbf{V}n$ can be found from equilibrium statistical mechanics. The physical meaning is that the electric field sets up a gradient of ion concentration, and at equilibrium the diffusion down this concentration gradient exactly balances the forced flow due to the electric field. At equilibrium the electric field produces an ion distribution given by a Boltzmann expression; in differential form this is

$$\frac{dn}{n} = \frac{eE\,dz}{kT},$$ (5-1-2)

or, in three dimensions,

$$\mathbf{V}n = \frac{ne}{kT}\,\mathbf{E}.$$ (5-1-3)

Substituting (5-1-3) into equation (5-1-1) and setting $\mathbf{J} = 0$, we obtain the Einstein relation

$$K = \frac{eD}{kT}.$$ (5-1-4)

The requirement of linear phenomenological equations means that this holds only in the low-field region ($E/p \ll 2$ V/cm-Torr, according to Section 1-2).

For dilute gases the Einstein relation will follow from the solution of the linearized Boltzmann equation in the low-field region, but the foregoing derivation shows the relation to be much more general than the Boltzmann equation, which describes only binary collision processes. Thus the Einstein relation should hold for all degrees of approximation in constructing solutions of the Boltzmann equation, not just in the first approximation.

B. DEPENDENCE ON DENSITY AND FIELD INTENSITY. The fact that diffusion and mobility are controlled by binary collisions in dilute gases allows further conclusions to be drawn without detailed calculations. In the first place both D and the zero-field K are inversely proportional to N (recall that $n \ll N$). The reason is that the number of acceleration-limiting events (collisions) increases directly with N; thus v_d is inversely proportional to N and so are D and K. This argument, despite its seeming casualness, gives an exact result, because only the counting of collisions is essential, and collision frequency is exactly proportional to N in the binary collision regime. The argument fails at higher fields for then collision frequency also depends on v_d itself when v_d is no longer negligible compared with thermal velocities, since fast ions make more collisions than slow ions;

v_d, however, must depend on E and N only through the ratio E/N even if it is no longer directly proportional to E/N, because of the dependence on it of collision frequency. Again, this result is exact even when obtained by apparently crude arguments, for the counting of collisions is the only crucial step. It is made more explicit by (5-2-5) in the next section. Indeed, v_d, D_L, and D_T all depend only on E/N at all fields, but there is no general relation among them like the Einstein relation at low fields.

Another general result can be obtained from symmetry considerations. As E/N increases from zero, K becomes a function of E/N, but because the gas is isotropic (since the field affects the gas only indirectly by collisions with ions of very low concentration), K must be an even function of E/N. Thus, if K is expanded as a power series in E/N, only even powers will appear:

$$K(E) = K(0)\left[1 + \alpha_1\left(\frac{E}{N}\right)^2 + \alpha_2\left(\frac{E}{N}\right)^4 + \cdots\right]. \qquad (5\text{-}1\text{-}5)$$

This result holds for any isotropic medium, not just dilute gases. The form of the expansion, however, limits its validity to medium fields.

5-2. ELEMENTARY THEORIES AND QUALITATIVE ARGUMENTS

By making the foregoing remarks a little more explicit we can obtain the dependence of the mobility on the masses of the ions and molecules and an interpolation formula to give the dependence of v_d on E/N for all values of field intensity. We can also obtain some rough results on D_L and D_T at high fields and the behavior of v_d in gas mixtures as a function of composition. The arguments introduce the collision cross section Q and a few comments about the magnitude and variation of Q can be made.

A. FREE-FLIGHT THEORY.   The arguments we use here are due largely to Wannier (1953). As in Section 1-2, we assume that the ion undergoes an acceleration eE/m for a mean free time τ but recognize that it loses only a fraction of its momentum on each collision, so that

$$v_d \propto \left(\frac{eE}{m}\right)\tau. \qquad (5\text{-}2\text{-}1)$$

The constant of proportionality must depend on the mass ratio m/M and on the law of force between ions and molecules. It is straightforward to determine the mass dependence of the momentum loss on collision from the equations of momentum and energy conservation. If we then average over all collisions and ignore subtleties about the average of a product and the product of the averages, we obtain

$$v_d = \xi\left(1 + \frac{m}{M}\right)\left(\frac{eE}{m}\right)\tau, \tag{5-2-2}$$

where ξ is a factor of order unity that may depend in a complicated way on the ion-molecule force law. It may also still contain a weak mass dependence because of the approximate averaging over collisions. The mean free time is related to the collision cross section Q and the mean relative speed $\bar{v}_r$ of ions and molecules by the expression

$$\frac{1}{\tau} = \bar{v}_r NQ. \tag{5-2-3}$$

It is reasonable to take $\bar{v}_r$ as the root-mean-square relative velocity

$$\bar{v}_r = (\overline{v^2} + \overline{V^2})^{1/2}, \tag{5-2-4}$$

where $\overline{v^2}$ is the mean-square ion velocity and $\overline{V^2}$ is the mean-square molecule velocity. Any inaccuracy in this equation can be absorbed eventually into the parameter ξ. By combining these results we obtain

$$v_d = \xi\left(\frac{1}{m} + \frac{1}{M}\right)\frac{eE}{(\overline{v^2} + \overline{V^2})^{1/2}NQ}. \tag{5-2-5}$$

The only remaining problem is to find $\overline{v^2}$, which has both thermal and field components.

For low fields $\overline{v^2}$ is entirely thermal, and energy equipartition gives

$$\overline{v^2} + \overline{V^2} = 3kT\left(\frac{1}{m} + \frac{1}{M}\right), \tag{5-2-6}$$

from which we obtain

$$v_d(0) = \frac{\xi}{3^{1/2}}\left(\frac{1}{m} + \frac{1}{M}\right)^{1/2}\frac{e}{(kT)^{1/2}Q}\left(\frac{E}{N}\right). \tag{5-2-7}$$

Here we see explicitly the dependence on the parameter E/N. Dividing by E to obtain $K(0)$ and applying the Einstein relation to obtain D, we also see explicitly the inverse dependence on N. Comparison with the accurate Chapman-Enskog first approximation for D (Chapman and Cowling, 1970, Section 9.81; Hirschfelder et al., 1964, Section 8.2) shows that all dimensional factors in (5-2-7) are correct and that

$$\xi = \frac{3}{16}(6\pi)^{1/2} = 0.814. \tag{5-2-8}$$

Thus in this limit ξ is fortuitously independent of both mass and force law, and even the numerical accuracy is reasonable in that 0.814 is not far from unity.

For high fields $\overline{v^2}$ has a negligible thermal component, but it would not be correct to set $\overline{v^2} = v_d^2$. The reason is that collisions do not merely absorb some of the ion energy, they also randomize some of it, so that the average total ion energy consists of a part visible as drift motion plus a random part. Using the equations of momentum and energy conservation and averaging over collisions in the same approximate way as before, we obtain (Wannier, 1953)

$$m\overline{v^2} = mv_d^2 + Mv_d^2, \tag{5-2-9}$$

the last term being the random part of the field energy. Thus light ions in a heavy gas $(m \ll M)$ have most of their field energy as random motion, since the heavy molecules are extremely effective in deflecting the ions but absorb little energy in recoil. Heavy ions in a light gas $(m \gg M)$, however, have most of their energy as drift motion, for the light molecules are ineffective either in deflecting the ions or in absorbing their energy. For $m = M$ the ion energy is equally divided between drift and random components. Substituting (5-2-9) back into (5-2-5) and remembering that $\overline{V^2} \ll v_d^2$ at high fields, we obtain

$$v_d(\infty) = \xi^{1/2} \left(\frac{1}{m} + \frac{1}{M}\right)^{1/4} \left(\frac{e}{M^{1/2}Q}\right)^{1/2} \left(\frac{E}{N}\right)^{1/2}. \tag{5-2-10}$$

This formula works remarkably well, although it may be complicated to use if Q depends strongly on $v_d(\infty)$. A few numerical comparisons with Q assumed constant will illustrate the agreement. For electrons $(m \ll M)$ it takes the same form as the accurate result of Druyvesteyn (Chapman and Cowling, 1970, p. 388), with

$$\xi = \frac{2\pi}{3(3)^{1/2}[\Gamma(\tfrac{3}{4})]^2} = 0.805. \tag{5-2-11}$$

For $m = M$ it checks the accurate calculations of Wannier (1953) for rigid spheres with

$$\xi = \frac{(1.1467)^2}{2^{1/2}} = 0.930. \tag{5-2-12}$$

For heavy ions in a light gas $(m \gg M)$ the results agree exactly with Wannier's calculations with $\xi = 1$. The 20% variation of ξ with mass ratio for high fields means that ξ still contains a weak mass dependence; this presumably entered through the approximate averaging used in obtaining (5-2-9).

The different field dependence of v_d at low and high fields is interesting:

$$v_d(0) \propto \frac{E}{NQ}, \tag{5-2-13}$$

$$v_d(\infty) \propto \left(\frac{E}{NQ}\right)^{1/2}. \tag{5-2-14}$$

Thus, if Q is independent of E/N, v_d varies as the first power of E/N at low fields and as the square root of E/N at high fields; Q, however, depends on v_r except in the case of rigid spheres. Thus at low fields Q is a function of temperature, since v_r is thermal and (5-2-13) is exact; but at high fields the thermal energy is negligible and Q is a function of v_d, hence of E/N. Thus the square-root prediction for high fields is only approximate, but the prediction that $v_d(\infty)$ is independent of temperature should be accurate. Experimental data for noble gas ions in their parent gases do show nearly the square-root dependence on E/N at high fields (Wannier, 1953). In any case, the dependence on field intensity is seen to be a function only of the ratio E/N.

For intermediate fields we take the thermal and field energies to be additive and combine (5-2-6) and (5-2-9) to obtain

$$\overline{v^2} + \overline{V^2} = 3kT\left(\frac{1}{m} + \frac{1}{M}\right) + v_d^2\left(\frac{m+M}{m}\right). \tag{5-2-15}$$

Substitution back into (5-2-5) yields

$$v_d = \xi\left(\frac{1}{m} + \frac{1}{M}\right)^{1/2} \frac{eE}{NQ}(3kT + Mv_d^2)^{-1/2}, \tag{5-2-16}$$

which is seen to be a quadratic in the variable v_d^2,

$$(v_d^2)^2 + \frac{3kT}{M}(v_d^2) - \frac{\xi^2}{M}\left(\frac{1}{m} + \frac{1}{M}\right)\left(\frac{eE}{NQ}\right)^2 = 0. \tag{5-2-17}$$

Solution of this quadratic gives a reasonable interpolation formula for all field intensities, but even without an explicit solution we can see several features.

In the first place, v_d will depend on E and N only through the ratio E/N, as mentioned before. In fact, v_d will be a function of the combination eE/NQ.

In the second place, for moderately low fields we can expand the right-hand side of (5-2-16) in powers of the small quantity $Mv_d^2/3kT$ and solve iteratively for v_d; this shows that v_d depends on higher powers of E/N only through $(E/N)^2$, as given on general grounds in (5-1-5). In particular,

$$v_d(E) = v_d(0)\left[1 + \alpha_1\left(\frac{E}{N}\right)^2 + \cdots\right], \tag{5-2-18}$$

$$\alpha_1 = -\frac{\xi^2}{18}\left(\frac{m+M}{m}\right)\left(\frac{e}{kTQ}\right)^2. \tag{5-2-19}$$

Comparison with accurate kinetic-theory calculations (Mason and Schamp, 1958) shows that α_1 is of this form but that it depends in a rather complicated way on the masses, the ion-molecule force law, and the temperature. For rigid spheres ξ is temperature independent and has the following values for different mass ratios: 0.71 for $m \ll M$, 0.82 for $m = M$, and 0.63 for $m \gg M$. These values are reasonably close to unity, especially for an expansion coefficient like α_1, but the deviations from unity will be greater for other than rigid-sphere interactions.

In the third place, for moderately strong fields we can expand in powers of the small quantity $3kT/Mv_d^2$ to obtain

$$v_d(E) = v_d(\infty)\left[1 + \beta_1\left(\frac{N}{E}\right) + \beta_2\left(\frac{N}{E}\right)^2 + \cdots\right], \tag{5-2-20}$$

$$\beta_1 = -\frac{3}{4\xi}\left(\frac{m}{m+M}\right)^{1/2}\left(\frac{kTQ}{e}\right). \tag{5-2-21}$$

Notice that all inverse powers of E/N occur here, whereas only even powers of E/N occurred in the expansion for moderately weak fields. There exists no accurate theory with which this result can be compared.

We can also obtain some approximate results on D by similar arguments. Free-flight or free-path treatments of diffusion lead to the well-known expression

$$D = \tfrac{1}{3}\xi \overline{v_r^2}\tau, \tag{5-2-22}$$

where we have included the factor ξ to be consistent with the Einstein relation. The usual arguments leading to this expression implicitly assume isotropy, hence do not apply to high fields, but we can reasonably extend the result to any field intensity by describing the diffusion as an isotropic random process superposed on the ordered drift velocity v_d. Then the transverse and longitudinal diffusion coefficients are

$$D_T = \xi\left(\frac{m}{\mu}\right)(\overline{v^2})_T\,\tau, \tag{5-2-23}$$

$$D_L = \xi\left(\frac{m}{\mu}\right)[(\overline{v^2})_L - v_d^2]\tau, \tag{5-2-24}$$

the factor m/μ occurring because we have changed from the relative velocity to the ion velocity. The drift energy needs to be subtracted only for the longitudinal motion. The factor $\frac{1}{3}$ has disappeared because we are dealing with velocity components only; it appeared in (5-2-22) in the first place only through the relation

$$\overline{v_x^2} = \overline{v_y^2} = \overline{v_z^2} = \tfrac{1}{3}\overline{v^2}. \tag{5-2-25}$$

We now proceed as before, taking thermal and field energies to be additive,

$$\frac{m}{\mu}(\overline{v^2})_T = kT\left(\frac{1}{m} + \frac{1}{M}\right) + \zeta_T v_d^2, \tag{5-2-26}$$

$$\frac{m}{\mu}[(\overline{v^2})_L - v_d^2] = kT\left(\frac{1}{m} + \frac{1}{M}\right) + \zeta_L v_d^2. \tag{5-2-27}$$

Here the argument has bogged down, as shown by the unknown factors ζ_T and ζ_L. The difficulty is as follows. We know that collisions tend to randomize part of the field energy and have already written down an expression for it in (5-2-9). What we do not know without a detailed calculation is how the random energy is divided between the transverse and longitudinal components. Considerations on the "persistence of velocities" after collision (Jeans, 1925, Sections 352–362, 416–418) suggest that

$$\zeta_L \ge \zeta_T, \tag{5-2-28}$$

and comparison of the thermal conductivity and viscosity of a pure monatomic gas suggests that for $m = M$ the persistence effect yields $\zeta_L/\zeta_T = \frac{5}{2}$ (Cowling, 1960, pp. 61–63). Comparison of (5-2-26) and (5-2-27) with (5-2-9) further shows that the distribution of total random field energy requires that

$$2\zeta_T + \zeta_L = \frac{m + M}{m}. \tag{5-2-29}$$

These results agree with the detailed calculations of Wannier (1953), but we cannot proceed by using only the sort of elementary theory given in this section.

Despite our inability to deal with ζ_T and ζ_L at this stage we can extract useful information from (5-2-26) and (5-2-27). First, because of the proportionality to $\overline{v^2}$, the diffusion coefficients are obtained by simply adding the low-field and high-field expressions. Second, the high-field portion is proportional to $v_d^2\tau$; since we have already found an interpolation formula for v_d at all fields, we now have similar formulas for the diffusion coefficients. Substituting back and eliminating τ in favor of v_d through (5-2-2), we obtain

$$D_T(E) = D(0) + \zeta_T\left(\frac{mM}{m + M}\right)\frac{v_d^3}{eE}, \tag{5-2-30}$$

$$D_L(E) = D(0) + \zeta_L\left(\frac{mM}{m + M}\right)\frac{v_d^3}{eE}, \tag{5-2-31}$$

where $D(0)$ is the zero-field value obeying the Einstein relation. We see, as expected, that $ND(E)$ depends only on the combination E/N; the explicit dependence is found by solution of (5-2-17) for v_d. The dimensionless parameters ζ_T and ζ_L are complicated functions of the mass ratio m/M and the ion-molecule force law. Thus the diffusion coefficients are expected to increase with increasing E/N, with $D_L \geq D_T$. Experimental results confirm this qualitative prediction (see Section 7-4). Detailed calculations by Wannier (1953) for a model with constant τ and an isotropic scattering pattern yield

$$\zeta_T = \frac{1}{3}\frac{(M + m)^2}{m(M + 2m)}, \tag{5-2-32}$$

$$\zeta_L = \frac{1}{3}\frac{(M + m)(M + 4m)}{m(M + 2m)}, \tag{5-2-33}$$

which are in agreement with the behavior we expected on general grounds. They are also in reasonable agreement with experiment (see Section 7-4).

We conclude with a few general remarks about free-flight (or free-path) theories. It would appear that we have obtained quite a number of reasonably accurate results with a minimum of calculation, but we must temper this success with two admissions. First, the success is to some extent fortuitous. Second, in the words of H. A. Kramers (1949), "... considerations along this line may well belong to those unimprovable speculations of which the kinetic theory of gases affords such ghastly examples." What is fortuitous is that mean-free-path theories of diffusion give good results only when one of the species is present in trace concentrations, as the ions are. When applied to binary mixtures of arbitrary composition, such theories predict that the diffusion coefficient should depend strongly on the relative composition of the mixture, a result in disagreement with both experiment and accurate theory (Chapman and Cowling, 1970, Section 14.31). Attempts to refine the theory by corrections for persistence of velocities soon meet the "unimprovable" difficulty; that is, the calculations are difficult and the convergence is slow. It is also difficult to see how the refinements would be just sufficient to nullify the composition dependence predicted by the first approximation.

In terms of the calculations presented in this section the unimprovable difficulty appears in the parameters ξ, ζ_T, and ζ_L. The explicit calculation of these quantities by the tracing of collision histories is clearly a difficult procedure—improved results have in fact usually been obtained by abandoning the free-path approach and starting over with an attempt to solve the Boltzmann equation or some similar integrodifferential equation. It is only recently that the complete mathematical connection was established between free-flight theories of diffusion and the accurate Chapman-Enskog theory (Monchick and Mason, 1967); even this connection applies to ion mobility only in the zero-field case. For higher fields the theoretical structure is still incomplete. These matters are discussed in more detail in subsequent sections.

B. COLLISION CROSS SECTIONS. The foregoing theory introduces an ill-defined quantity, the collision cross section Q. This fundamental defect is tolerated because the simplicity and physical appeal of the theory make it so convenient for thinking and discussion, but some extra information about Q must be grafted on if the theory is to be semiquantitatively useful. Here we give some rough-and-ready methods for obtaining reasonable estimates of the magnitude and energy dependence of Q.

As shown subsequently, Q really is a well-defined quantity, but to establish this requires either a momentum-transfer type of theory or an accurate solution of the Boltzmann equation. It turns out that Q is a diffusion or momentum-transfer cross section, defined as

$$
Q_D = \int (1 - \cos \theta) I_s(\theta) \, d\Omega_{CM}
$$
$$
= 2\pi \int_0^\pi (1 - \cos \theta) I_s(\theta) \sin \theta \, d\theta, \qquad (5\text{-}2\text{-}34)
$$

where θ is the scattering angle in the center-of-mass system and $I_s(\theta)$ is the differential-scattering cross section. This expression is limited to elastic collisions. Classical mechanics often suffices to describe the collisions, in which case (5-2-34) can be written as

$$
Q_D = 2\pi \int_0^\infty (1 - \cos \theta) b \, db, \qquad (5\text{-}2\text{-}35)
$$

where b is the impact parameter and θ is given by

$$
\theta = \pi - 2b \int_{r_a}^\infty \left[1 - \frac{b^2}{r^2} - \frac{V(r)}{E} \right]^{-1/2} \frac{dr}{r^2}; \qquad (5\text{-}2\text{-}36)
$$

in this equation r_a is the distance of closest approach (the outermost zero of the bracketed expression), $V(r)$ is the ion-molecule potential energy, and $E = \frac{1}{2}\mu v_r^2$ is the initial relative kinetic energy (not to be confused with electric field). Although classical theory always fails to describe $I_s(\theta)$ at small angles, these classical expressions are adequate because the angular weighting factor $(1 - \cos\theta)$ suppresses the small-angle contribution.

The accurate evaluation of Q_D for any but trivially simple forms of $V(r)$ requires extensive numerical integration, as discussed in Chapter 6, but often fairly good estimates can be obtained from simple physical arguments, and this will suffice for present purposes. One accurate result that can be obtained by dimensional arguments alone goes back at least to Lord Rayleigh (1900): for a potential of the form

$$V(r) = \pm \frac{C}{r^n}, \tag{5-2-37}$$

where C and n are positive constants, the energy dependence of any cross section is

$$Q \propto \left(\frac{C}{E}\right)^{2/n}. \tag{5-2-38}$$

Of course, nothing is known about the constant of proportionality from this argument. One quick way to verify this result is to note from (5-2-36) that θ is a function of the single dimensionless variable Eb^n/C; then (5-2-38) follows because any Q is proportional to $b\,db$ in a three-dimensional world.

When the scattering is dominated by an attractive potential of the form $-C/r^n$, the magnitude of Q can be estimated by an argument given by Wannier (1953) and elaborated somewhat by Dalgarno et al. (1958). At a given energy there is a certain impact parameter b_0 at which the particles take up an unstable orbiting motion (Hirschfelder et al., 1964, Section 8.4); for $b < b_0$ the particles are drawn together until forced apart again by a repulsive core in the potentials. The diffusion cross section is taken to be equal to the cross section for orbiting or "capture"

$$Q_D(\text{orbit}) = \pi b_0^2 = \frac{n\pi}{n-2}\left[\left(\frac{n-2}{2}\right)\frac{C}{E}\right]^{2/n}. \tag{5-2-39}$$

For a more careful derivation that usually yields a better numerical estimate we rewrite (5-2-35) as

$$Q_D = 4\pi \int_0^\infty \sin^2\left(\frac{\theta}{2}\right)b\,db. \tag{5-2-40}$$

If orbiting occurs, θ varies rapidly with b out to large b and then slowly tails off. We therefore replace $\sin^2(\theta/2)$ with its mean value of $\frac{1}{2}$ out to some $b*$ and with zero for larger b. This can be called a "random-angle" approximation. The precise criterion for the choice of $b*$ is not critical; analogous calculations on charge transfer cross sections suggest choosing $b*$ as that value of θ for which (Firsov, 1951; Bates and Boyd, 1962)

$$\left|\frac{\theta}{2}\right| = \frac{1}{\pi}. \qquad (5\text{-}2\text{-}41)$$

This is an implicit equation for $b*$; for an explicit equation we expand the square root in (5-2-36) for θ and obtain a small-angle approximation (Kennard, 1938, Section 70),

$$\theta \approx \frac{b}{E} \int_b^\infty \frac{V(b) - V(r)}{(r^2 - b^2)^{3/2}} r \, dr, \qquad (5\text{-}2\text{-}42)$$

with $r_a \approx b$. This integral can be evaluated for inverse-power potentials

$$\theta \approx \pm A_n\left(\frac{C}{Eb^n}\right), \qquad (5\text{-}2\text{-}43)$$

where

$$A_n = \frac{\pi^{1/2}\Gamma[(n + 1)/2]}{\Gamma(n/2)}, \qquad (5\text{-}2\text{-}44)$$

from which we can solve for $b*$ and obtain

$$Q_D(\text{random}) = \pi b*^2 = \pi\left(\frac{\pi A_n}{2}\frac{C}{E}\right)^{2/n}. \qquad (5\text{-}2\text{-}45)$$

A comparison of these two approximations with accurate value of Q_D calculated by numerical integration (Higgins and Smith, 1968) is shown in Table 5-2-1. The agreement is not earthshaking, but it is good enough for many purposes.

Many of the values of n in Table 5-2-1 have physical significance: $n = 2$ is the ion-dipole interaction, $n = 3$, the ion-quadrupole, $n = 4$, the ion-induced dipole, $n = 6$, the ion-induced quadrupole and the dispersion energy, and so on.

The common feature of both approximations is that Q_D is proportional to the square of some characteristic distance that is a function of energy. This observation suggests that we might extend the results to arbitrary potentials, both attractive and repulsive, by picking a more general characteristic distance that could be taken as proportional to an effective collision diameter. A simple choice is the distance at which the potential

TABLE 5-2-1. Accuracy of approximate diffusion cross sections for the potential $V(r) = -C/r^n$

n	$Q_D(\text{approx})/Q_D$	
	Orbiting Approx (5-2-39)	Random-Angle Approx (5-2-45)
2	0.310	0.764
3	0.709	0.804
4	0.905	0.871
6	1.197	1.056
8	1.349	1.172
10	1.405	1.222
25	1.302	1.182
50	1.180	1.108
∞	1.000	1.000

energy is equal to the initial relative kinetic energy (i.e., the distance of closest approach in a head-on collision); this yields the result

$$Q_D \propto \pi d^2 \qquad (5\text{-}2\text{-}46)$$

$$|V(d)| = E. \qquad (5\text{-}2\text{-}47)$$

For inverse-power potentials (5-2-37) we then find

$$\pi d^2 = \pi \left(\frac{C}{E}\right)^{2/n}. \qquad (5\text{-}2\text{-}48)$$

The ratio $Q_D/\pi d^2$ is given in Table 5-2-2 for a number of attractive and repulsive potentials. Although the ratio is not constant, we would never be off by more than a factor of 2 by taking $Q_D \approx 2\pi d^2$, and if we have any idea of the shape of the potential a much closer estimate of Q_D can be obtained by interpolation in this table.

The effective collision diameter approximation has been used by Hirschfelder and Eliason (1957) as a basis for the rapid estimation of transport properties of gases.

Short-range repulsion energy can often be represented by an exponential potential

$$V(r) = V_0 e^{-r/a}, \qquad (5\text{-}2\text{-}49)$$

where V_0 and a are positive constants. The effective collision diameter approximation yields for this potential

$$\pi d^2 = \pi a^2 \left[\ln\left(\frac{V_0}{E}\right)\right]^2. \qquad (5\text{-}2\text{-}50)$$

TABLE   5-2-2.   Accuracy of the
effective collision diameter approximation to the diffusion cross section
for the potentials $V(r) = \pm C/r^n$

n	$\dfrac{Q_D}{\pi d^2}$	
	Attraction	Repulsion
2	3.23	1.59
3	2.67	1.30
4	2.21	1.19
6	1.58	1.11
8	1.30	1.08
10	1.17	1.06
25	1.02	1.02
50	1.00	1.01
∞	1.00	1.00

The ratio $Q_D/\pi d^2$ for this potential is a function of energy, unlike that
for the inverse-power potential. Some values based on accurate Q_D obtained
by numerical integration (Monchick, 1959) are given in Table 5-2-3; the
agreement is again reasonable.

Thus reasonable estimates of both the energy dependence and magnitude
of the cross section can be obtained without undue difficulty even for
complicated potentials. Even if the potential is so complicated that (5-2-47)

TABLE 5-2-3.   Accuracy of
the effective collision diameter
approximation to the diffusion
cross section for the potential
$V(r) = V_0 e^{-r/a}$

$\dfrac{E}{V_0}$	$\dfrac{Q_D}{\pi d^2}$
0	1.00
10^{-4}	1.07
10^{-3}	1.10
10^{-2}	1.16
10^{-1}	1.38

cannot be solved explicitly for d, results can be obtained by treating d parametrically; that is, an arbitrary value of d is chosen, the value of Q_D calculated from (5-2-46), and the corresponding value of E from (5-2-47).

Finally, a few words must be given about the estimate of Q_D when an ion moves in its parent gas and resonant charge transfer is possible. Except at very low energies, the diffusion of an ion is then dominated by the charge transfer process. The reasons for this are discussed in some detail later in this chapter; all we need to do at this point is to mention that Q_D is related to the charge transfer cross section Q_T and that a sufficiently accurate relation for the present purposes (Dalgarno, 1958) is

$$Q_D \approx 2Q_T. \qquad (5\text{-}2\text{-}51)$$

The value of this equation is that Q_T is often known experimentally from measurements with ion beams. Only a few measurements may be needed, for Q_T can often be accurately represented over a large energy range by an expression of the form

$$Q_T^{1/2} = a_1 - a_2 \ln E. \qquad (5\text{-}2\text{-}52)$$

This gives the same energy dependence for Q_D as (5-2-50) for an exponential potential; this similarity is not accidental for reasons discussed in Section 5-7-E.

C. MOMENTUM TRANSFER THEORY.   Our main purpose in discussing the momentum transfer theory of diffusion here is to show how the explicit expression (5-2-35) for Q_D can be obtained by relatively simple physical arguments. Other than that, the results for ion mobilities are much the same by free-flight and momentum transfer arguments. This similarity, however, is rather accidental and is a consequence of the fact that the ions are present only in trace concentration. The results of the two approaches for mixtures of arbitrary composition are quite different, at least in first approximation. The discussion of ion mobility in gas mixtures is especially simple by the momentum transfer approach and is also included in this section.

Free-path or free-flight arguments go back to Maxwell (1860). Interestingly enough, in the same paper, Maxwell also first devised a momentum transfer theory of diffusion but was apparently not too happy with it (because of a mistake, corrected by Clausius). It was independently invented by Stefan (1871, 1872) and later used by Langevin (1905) to give the first quantitatively accurate theory of low-field ion mobility. The theory was then apparently forgotten for many years, until it was discovered still one more time (apparently independently) by Frankel (1940) and by Present and de

Bethune (1949). One impetus for this revival was the problem of isotope separation in World War II (Present, 1958, p. x).

The momentum transfer approach is particularly simple for ion mobilities. By Newton's second law of motion the electric field transfers momentum to the ions, but since the ions are not accelerated on the average the momentum must be transferred by collisions to the gas molecules. It is easy to show that in one collision the momentum transferred to a gas molecule has a component parallel to v_r of (Present, 1958, p. 136)

$$\delta \mathbf{p}_\| = \mu \mathbf{v}_r (1 - \cos \theta). \tag{5-2-53}$$

If we average over many collisions, we expect all the random components to average to zero and only the drift velocity to contribute; the average momentum communicated to the gas per collision is therefore $\mu v_d (1 - \cos \theta)$. But the average number of collisions an ion makes per unit time per unit volume with impact parameter in the range db is

$$N\bar{v}_r (2\pi b\, db). \tag{5-2-54}$$

The total momentum transferred per ion to the gas, which must be equal to eE, is therefore

$$\mu v_d N\bar{v}_r \int_0^\infty (1 - \cos \theta) 2\pi b\, db = eE. \tag{5-2-55}$$

Within a factor ξ of order unity this is the same as (5-2-2) and (5-2-3) of the free-path theory, with the cross section identified as

$$Q_D = 2\pi \int_0^\infty (1 - \cos \theta) b\, db. \tag{5-2-56}$$

Since we only want to identify Q_D, we can be satisfied with the above rather rough averaging over collisions, but it should be mentioned that a more careful calculation gives the correct numerical value of ξ as well (Present, 1958, Section 8-3).

The rest of the development clearly parallels that of the free-flight theory and need not be elaborated here. The extension to multicomponent mixtures by the momentum transfer method, however, is so straightforward that it is worth presenting at this point (Mason and Hahn, 1972). The total momentum transferred per ion to a gas mixture is just the sum of the momenta transferred to each species in the mixture if only binary collisions are considered; the generalization of (5-2-55) to a mixture is therefore

$$v_d \sum_j \mu_j N_j \bar{v}_{r(j)} \int_0^\infty (1 - \cos \theta_j) 2\pi b\, db = eE \tag{5-2-57}$$

or, in more detail,

$$v_d \sum_j \left(\frac{mM_j}{m + M_j}\right)(\overline{v^2} + \overline{V_j^2})^{1/2} N_j Q_{D(j)} = eE. \tag{5-2-58}$$

This result is valid at all field strengths. We can safely ignore the factor ξ in this calculation because we eventually express the results in terms of the drift velocities in the pure gases. For low fields the energies are entirely thermal,

$$\overline{v^2} + V_j^2 = 3kT\left(\frac{1}{m} + \frac{1}{M_j}\right), \tag{5-2-59}$$

and (5-2-58) becomes

$$v_d(0) \sum_j \left(\frac{mM_j}{m + M_j}\right)^{1/2} (3kT)^{1/2} N_j Q_{D(j)} = eE. \tag{5-2-60}$$

The drift velocity $v_{d(j)}(0)$ in a pure gas j of density N equal to the total density of the mixture is given by òne term from the above sum,

$$v_{d(j)}(0)\left(\frac{mM_j}{m + M_j}\right)^{1/2} (3kT)^{1/2} N Q_{D(j)} = eE, \tag{5-2-61}$$

from which we reduce (5-2-60) to the form

$$v_d(0) \sum_j \frac{x_j}{v_{d(j)}(0)} = 1, \tag{5-2-62}$$

where $x_j = N_j/N$ is the mole fraction of species j in the mixture. This is more conventionally written as

$$\frac{1}{v_d(0)} = \sum_j \frac{x_j}{v_{d(j)}(0)} \tag{5-2-63}$$

or, in terms of mobilities,

$$\frac{1}{K_{mix}(0)} = \sum_j \frac{x_j}{K_j(0)}, \tag{5-2-64}$$

a result known as Blanc's law (Blanc, 1908).

At high fields we face the problem of determining the partitioning of the ion energy between drift and random field components. The extension of the single gas result $m\overline{v^2} = mv_d^2 + Mv_d^2$ to mixtures turns out to be not so simple, but we can always write the result formally as

$$m\overline{v^2} = mv_d^2 + \overline{M}v_d^2, \tag{5-2-65}$$

where $\overline{M}$ is some sort of average mass of the gas mixture whose precise form requires a more elaborate calculation. Then (5-2-58) becomes

$$[v_d(\infty)]^2 \sum_j \left(\frac{mM_j}{m + M_j}\right)\left(\frac{m + \overline{M}}{m}\right)^{1/2} N_j Q_{D(j)} = eE. \qquad (5\text{-}2\text{-}66)$$

The drift velocity $v_{d(j)}(\infty)$ in a pure gas j of density N is, since $\overline{M} = M_j$ for a single gas,

$$[v_{d(j)}(\infty)]^2 \left(\frac{mM_j}{m + M_j}\right)\left(\frac{m + M_j}{m}\right)^{1/2} N Q_{D(j)} = eE, \qquad (5\text{-}2\text{-}67)$$

from which we reduce (5-2-66) to the form, assuming constant $Q_{D(j)}$ (Whealton and Mason, 1972),

$$\left(\frac{1}{v_d(\infty)}\right)^2 = \sum_j x_j \left(\frac{m + \overline{M}}{m + M_j}\right)^{1/2} \left(\frac{1}{v_{d(j)}(\infty)}\right)^2. \qquad (5\text{-}2\text{-}68)$$

This result is quite unlike Blanc's law, in which reciprocals of drift velocities are added linearly with respect to mole fraction. Here the reciprocals of the *squares* of drift velocities are added with respect to a peculiar sort of mass fraction.

For intermediate fields we assume that thermal and field energies are additive:

$$\overline{v^2} + \overline{V_j^2} = 3kT\left(\frac{1}{m} + \frac{1}{M_j}\right) + v_d^2\left(\frac{m + \overline{M}}{m}\right). \qquad (5\text{-}2\text{-}69)$$

From this we readily obtain the result

$$\frac{1}{v_d^2(E)} = \sum_j \frac{x_j}{v_{d(j)}^2(E)} \left[\left(\frac{m + \overline{M}}{m + M_j}\right) + \frac{3kT}{M_j v_d^2(E)}\right]^{1/2} \left[1 + \frac{3kT}{M_j v_{d(j)}^2(E)}\right]^{-1/2}. \qquad (5\text{-}2\text{-}70)$$

This reduces to Blanc's law at low fields and to (5-2-68) at high fields.

The calculation of $\overline{M}$ is beyond the scope of any simple theory. A velocity-independent scattering model with constant τ yields (Mason and Hahn, 1972; Hahn and Mason, 1973)

$$\overline{M} = \frac{\sum_j \omega_j M_j}{\sum_j \omega_j}, \qquad (5\text{-}2\text{-}71)$$

where the weight factors ω_j are, for high fields,

$$\omega_j = \frac{x_j M_j Q_{D(j)}}{(m + M_j)^2}. \qquad (5\text{-}2\text{-}72)$$

The derivation for $\overline{M}$ is given in Section 5-5-F and numerical results are discussed in Section 5-6-B.

Similar results hold for the transverse and longitudinal diffusion coefficients (Whealton and Mason, 1972).

5-3. LOW-FIELD THEORY

Because of the Einstein relation, the theory of ion mobility at low fields is equivalent to the classical kinetic theory of diffusion. The original formulation of a rigorous kinetic theory goes back to Maxwell and to Boltzmann in the latter part of the nineteenth century, but they were unable to overcome many of the mathematical problems involved in the computation of transport properties. The solutions came in the second decade of the twentieth century from two independent workers—Chapman, who started with Maxwell's equation of transfer, and Enskog, who started with Boltzmann's equation for the distribution function. The results are mathematically equivalent. A historical summary of these developments appears in the first (1939) and second (1952) editions of the well-known treatise by Chapman and Cowling (1970). A delightful personal account has been given by Chapman (1967). In this section we give a brief sketch of the Chapman-Enskog treatment of diffusion as applied to ion mobility and then discuss some of its consequences and extensions.

A. CHAPMAN-ENSKOG THEORY. All transport phenomena arise by deviations from the Maxwell equilibrium distribution function. The basic problem of rigorous kinetic theory is to find the nonequilibrium distribution function from which the various fluxes can be found by integration. The diffusive flux, for instance, is the integral of molecular velocity over the distribution function. Comparison with the phenomenological equation defining D (Section 1-1) then identifies D in terms of expressions involving molecular collisions.

The distribution function is described by the Boltzmann equation, which has the form of an equation of continuity. The Boltzmann equation for the ions in a gas mixture is

$$\frac{\partial f_i}{\partial t} + \mathbf{v} \cdot \mathbf{V}_r f_i + \frac{e}{m} \mathbf{E} \cdot \mathbf{V}_v f_i$$

$$= \sum_j \int \cdots \int [f_i(\mathbf{v}')f_j(\mathbf{V}_j') - f_i(\mathbf{v})f_j(\mathbf{V}_j)]v_r I_s(\theta) \, d\Omega_{CM} \, d^3V_j, \qquad (5\text{-}3\text{-}1)$$

where the right-hand side of the equation represents the source-sink terms due to collisions. The summation extends over all the species of neutral molecules in the mixture; no term for ion-ion collisions occurs because we assume that ions are present only in trace quantities. The primes refer to velocities after collision. The distribution functions are normalized to the densities

$$\int f_i(\mathbf{v}) \, d^3 v = n, \tag{5-3-2}$$

$$\int f_j(\mathbf{V}_j) \, d^3 V_j = N_j. \tag{5-3-3}$$

If we could solve (5-3-1), the computation of the drift velocity would be straightforward:

$$\mathbf{v}_d = \frac{1}{n} \int \mathbf{v} f_i(\mathbf{v}) \, d^3 v. \tag{5-3-4}$$

At low fields the ion energy is related to the temperature by

$$\tfrac{3}{2}kT = \frac{1}{n} \int (\tfrac{1}{2}mv^2) f_i(\mathbf{v}) \, d^3 v, \tag{5-3-5}$$

since the drift energy is negligible compared with the thermal energy.

The Boltzmann equation assumes that only binary collisions are frequent enough to affect the distribution function. In addition, the form (5-3-1) assumes that all collisions are elastic.

To find an acceptable solution of the Boltzmann equation the distribution function is written as

$$f_i = f_i^{(0)}(1 + \phi_i), \tag{5-3-6}$$

where $f_i^{(0)}$ is the equilibrium distribution

$$f_i^{(0)} = n \left(\frac{m}{2\pi k T} \right)^{3/2} \exp \left(-\frac{mv^2}{2kT} \right) \tag{5-3-7}$$

and ϕ_i is a deviation term. On the assumption that all the ϕ's are small the Boltzmann equation can be linearized to yield

$$\frac{\partial f_i^{(0)}}{\partial t} + \mathbf{v} \cdot \nabla_r f_i^{(0)} + \frac{e}{m} \mathbf{E} \cdot \nabla_v f_i^{(0)}$$

$$= \sum_j \int \cdots \int f_i^{(0)} f_j^{(0)} [\phi_i(\mathbf{v}') + \phi_j(\mathbf{V}_j') - \phi_i(\mathbf{v}) - \phi_j(\mathbf{V}_j)] v_r I_s(\theta) \, d\Omega_{CM} \, d^3 V_j. \tag{5-3-8}$$

We could drop the term involving $\mathbf{E}$ and still find the mobility by virtue of the Einstein relation, but we shall keep this term and thereby demonstrate directly that the linearized Boltzmann equation leads to the Einstein relation. To keep the following algebra simple, however, we assume that only one neutral gas component is present; the algebraic details for multicomponent mixtures can be found in standard treatises (Chapman and Cowling, 1970, Chapter 18; Hirschfelder et al., 1964, Chapter 7).

The differentiations of $f_i^{(0)}$ in (5-3-8) can be carried out. Space and time derivatives of n occur; it is consistent with the linear approximation to drop the time derivative, but the space derivative survives. To put it another way, we need seek only a stationary-state solution to identify D and K by comparison with the phenomenological equation (5-1-1). The result is

$$\frac{1}{n} f_i^{(0)} [(\mathbf{v} \cdot \nabla_r n) - \frac{en}{kT} (\mathbf{v} \cdot \mathbf{E})]$$

$$= \int \cdots \int f_i^{(0)} f_j^{(0)} (\phi_i' + \phi_j' - \phi_i - \phi_j) v_r I_s(\theta) \, d\Omega_{CM} \, d^3 V_j. \qquad (5\text{-}3\text{-}9)$$

The form of this equation suggests that we seek a solution of the type

$$\phi_i = -n \left[(\mathbf{v} \cdot \nabla n) - \frac{en}{kT} (\mathbf{v} \cdot \mathbf{E}) \right] C_i(v^2), \qquad (5\text{-}3\text{-}10)$$

where $C_i(v^2)$ is an unknown scalar function of ion velocity. (From here on there is no ambiguity in dropping the subscript r from the gradient operator.) This is also a physically reasonable form for ϕ_i. Since it is the gradient and the field that cause the deviations from an equilibrium distribution, it is plausible to take the deviations as directly proportional to the gradient and the field. An analogous result holds for the ϕ_j describing the neutral molecules.

At this point we can already obtain the Einstein relation and an expression for D and K in terms of the unknown function $C_i(v^2)$. The ion flux density is

$$\mathbf{J} = n\mathbf{v}_d = \int \mathbf{v} f_i \, d^3 v = \int \mathbf{v} f_i^{(0)} \phi_i \, d^3 v, \qquad (5\text{-}3\text{-}11)$$

and, after a little manipulation, substitution of (5-3-10) into this expression yields

$$\mathbf{J} = -\tfrac{1}{3} n \left(\nabla n - \frac{en}{kT} \mathbf{E} \right) \int f_i^{(0)} C_i(v^2) v^2 \, d^3 v. \qquad (5\text{-}3\text{-}12)$$

Comparison with the phenomenological equation $\mathbf{J} = -D \nabla n + n K \mathbf{E}$ shows that

$$D = \left(\frac{kT}{e} \right) K = \tfrac{1}{3} n \int f_i^{(0)} C_i(v^2) v^2 \, d^3 v. \qquad (5\text{-}3\text{-}13)$$

This demonstrates that the assumption (5-3-10) for the form of ϕ_i leads to a transport equation that agrees with experiment and yields the Einstein relation.

A little progress in evaluating (5-3-13) can be made even without first finding the unknown function $C_i(v^2)$. The form of the integrand suggests expanding $C_i(v^2)$ in terms of some complete set of orthogonal functions whose orthogonality condition will make most of the integrations in (5-3-13) go to zero. Sonine polynomials are convenient for this purpose:

$$C_i(v^2) = \sum_{s=0}^{\infty} c_{is} S_{3/2}^{(s)}(w^2), \qquad (5\text{-}3\text{-}14)$$

where the c_{is} are expansion coefficients and the argument of the polynomials is $w^2 = mv^2/2kT$. Then (5-3-13) for D becomes

$$D = \frac{nkT}{m} c_{i0}. \qquad (5\text{-}3\text{-}15)$$

Thus we need find only the first expansion coefficient rather than the complete function $C_i(v^2)$.

The remainder of the manipulations are now straightforward but tedious. Substitution of (5-3-10) back into (5-3-9) yields a linear integral equation for $C_i(v^2)$. The expansion (5-3-14) is then substituted and the resulting equation solved by a moment method in which the equation is multiplied by successive Sonine polynomials and integrated. The result is an infinite set of linear algebraic equations for the c_{is} as unknowns, and the coefficients of these unknowns are complicated multiple integrals over d^3v and d^3V_j for a collision. These integrals result from the moment formation and most of the integrations can be carried out explicitly. The final two integrations, however, cannot be performed until the ion-neutral potential is specified. Thus the coefficients of the c_{is} can be reduced to linear combinations of irreducible "collision integrals" of the form

$$\overline{\Omega}^{(l,\,s)}(T) = [(s+1)!\,(kT)^{s+2}]^{-1} \int_0^{\infty} e^{-E/kT} E^{s+1} Q^{(l)}(E)\, dE, \qquad (5\text{-}3\text{-}16)$$

where the generalized transport cross sections are

$$Q^{(l)}(E) = 2\pi \left[1 - \frac{1 + (-1)^l}{2(1+l)}\right]^{-1} \int_0^{\pi} (1 - \cos^l \theta) I_s(\theta) \sin \theta \, d\theta. \qquad (5\text{-}3\text{-}17)$$

The normalization factors here have been chosen so that both the collision integrals and the transport cross sections are equal to πd^2 for the collision of classical rigid spheres of diameter d. Given the ion-neutral potential, the $Q^{(l)}$ and $\overline{\Omega}^{(l,\,s)}$ can be computed, but numerical integration is usually necessary.

Thus D is given by the value of a single unknown in an infinite set of algebraic equations that cannot be solved exactly except in very special cases. A remarkably good approximation is obtained by ignoring everything but the coefficient of interest, c_{i0}; this is equivalent to truncating the expansion (5-3-14) for $C_i(v^2)$ after the first term. This first approximation yields (Chapman and Cowling, 1970, Chapter 9; Hirschfelder et al. 1964, Chapter 7)

$$[D]_1 = \left(\frac{kT}{e}\right)[K]_1 = \frac{3}{16}\left(\frac{1}{m} + \frac{1}{M}\right)^{1/2} \frac{(2\pi kT)^{1/2}}{N\overline{\Omega}^{(1,\,1)}}. \qquad (5\text{-}3\text{-}18)$$

This is equivalent to the free-flight result given in (5-2-7), for $\overline{\Omega}^{(1,\,1)}$ is a weighted average of the cross section $Q^{(1)}$, which is seen to be exactly the same as Q_D. The higher approximations are considered in a little more detail in the next section.

To summarize, these results are valid only in the limit of zero field, binary collision mechanism, elastic collisions, and small enough deviations from equilibrium that the flux is linear in the gradient and the field. As written, the results are correct in quantum mechanics as well as classical mechanics, provided only that the Boltzmann equation is valid.

B. CONVERGENCE OF APPROXIMATIONS.   For higher approximations to D the infinite set of algebraic equations must be systematically truncated in some plausible way. Two truncation schemes are commonly used. The Chapman-Cowling scheme carves out successively larger square blocks around c_{i0}; it is equivalent to including more and more terms in the expansion (5-3-14) for $C_i(v^2)$. The Kihara scheme notes that the elements of the main diagonals of the infinite set of equations are usually much larger than the other elements; it is then assumed that the elements fall off in magnitude in a systematic way with distance from a main diagonal, and a successive approximation scheme is set up in which more and more off-diagonal elements are included. A fairly detailed exposition of these two truncation schemes and comparisons of their numerical accuracy are available (Mason, 1957a,b). Here we use only the Kihara result, which is slightly simpler than the Chapman-Cowling result, at least in the lower approximations, and is often somewhat more accurate:

$$\frac{D}{[D]_1} = \frac{K}{[K]_1} = 1 + \frac{M^2(6C^* - 5)^2}{30m^2 + 10M^2 + 16mMA^*} + \cdots, \qquad (5\text{-}3\text{-}19)$$

where

$$A^* = \frac{\overline{\Omega}^{(2,\,2)}}{\overline{\Omega}^{(1,\,1)}}, \qquad (5\text{-}3\text{-}20)$$

$$C^* = \frac{\overline{\Omega}^{(1,\,2)}}{\overline{\Omega}^{(1,\,1)}}. \qquad (5\text{-}3\text{-}21)$$

There are three known cases for which the exact value of D can be found. The first is an ion-neutral potential varying as r^{-4}, discovered by Maxwell and therefore often known as the Maxwellian model (Chapman and Cowling, 1970, Chapter 10). For this model the first approximation is exact, regardless of the relative magnitude of m and M. In terms of the infinite set of algebraic equation that must be solved in the Chapman-Enskog theory, all the elements off the main diagonals are exactly zero for the Maxwellian model; in terms of (5-3-19), the value of C^* is $\frac{5}{6}$ for this model. A similar result occurs when $m \gg M$, regardless of the ion-neutral potential; this is called a quasi-Lorentzian mixture (Mason, 1957b). Finally, for $m \ll M$ we have the special case of a Lorentzian mixture in which a few light particles move in a bed of fixed scatterers. This model was invented by Lorentz to discuss the motion of electrons in metals and can be treated exactly by different methods to give (Chapman and Cowling, 1970, Section 10.5)

$$\frac{D}{[D]_1} = \frac{K}{[K]_1} = \frac{32\overline{\Omega}^{(1,1)}}{9\pi(kT)^2} \int_0^\infty \frac{e^{-E/kT}E\,dE}{Q^{(1)}(E)}. \tag{5-3-22}$$

The Chapman-Enskog procedure converges comparatively slowly for a Lorentzian mixture.

To test the numerical convergence of the Chapman-Enskog result we can compare the approximate expression (5-3-19) with the exact results for a Lorentzian mixture, which is probably the most unfavorable case. For a potential varying as r^{-n} (5-3-19) becomes, with $m \ll M$,

$$\frac{D}{[D]_1} = \frac{K}{[K]_1} = 1 + \frac{1}{10}\left(1 - \frac{4}{n}\right)^2 + \cdots \tag{5-3-23}$$

and the exact (5-3-22) becomes

$$\frac{D}{[D]_1} = \frac{K}{[K]_1} = \frac{16}{9\pi}\Gamma\left(3 - \frac{2}{n}\right)\Gamma\left(2 + \frac{2}{n}\right). \tag{5-3-24}$$

Some numerical results are given in Table 5-3-1 for the inverse power potential $V(r) = \pm C/r^n$. From these it is apparent that the first approximation may not always be sufficiently accurate but that the second will usually be satisfactory.

It is worth examining the Chapman-Enskog convergence for an ion-molecule potential that is physically more realistic than a single inverse power. A widely used potential model includes the long-range attractive polarization energy and represents the short-range repulsion energy as an inverse power. It is convenient to write such an $(n-4)$ potential in the form

$$V(r) = \frac{n\varepsilon}{n-4}\left[\frac{4}{n}\left(\frac{r_m}{r}\right)^n - \left(\frac{r_m}{r}\right)^4\right], \tag{5-3-25}$$

TABLE 5-3-1. Convergence of Chapman-Enskog calculation for the potential $V(r) = \pm C/r^n$: values of $K/[K]_1$ at zero field for the unfavorable case of $m \ll M$ (Lorentzian mixture)

n	Approximate (5-3-23)	Exact (5-3-24)
2	1.100	1.132
3	1.011	1.014
4	1.000	1.000
6	1.011	1.014
8	1.025	1.031
10	1.036	1.045
25	1.071	1.091
50	1.085	1.110
∞	1.100	1.132

where ε is the depth of the potential minimum and r_m is the position of the minimum. For the unfavorable case of $m \ll M$ we need only the dimensionless ratio C^*, according to (5-3-19). This has been calculated by numerical integration for $n = 12$ and is tabulated (Mason and Schamp, 1958). Results equivalent to $\overline{\Omega}^{(1,\,1)}$ have been tabulated for $n = \infty$ (Hassé, 1926) and for $n = 8$ (Hassé and Cook, 1931); values of C^* can be obtained from them by numerical differentiation according to the recursion formula

$$C^* = 1 + \frac{1}{3} \frac{d \ln \overline{\Omega}^{(1,\,1)}}{d \ln T}, \qquad (5\text{-}3\text{-}26)$$

which can be verified by differentiation of (5-3-16). Some values of $K/[K]_1$ for $m \ll M$ according to (5-3-19) are given in Table 5-3-2 as a function of temperature for $n = 8$, 12, ∞. At low temperatures the Chapman-Enskog results are exact in the first approximation, since the potential (5-3-25) behaves like the r^{-4} Maxwellian model at low energies. At high temperatures the behavior approaches that of an r^{-n} repulsive potential for which the exact results have already been given in Table 5-3-1. From the results shown we may again conclude that if the first approximation is not sufficiently accurate the second approximation usually will be.

Finally we can determine whether the addition of an additonal attraction energy term to (5-3-25) is likely to affect the convergence appreciably.

TABLE 5-3-2. Convergence of Chapman-Enskog calculation for $(n\text{-}4)$ potentials: values of $K/[K]_1$ at zero field for the unfavorable case of $m \ll M$ (Lorentzian mixture), according to (5-3-19)

$\dfrac{kT}{\varepsilon}$	$\dfrac{K}{[K]_1}$		
	$n = 8$	$n = 12$	$n = \infty$
0	1.000	1.000	1.000
0.5	1.044	1.020	1.005
1	1.017	1.005	1.012
2	1.000	1.004	1.043
4	1.013	1.024	1.072
9	1.024	1.039	1.088
∞	1.025	1.044	1.100

In particular, it is believed that an r^{-6} attraction energy may be important, corresponding physically to the charge-induced quadrupole energy plus the London disperson energy (Margenau, 1941; Mason and Schamp, 1958). A model embodying these characteristics is a (12-6-4) potential, which can be written as

$$V(r) = \frac{\varepsilon}{2}\left[(1 + \gamma)\left(\frac{r_m}{r}\right)^{12} - 4\gamma\left(\frac{r_m}{r}\right)^{6} - 3(1 - \gamma)\left(\frac{r_m}{r}\right)^{4} \right], \quad (5\text{-}3\text{-}27)$$

where ε and r_m are the depth and position of the potential minimum and γ is a dimensionless third parameter that measures the relative strengths of the r^{-6} and r^{-4} energies. For $\gamma = 0$ the r^{-4} energy dominates the attraction and for $\gamma = 1$ the r^{-6} energy dominates. Some values of $K/[K]_1$ for $m \ll M$, according to (5-3-19), are shown in Table 5-3-3 as a function of temperature for $\gamma = 0, 0.5, 1$, calculated from existing tabulations (Mason and Schamp, 1958). It is apparent that the addition of the r^{-6} energy does not affect the convergence in any important way.

In short, it appears that the first two terms of (5-3-19) should give the zero-field mobility with a convergence error of not more than 3%. Although the foregoing results have ignored quantum effects and the possibility of resonant charge transfer, inclusion of these effects does not seem to alter the convergence appreciably (Dalgarno and Williams, 1958).

TABLE 5-3-3. Effect of additional r^{-6} energy on the convergence of the Chapman-Enskog calculation: values of $K/[K]_1$ at zero field for (12-6-4) potentials according to (5-3-19) for the unfavorable case of $m \ll M$ (Lorentzian mixture)

$\dfrac{kT}{\varepsilon}$	$\dfrac{K}{[K]_1}$		
	$\gamma = 0$	$\gamma = 0.5$	$\gamma = 1$
0	1.000	1.000	1.011
0.5	1.020	1.006	1.000
1	1.005	1.001	1.000
2	1.004	1.007	1.009
4	1.024	1.027	1.029
9	1.039	1.041	1.043
∞	1.044	1.044	1.044

C. CONNECTION WITH ELEMENTARY THEORIES. For many years the accurate Chapman-Enskog theory coexisted with the earlier free-path or free-flight theories, seemingly inconsistent with or at least disconnected from them, although there was an intuitive feeling that a connection had to exist. The connection was established by Monchick (1962) for a single gas and by Monchick and Mason (1967) for mixtures. In this procedure an expression is written down for the distribution function (or for any flux) in which the history of a particle is followed back in time through N collisions; this is the formal generalization of simple free-flight theory to include N collision histories instead of just one. This expression has the appearance of a Liouville-Neumann iteration solution to an integral equation, and it is not difficult to tell by inspection what the integral equation must be. This integral equation turns out to be equivalent to the integral equation (5-3-9) which was solved by the Chapman-Enskog method. Thus the connection between the two types of theory becomes clear: one corresponds to solution of an integral equation by iteration, the other to solution by expansion in orthogonal functions.

The relation between the momentum-transfer theory of diffusion and the Chapman-Enskog theory has been apparent for some time, although it does not seem to be widely known. A rigorous treatment of diffusion can be based on Maxwell's equation of change, which is obtained from the

Boltzmann equation by multiplication by molecular mass and integration over velocities. This was, in fact, the procedure originally used by Chapman. The calculation is reduced to finding the effect of collisions on $\sum v_x$ and then integrating over all collisions, expanding the distribution function in a power series in the molecular velocity components (Kennard, 1938, Section 109). The results are identical to those obtained by the Chapman-Enskog method, but the computation of the effect of collisions on $\sum v_x$ is the same as the computation of momentum transfer, which is $m \sum v_x$. The only difference is that the usual momentum transfer calculation takes account of the distribution function only in a simple approximate way (Present, 1958, Sections 4-2 and 8-3); a more systematic treatment of the distribution function makes the momentum transfer calculation mathematically identical to the Maxwell-Chapman calculation.

Thus the connections among the various theories can be summarized by the following diagram:

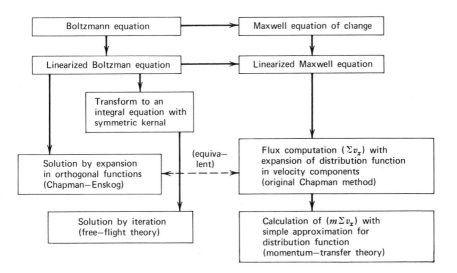

D. TEMPERATURE DEPENDENCE OF MOBILITY.   As we can see from (5-3-18), the temperature dependence of the mobility is given essentially by the temperature dependence of the quantity $[T^{1/2}\overline{\Omega}^{(1, 1)}]$, except for a small contribution from the higher Chapman-Enskog corrections. The effect of the ion-neutral potential is contained in $\overline{\Omega}^{(1, 1)}$, as shown by (5-3-16) and (5-3-17). The temperature dependence of the mobility is thus closely connected with the nature of the ion-neutral potential; as we show in this section, the connection is sensitive enough that measurements of the temperature dependence of mobility can supply information on the ion-neutral potential.

We proceed by means of model calculations. Since all ion-neutral inter-actions must involve a polarization contribution, we always include an r^{-4} energy in the models. Unless the neutral molecule possesses an appreciable permanent dipole or quadrupole moment, this polarization energy is the longest-ranged component of the potential, and therefore dominates the scattering at low energies. Thus in the limit of low temperatures the mobility is dominated by the polarization and becomes independent of the other components of the potential. Since $\overline{\Omega}^{(1,1)}$ varies as $T^{-2/n}$ for an r^{-n} potential, we see that the polarization limit of the mobility is a nonzero constant. This constant must be found by numerical integration; the first evaluation was carried out by Langevin (1905) and improved by Hassé (1926), and the most accurate recent value has been reported by Heiche and Mason (1970). In particular, the polarization potential is

$$V_p(r) = -\frac{e^2\alpha}{2r^4},\qquad(5\text{-}3\text{-}28)$$

where α is the polarizability of the neutral, and the polarization limit of the mobility at standard gas density is

$$K_p = \frac{13.876\ \text{cm}^2}{(\alpha\mu)^{1/2}\ \text{V-sec}},\qquad T\to 0°\text{K},\qquad(5\text{-}3\text{-}29)$$

where α is in Å^3 and μ is the reduced mass in g/mole. In this case the first Chapman-Enskog approximation is exact. For ion-dipole (r^{-2}) and ion-quadrupole (r^{-3}) potentials the mobility should continue to decrease as T approaches zero. But these potentials are orientation-dependent and are zero for the lowest rotational state of the molecule so that the mobility must ultimately pass through a minimum at some very low temperature and rise back to the polarization limit (Arthurs and Dalgarno, 1960).

First we investigate the effect of including a short-range repulsion energy in the potential, using the $(n\text{-}4)$ potential given by (5-3-25). Variation of n then shows how the steepness of the repulsion affects the mobility. In Fig. 5-3-1 the ratio $[K]_1/K_p$ is plotted as a function of temperature for $n = 8, 12, \infty$. These results all required extensive numerical integration and are taken from published tabulations (Hassé, 1926; Hassé and Cook, 1931; Mason and Schamp, 1958). The interesting feature is that the mobility rises to a maximum and then decreases slowly, with the height of the maximum strongly dependent on the steepness of the repulsion energy. It is as if the addition of repulsion energy tended to cancel the polarization attraction to some extent, thus reducing the magnitude of the

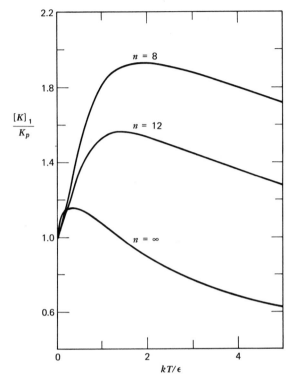

FIG. 5-3-1. Effect of repulsion energy on mobility, as illustrated for a series of $(n$-4$)$ potentials.

cross section and increasing the mobility at intermediate temperatures. At high temperatures the repulsion dominates and the value of $\overline{\Omega}^{(1,\,1)}$ decreases as $T^{-2/n}$; the eventual decrease of mobility at high temperatures therefore is as $T^{-(1/2)+(2/n)}$.

We next consider the effect of adding some r^{-6} attraction energy to a (12-4) potential; this yields the (12-6-4) potential given by (5-3-27). The ratio $[K]_1/K_p$ is shown in Fig. 5-3-2 as a function of temperature for $\gamma = 0$, 0.25, 0.50, and 0.75, as obtained from published tabulations (Mason and Schamp, 1958). The curve for $\gamma = 0$ here is the same as the curve for $n = 12$ in Fig. 5-3-1. The addition of r^{-6} attraction energy is seen to have the effect of decreasing the height of the maximum in the mobility versus temperature curve and in that sense simulates the effect of increasing the steepness of the repulsion. More surprising is the appearance of a minimum in the curve for sufficiently large values of γ. Thus the addition of sufficient r^{-6} attraction energy has the effect of permitting the mobility to fall lower

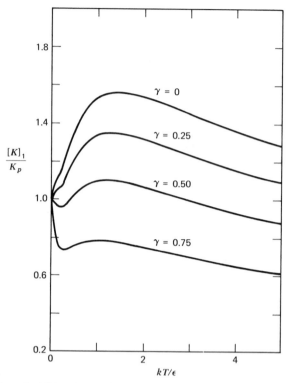

FIG. 5-3-2. Effect of additional attraction energy on mobility, as illustrated for a series of (12-6-4) potentials.

than the polarization limit and greatly reduces the height of the maximum. The (12-6-4) model gives a good account of the mobility of alkali ions in noble gases, with well depths ε in the range of 0.014 to 0.12 eV (Mason and Schamp, 1958).

The initial decrease of the mobility with the addition of sufficient r^{-6} attraction energy suggests that the temperature variation of mobility can be very different if the short-range potential is attractive rather than repulsive. This can occur when valence forces come into play. Some examples would be H^+ in H_2, H^+ in He, N^+ in N_2, and O^+ in O_2, since H_3^+, HeH^+, N_3^+, and O_3^+ are all known to be stable species with rather large binding energies. For such cases we might expect the mobility to decrease from the polarization limit with increasing temperature. Eventually, of course, the interaction energy must be repulsive if the particles are close enough together, but this would not cause the mobility to rise until very high temperatures were reached.

To illustrate the foregoing points, the calculated and experimental mobilities of Li^+ in He and of H^+ in H_2 are shown in Fig. 5-3-3. The Li^+ in He calculations were obtained by fitting a (12-6-4) potential to the experimental measurements (Mason and Schamp, 1958). The H^+ in H_2 calculations were based on a determination of the potential from a combination of quantum-mechanical calculations and analysis of

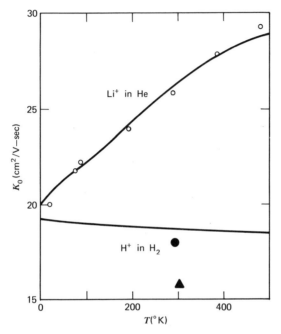

FIG. 5-3-3. Differing effects of short-range repulsion (Li^+ in He) and valence attraction (H^+ in H_2) on the temperature dependence of mobility. Experimental results for Li^+ in He are from Hoselitz (1941) and for H^+ in H_2 from Persson and Brown (●, 1955) and Miller et al. (▲, 1968). (The results for Li^+ in He refer to a standard gas density at 18°C and 1 atm rather than at 0°C and 1 atm.)

measurements of the scattering of proton beams by H_2 gas (Mason and Vanderslice, 1959). The difference in the effects of short-range repulsion (Li^+ in He) and of valence attraction (H^+ in H_2) is striking. (Possible inelastic effects have been neglected in these calculations.)

As a final example we consider a core model suggested as a representation of polyatomic ions, which are bulky and whose charges are often not located near their geometric centers. This model consists of a (12-4) central potential displaced from the origin,

$$V(r) = \frac{\varepsilon}{2}\left[\left(\frac{r_m - a}{r - a}\right)^{12} - 3\left(\frac{r_m - a}{r - a}\right)^4\right],\qquad (5\text{-}3\text{-}30)$$

where the ion-neutral separation r is measured between geometric centers, a is the rigid core diameter, and ε and r_m have their usual meaning. The ratio $[K]_1/K_p$ is shown in Fig. 5-3-4 as a function of temperature for several values of $a^* = a/r_m$ (Mason et al., 1972). It is clear that even a small core size greatly diminishes the height of the mobility maximum, and a moderate core size is sufficient to suppress the maximum and reduce the whole curve below the polarization limit. The effect, however, is *not* the same as the simple addition of a rigid core to a polarization potential,

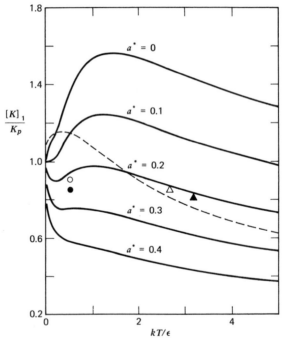

FIG. 5-3-4. Effect of core size on mobility, as illustrated for a series of (12-4) core potentials. The added core suppresses the maximum and eventually reduces the mobility everywhere below the polarization limit. This is not true for the (∞-4) model of a rigid sphere with an added polarization potential, shown as the dashed curve. The points correspond to the following measurements: $\bigcirc$ SF$_5^-$, $\bullet$ SF$_6^-$, $\triangle$ (SF$_6$)SF$_6^-$, $\blacktriangle$ (SF$_6$)$_2$SF$_6^-$, all in SF$_6$ (Patterson, 1970a).

which yields the dashed curve in Fig. 5-3-4 (this is the same as the $n = \infty$ curve of Fig. 5-3-1). The reason is that the core model of (5-3-30) corresponds roughly to the charge being located on the surface of the sphere, whereas the $(\infty\text{-}4)$ model corresponds to the charge located at the center of the sphere. Also shown in Fig. 5-3-4 are mobilities measured at 300°K in SF_6 for $SF_5{}^-$, $SF_6{}^-$, $(SF_6)SF_6{}^-$, and $(SF_6)_2SF_6{}^-$ (Patterson, 1970a). It is apparent that either the core model or the $(\infty\text{-}4)$ model can account for the mobilities of the clustered ions $(SF_6)SF_6{}^-$ and $(SF_6)_2SF_6{}^-$, but the $(\infty\text{-}4)$ model cannot account for the low mobilities of $SF_5{}^-$ and $SF_6{}^-$.

Comparison of Figs. 5-3-2 and 5-3-4 shows that the effects of adding r^{-6} energy and adding a core are rather similar, although on close inspection the curves are found to have somewhat different shapes.

There are two important restrictions on the results presented in this section. The first is that all the calculations are based on classical mechanics. There will be appreciable quantum effects for light particles at low temperatures. The second is that the possibility of resonant charge transfer has been ignored. Charge transfer has a strong effect on the temperature dependence of mobility, causing it to fall smoothly and slowly with increasing temperature. Both effects are discussed in detail later.

E. QUANTUM EFFECTS.   The expressions (5-3-16), (5-3-17), and (5-3-18) for the mobility are valid in both classical and quantum mechanics. The entire difference between classical and quantal results resides in the differential scattering cross section $I_s(\theta)$ and all quantum effects originate there. The entire dependence of the mobility on the ion-neutral potential is also contained in $I_s(\theta)$.

Given the potential, we could first calculate $I_s(\theta)$ and then obtain the transport cross sections and collision integrals by integration, an inefficient procedure from a computational standpoint. In a classical calculation it is better to convert to the impact parameter formulation,

$$I_s(\theta) \sin \theta \, d\theta = b \, db,$$

calculate $\theta(b)$ from Newton's second law of motion according to (5-2-36), and then obtain the transport cross sections by integration over b. In a quantal calculation it is better to perform a phase-shift analysis and then obtain the transport cross sections directly from the phase shifts without the intermediate use of $I_s(\theta)$.

In this section we first show how a phase-shift calculation proceeds and then indicate the typical quantum effects that can occur. More detailed discussion is reserved for Chapter 6.

In a phase-shift calculation a plane wave incident on a scattering center is first decomposed into component partial waves, each partial wave corresponding to a different value of the angular-momentum quantum number l. Each partial wave is then followed through the scattering region by integration of the corresponding radial wave equation; since energy and angular momentum are conserved, the only effect of the scattering potential is to shift the phase of the partial wave by an amount δ_l. The partial waves are then recombined to form the scattered wave, from which the differential cross section is identified. In particular, the scattering amplitude $f_s(\theta)$ in terms of the phase shifts is

$$f_s(\theta) = \frac{1}{2i\kappa} \sum_{l=0}^{\infty} (2l + 1)(e^{2i\delta_l} - 1)P_l(\cos \theta), \qquad (5\text{-}3\text{-}31)$$

where $\kappa = \mu v/\hbar$ is the wavenumber of relative motion, and the differential cross section is

$$I_s(\theta) = |f_s(\theta)|^2. \qquad (5\text{-}3\text{-}32)$$

On substituting (5-3-31) and (5-3-32) back into the expression (5-3-17) for the transport cross sections, we find that the integration over θ can be carried out explicitly. After considerable trigonometric transformation the expressions for the first four transport cross sections can be put in the following neat forms:

$$Q^{(1)} = \frac{4\pi}{\kappa^2} \sum_{l=0}^{\infty} (l + 1) \sin^2 (\delta_l - \delta_{l+1}), \qquad (5\text{-}3\text{-}33)$$

$$Q^{(2)} = \frac{4\pi}{\kappa^2} \left(\frac{3}{2}\right) \sum_{l=0}^{\infty} \frac{(l + 1)(l + 2)}{(2l + 3)} \sin^2 (\delta_l - \delta_{l+2}), \qquad (5\text{-}3\text{-}34)$$

$$Q^{(3)} = \frac{4\pi}{\kappa^2} \sum_{l=1}^{\infty} \left(\frac{l + 1}{2l + 5}\right) \left[\frac{(l + 2)(l + 3)}{(2l + 3)} \sin^2 (\delta_l - \delta_{l+3}) \right.$$
$$\left. + \frac{3(l^2 + 2l - 1)}{(2l - 1)} \sin^2 (\delta_l - \delta_{l+1}) \right], \qquad (5\text{-}3\text{-}35)$$

$$Q^{(4)} = \frac{4\pi}{\kappa^2} \left(\frac{5}{4}\right) \sum_{l=0}^{\infty} \frac{(l + 1)(l + 2)}{(2l + 3)(2l + 7)} \left[\frac{(l + 3)(l + 4)}{(2l + 5)} \sin^2 (\delta_l - \delta_{l+4}) \right.$$
$$\left. + \frac{2(2l^2 + 6l - 3)}{(2l - 1)} \sin^2 (\delta_l - \delta_{l+2}) \right]. \qquad (5\text{-}3\text{-}36)$$

The factors 3/2 and 5/4 come from the normalization constant in the definition of the generalized transport cross sections (5-3-17) and have no special significance. The general formulation of all the transport cross sections in terms of phase shifts is rather complicated and has been given by Wood (1971).

Before indicating specific quantum effects we should see how the foregoing expressions reduce to the corresponding classical ones. Two conditions are sufficient to obtain the classical limit:

1. The phase shifts can be calculated by the semiclassical Jeffreys-Wentzel-Kramers-Brillouin (JWKB) approximation.
2. The quantum number l can be taken as a continuous variable, so that differences can be replaced by differentials and sums replaced by integrals.

From the first condition we make the identification between l and b:

$$l + \tfrac{1}{2} = \kappa b; \tag{5-3-37}$$

the phase shift is given by the JWKB expression

$$\delta_l = \delta(b) = \kappa \int_{r_a}^{\infty} \left[1 - \frac{b^2}{r^2} - \frac{V(r)}{E} \right]^{1/2} dr - \kappa \int_{b}^{\infty} \left(1 - \frac{b^2}{r^2} \right)^{1/2} dr. \tag{5-3-38}$$

Differentiation of (5-3-38) and comparison with the expression (5-2-36) for the deflection angle establishes the semiclassical connection

$$\theta(b) = 2\frac{d\delta_l}{dl} = \frac{2}{\kappa}\frac{d\delta(b)}{db}. \tag{5-3-39}$$

Phase shift differences are thus directly related to the deflection angle

$$\delta_{l+n} - \delta_l = \frac{n}{2}\theta. \tag{5-3-40}$$

Substitution of (5-3-40) into the expressions for the transport cross sections and replacement of the sums by integrations yields

$$Q^{(1)} \approx \frac{4\pi}{\kappa^2} \int_0^{\infty} (l + 1) \sin^2 \frac{\theta}{2} \, dl \rightarrow 2\pi \int_0^{\infty} (1 - \cos \theta) b \, db, \tag{5-3-41}$$

with similar expressions for the other cross sections. Thus the two conditions lead *exactly* to the classical limit. This is a special result for transport cross sections and is due to the occurrence of phase shift differences. For other types of cross section these two conditions lead to semiclassical results, which still exhibit many quantum effects, and further approximations must be made to obtain classical limits.

Quantum effects can usually be ascribed to either of two causes: (a) inaccuracy of the JWKB approximation or (b) the discrete nature of l. It is the latter that produces the most striking effects in the form of maxima and minima in the diffusion cross section, caused by resonances in the phase shifts (i.e., rapid changes in δ_l through multiples of $\pi/2$), which are associated with the phenomenon of classical orbiting occurring at a maximum in the effective potential (actual potential plus centrifugal potential). Failure of the JWKB approximation usually modifies quantitative features but does not produce qualitative changes. The orbiting resonances, however, are often washed out by the averaging of the cross section to form the collision integrals; only for light particles do the effects persist through to the mobility.

The foregoing effects are illustrated in Fig. 5-3-5, which shows the diffusion cross section as a function of energy calculated by Munn et al. (1964) for a (12-6) potential,

$$V(r) = 4\varepsilon\left[\left(\frac{\sigma}{r}\right)^{12} - \left(\frac{\sigma}{r}\right)^{6}\right],$$ (5-3-42)

where σ is the value of r such that $V(\sigma) = 0$. This potential is not quantitatively suitable as a model for ion-neutral interactions because it lacks the asymptotic r^{-4} form but it should show the same general behavior. A measure of the quantum behavior of the system is the ratio $\lambda/\sigma = 2\pi/\kappa\sigma$, where λ is the de Broglie wavelength; for potentials like (5-3-42) it is common practice to use the de Boer parameter Λ^*,

$$\Lambda^* = \frac{h}{\sigma(2\mu\varepsilon)^{1/2}},$$ (5-3-43)

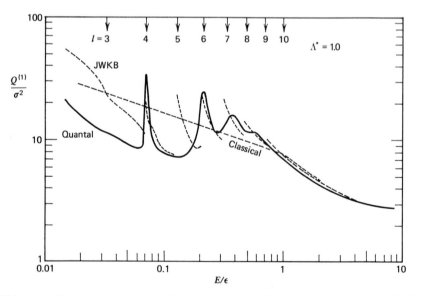

FIG. 5-3-5. Quantal, semiclassical JWKB, and classical diffusion cross sections for a (12-6) potential, showing orbiting resonances and other quantum effects discussed in the text.

which is the value of λ/σ for collisions of energy ε. The results in Fig. 5-3-5 are for $\Lambda^* = 1$. The correct quantal curve shows the orbiting resonances clearly; they are seen to occur at those energies for which one of the phase shifts shows a rapid change, as indicated by the corresponding value of l at the top of the figure. The curve marked JWKB was calculated by properly summing (not integrating) phase shifts calculated by the JWKB approximation. It also shows structure due to orbiting resonances but differs from the exact quantal curve, especially at low energies, because of the inaccuracies of the JWKB approximation for Λ^* as large as 1.0. The classical curve shows no trace of the orbiting resonances and roughly represents an average drawn through the resonances; this classical average representation becomes more accurate as Λ^* is decreased. Although for the (12-6) model classical orbiting is possible only for $E/\varepsilon \leq 0.8$, close inspection shows a small undulation on the classical curve at $E/\varepsilon \approx 1$, due to rainbow scattering, which in this sense is a sort of classical resonance behavior. There is no sharp distinction between orbiting and rainbows from a quantum-mechanical viewpoint because of the possibility of tunneling near the top of the centrifugal barrier. These effects are discussed in more detail in Chapter 6.

To give some idea of quantum effects on mobilities the mobility of H^+ in He is shown in Fig. 5-3-6, as based on accurately calculated values of

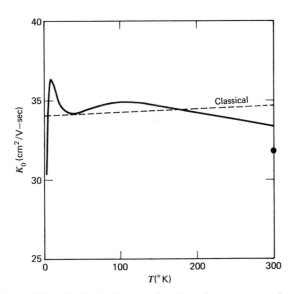

FIG. 5-3-6. The mobility of H^+ in He as a function of temperature, showing quantum effects. The experimental point is from Orient (1971).

the potential (Peyerimhoff, 1965; Wolniewicz, 1965). The quantal calculations of the mobility were made by Dickinson (1968b). The classical curve shown in the figure is a rough calculation based on fitting a (12-6-4) model to the accurate potential. The well depth is so great ($\varepsilon = 2.04\,\mathrm{eV}$) that the classical mobility is little different from the polarization limit even up to 500°K. Even the quantal mobility remains close to the polarization limit. The maximum at 110°K is attributed by Dickinson to a minimum in $Q^{(1)}$ caused by a resonance in the $l = 11$ phase shift.

The quantal mobility does not approach the classical polarization limit at $T = 0°K$ because at low energies only a few phase shifts contribute to the cross section. It can be shown (Landau and Lifshitz, 1958, Section 108) that, for potentials falling off faster than r^{-3}, the phase shifts for $l \neq 0$ vanish as a power of κ but δ_0 is linear in κ. In particular, for a potential vanishing asymptotically as r^{-4}

$$\delta_l \propto \kappa^2 \qquad (l \neq 0), \qquad (5\text{-}3\text{-}44)$$

$$\tan \delta_0 = -\kappa a, \qquad (5\text{-}3\text{-}45)$$

where a is called the scattering length. Moreover, in the limit of zero energy

$$\delta_0(0) = n_0\,\pi, \qquad (5\text{-}3\text{-}46)$$

where n_0 is the number of bound states in the potential well (Levinson's theorem). Thus at low energies all the δ_l are negligible compared with δ_0, which has the form

$$\delta_0 = n_0\,\pi + \kappa a + \cdots,$$

and all the transport cross sections take on the same limiting form

$$Q^{(l)} \to \frac{4\pi}{\kappa^2} \sin^2 \delta_0 = 4\pi a^2, \qquad (5\text{-}3\text{-}47)$$

as for rigid spheres. The mobility at low temperatures must therefore ultimately diverge as $T^{-1/2}$, but it happens at a temperature so low that it is experimentally unimportant. For the rather extreme case of H^+ in He there is no trace of such behavior even at 1°K (Dickinson, 1968b).

F. INELASTIC COLLISIONS. All the foregoing results apply strictly only to monatomic ions and neutrals; a proper account of polyatomic ions and neutrals must allow for inelastic collisions involving rotational and vibrational degrees of freedom. This requires starting over again at the very beginning with a reformulation of the Boltzmann equation, but we can anticipate two results without any calculations at all. First, the Einstein relation will continue to hold rigorously, since it is much more general than any Boltzmann equation. Second, since diffusion involves mass transfer and mass is conserved in an inelastic collision as well as in an elastic collision, we should not expect inelastic collisions to modify diffusion very much. Inelastic collisions would be expected to modify energy transfer strongly, however.

The usual procedure treats the translational degrees of freedom classically and the internal degrees of freedom quantum mechanically. Each internal state of ion and neutral is considered a separate chemical species, and inelastic collisions are considered as special chemical reactions (Wang Chang et al., 1964). Thus there is a Boltzmann equation for each internal state, just as there is for a multicomponent mixture. Moreover, the differential cross section for elastic scattering $I(\theta, E)$ must be replaced by a differential cross section $I_{ab}^{cd}(\theta, \phi, E)$ that describes collisions between two particles initially in internal states a and b which emerge from the collision in final states c and d at the angle θ, ϕ. Procedures in which the internal degrees of freedom are also treated classically have been given by Taxman (1958), by She and Sather (1967), and by Hoffman and Dahler (1969). Completely quantal treatments have been given by Waldmann (1957, 1958) and by Snider (1960).

The set of Boltzmann equations can be solved by the Chapman-Enskog method, with the following modifications. The expansion for the perturbation ϕ_i of the distribution function must include terms that involve angular velocities as well as translational velocities (Kagan and Afanas'ev, 1962; Waldmann, 1963). These extra "spin polarization" terms appear to have only a slight effect on the numerical value of the diffusion coefficient, as judged by calculations for the model of loaded spheres (Sandler and Dahler, 1967; Sandler and Mason, 1967). The expansion (5-3-14) for the function C_i must also include terms relating to internal energy as well as translational energy, even if spin polarization is ignored, and (5-3-15) for D will then contain two expansion coefficients instead of one.

Detailed formal calculations for diffusion coefficients according to the above procedure have been carried through by Monchick et al. (1963, 1966, 1968) and by Alievskiĭ and Zhdanov (1969). The external appearances of the expressions for D and K, including higher approximations, remain the same as for elastic collisions, but the explicit expressions for the collision integrals and transport cross sections are much more complicated. The first few are, without spin polarization,

$$\overline{\Omega}^{(l,\,s)}(T) = \frac{2}{(s+1)!Z_i Z_j} \sum_{abcd} \exp\left(-\varepsilon_{ia} - \varepsilon_{jb}\right) \int_0^\infty \gamma^{2s+3} e^{-\gamma^2} Q_{ab}^{(l)cd} \, d\gamma,$$

$$\text{for } l, s \leq 2, \qquad (5\text{-}3\text{-}48)$$

$$\gamma^2 Q_{ab}^{(1)cd}(E) = \int_0^{2\pi} d\phi \int_0^{\pi} (\gamma^2 - \gamma\gamma' \cos\theta) I_{ab}^{cd} \sin\theta \, d\theta, \qquad (5\text{-}3\text{-}49)$$

$$\gamma^4 Q_{ab}^{(2)cd}(E) = \tfrac{3}{2} \int_0^{2\pi} d\phi \int_0^{\pi} [\gamma^2(\gamma^2 - \gamma'^2 \cos^2\theta) - \tfrac{1}{6}(\gamma^2 - \gamma'^2)^2] I_{ab}^{cd} \sin\theta \, d\theta,$$

$$(5\text{-}3\text{-}50)$$

where

$$\gamma^2 = \frac{E}{kT}, \qquad \gamma'^2 = \frac{E'}{kT}, \qquad (5\text{-}3\text{-}51)$$

$$\gamma^2 - \gamma'^2 = \varepsilon_{ic} + \varepsilon_{jd} - (\varepsilon_{ia} + \varepsilon_{jb}), \qquad (5\text{-}3\text{-}52)$$

in which primes denote quantities after collision and the ε's are the energies of the internal quantum states divided by kT. The normalization factors Z_i and Z_j are the internal partition functions

$$Z_i = \sum_a e^{-\varepsilon_{ia}}, \qquad Z_j = \sum_b e^{-\varepsilon_{jb}}. \qquad (5\text{-}3\text{-}53)$$

No evaluations of any inelastic collision integrals or transport cross sections have yet been made for realistic potential models, but the following argument suggests that inelastic effects on diffusion coefficients are small. Inelastic collisions enter $\overline{\Omega}^{(1,\,1)}$ mainly through the term $\gamma\gamma' \cos\theta$; to a first approximation $\gamma' \approx \gamma$ and there is no effect from inelastic collisions. For a second approximation γ' can be written as γ plus some terms in $\Delta\varepsilon = \gamma^2 - \gamma'^2$; inelastic correction terms are then of the form of integrals of $\gamma(\Delta\varepsilon) \cos\theta$ over angles and energies. Such terms would vanish for isotropic scattering and even for nonisotropic scattering would be expected to be small unless there is some special correlation between θ and $\Delta\varepsilon$. However, even if these extra terms are ignored, so that the formulas have the same appearance as those for elastic collisions only, the differential cross sections I_{ab}^{cd} still include all the inelastic scattering. Thus the use of an effective spherically symmetric potential can be justified, but the final collision integrals should be regarded as made up of elastic plus inelastic excitation and de-excitation contributions. At any rate it is an empirical fact that diffusion coefficients of polyatomic gases can usually be correlated by potential models in which inelastic collisions are ignored. This argument probably does not apply at very low temperatures, however.

As far as mobilities are concerned, the most important results of all these elaborate formal kinetic-theory calculations are that the Einstein relation remains valid, that the external forms of the transport equations and transport coefficients remain unchanged, and that numerical values of mobilities are probably only slightly affected by inelastic collisions (except possibly at very low temperatures).

G. MULTICOMPONENT DIFFUSION. BLANC'S LAW.   In the first Chapman-Enskog approximation multicomponent diffusion is described by the same set of diffusion coefficients that would describe diffusion in all the possible binary mixtures contained in the multicomponent mixture. The results take a particularly simple form for diffusion of a trace species through a stagnant mixture of uniform composition,

$$\frac{1}{[D_{i\,\text{mix}}]_1} = \frac{x_1}{[D_{i1}]_1} + \frac{x_2}{[D_{i2}]_1} + \cdots, \tag{5-3-54}$$

where i denotes the trace species (ions) and the x's are the mole fractions of the mixture. The corresponding expression for mobilities is

$$\frac{1}{[K_{\text{mix}}]_1} = \frac{x_1}{[K_1]_1} + \frac{x_2}{[K_2]_1} + \cdots, \tag{5-3-55}$$

where K_1, K_2, etc., are the mobilities of the ion in the various pure gases. This expression is known as Blanc's law (Blanc, 1908). Physically, (5-3-54) and (5-3-55) are simply statements about additivity of cross sections in a binary collision limit and as such can be obtained from almost any elementary theory, as in Section 5-2-C.

Blanc's law does not hold in the higher Chapman-Enskog approximations. The correction terms for each $[D_{ij}]_1$ do not depend solely on the species i and j but involve all the species in the mixture. A general variational proof, known for some time (Holstein, 1955; Biondi and Chanin, 1961), shows that the deviations from Blanc's law are always positive and of the same order of magnitude as the higher order corrections to the pure component mobilities, but explicit expressions for the deviations from Blanc's law were not worked out until much later (Sandler and Mason, 1968).

The reason for worrying enough about the kinetic-theory deviations from Blanc's law to bother working through the complicated and tedious calculations involved is that experimental deviations from this law are often taken as evidence that the ion changes its chemical identity through clustering or complex formation as the gas composition is varied. It is important therefore to know how much deviation can be attributed to purely kinetic-theory effects in which the ion remains unchanged.

The Einstein relation remains valid in multicomponent mixtures to all degrees of Chapman-Enskog approximation.

Because of the algebraic complexity involved, expressions for kinetic-theory deviations from Blanc's law have been given explicitly only for binary gas mixtures and for the second order of Chapman-Enskog approximation (Sandler and Mason, 1968). The following expression is valid when only elastic collisions occur:

$$\frac{1}{[K_{\mathrm{mix}}]_2} = \frac{x_1}{[K_1]_2} + \frac{x_2}{[K_2]_2} + \Delta(\mathrm{Blanc}), \qquad (5\text{-}3\text{-}56)$$

$$\Delta(\mathrm{Blanc}) = \frac{x_1 x_2}{4} \frac{q_1 q_2}{[K_1]_1 [K_2]_1} \left[\frac{x_1 q_1}{[K_1]_1} + \frac{x_2 q_2}{[K_2]_1} \right]^{-1}$$

$$\times \left[\frac{M_1}{m + M_1} \left(\frac{6C_{i1}^* - 5}{q_1} \right) - \frac{M_2}{m + M_2} \left(\frac{6C_{i2}^* - 5}{q_2} \right) \right]^2, \qquad (5\text{-}3\text{-}57)$$

where

$$q_j = \frac{1}{(m + M_j)^2} \left(\frac{15}{2} m^2 + \frac{5}{2} M_j^2 + 4mM_j A_{ij}^* \right), \qquad (5\text{-}3\text{-}58)$$

and the C_{ij}^* and A_{ij}^* are the dimensionless ratios of collision integrals already defined. Since the q_j are always positive, it is obvious from the form of (5-3-57) that $\Delta(\mathrm{Blanc})$ is always positive, in accord with Holstein's theorem. The maximum value of $\Delta(\mathrm{Blanc})$ is

$$\max \Delta(\mathrm{Blanc}) = \frac{q_1 q_2}{4[K_1]_1 [K_2]_1} \left[\left(\frac{q_1}{[K_1]_1} \right)^{1/2} + \left(\frac{q_2}{[K_2]_1} \right)^{1/2} \right]^{-2}$$

$$\times \left[\frac{M_1}{m + M_1} \left(\frac{6C_{i1}^* - 5}{q_1} \right) - \frac{M_2}{m + M_2} \left(\frac{6C_{i2}^* - 5}{q_2} \right) \right]^2, \qquad (5\text{-}3\text{-}59)$$

and occurs at a composition

$$(x_1)_{\mathrm{max}} = \left(\frac{q_2}{[K_2]_1} \right)^{1/2} \left[\left(\frac{q_1}{[K_1]_1} \right)^{1/2} + \left(\frac{q_2}{[K_2]_1} \right)^{1/2} \right]^{-1}. \qquad (5\text{-}3\text{-}60)$$

We can immediately note several cases for which Blanc's law is exact. The first is the Maxwellian r^{-4} potential for which $(6C_{ij}^* - 5) = 0$. The second is that of heavy ions in which $m \gg M_1, M_2$. These are just the cases for which it was previously pointed out that the Chapman-Enskog first approximation is exact; therefore it is no surprise that Blanc's law is also exact. The third case occurs when $M_1 = M_2$ and the ion-neutral interaction is the same for both neutral species. The factor containing the $(6C_{ij}^* - 5)$ is then zero because the two terms composing it cancel each other.

The major percentage variation of $\Delta(\mathrm{Blanc})$ occurs through the factor containing the $(6C_{ij}^* - 5)$, as can be seen by writing G^2 for this factor and rearranging (5-3-59) to yield (here we drop the subscripts denoting the order of the approximation)

$$\frac{\max \Delta(\text{Blanc})}{1/K_{\text{mix}}(\text{Blanc})} = \frac{\frac{1}{4}q_1 q_2 G^2}{q_1 + q_2 + (K_1 + K_2)(q_1 q_2/K_1 K_2)^{1/2}} \leq \frac{\frac{1}{4}q_1 q_2 G^2}{(q_1^{1/2} + q_2^{1/2})^2}$$

(5-3-61)

The last step follows because $\frac{1}{2}(K_1 + K_2) \geq (K_1 K_2)^{1/2}$. Since the q_j lie between 2.5 and 7.5, the contribution of the factors other than G^2 lies between about $\frac{5}{32}$ and $\frac{15}{32}$. It is possible to concoct fairly large deviations from Blanc's law by suitable choices of mass ratios and ion-neutral potentials; for instance, $(6C_{ij}^* - 5)$ is $+1$ for a rigid-sphere potential and -1 for an r^{-2} potential. Taking these values and taking $m \ll M_1$, M_2, we find a maximum deviation of about 10%. An even greater deviation can be obtained by use of a screened Coulomb potential in place of the r^{-2} potential, which leads to a maximum deviation of more than 50%; this corresponds physically to the mobility of foreign ions through a partially ionized gas. A more realistic case would be ions of mass intermediate between M_1 and M_2, with an r^{-2} potential for the ion and light neutrals. Here the maximum deviation is only about 5%. Most other combinations of masses and potentials lead to much smaller deviations. Some values of $(6C^* - 5)$ are shown as a function of temperature in Fig. 5-3-7 for $(n\text{-}4)$ potentials and in Fig. 5-3-8 for $(12\text{-}6\text{-}4)$ potentials.

In short, kinetic-theory deviations from Blanc's law are usually small, although it is possible to imagine systems with rather large deviations. In any case, the deviation is given explicitly by (5-3-57).

H. EFFECT OF TEMPERATURE GRADIENTS. THERMAL DIFFUSION. In all the discussion so far it has been assumed that the temperature is uniform. Temperature gradients can, however, cause diffusion in mixtures; this phenomenon, known as thermal diffusion, requires the addition of another term to the diffusion equation (5-1-1), which becomes

$$\mathbf{J} = -D \, \nabla n + nK\mathbf{E} - nD\alpha_T \, \nabla \ln T, \qquad (5\text{-}3\text{-}62)$$

where α_T is a dimensionless quantity known as the thermal diffusion factor. The dependence of α_T on mass and ion-neutral interaction is fairly complicated. We shall not go into much detail, since our main concern here is only to assess the importance of thermal diffusion as a source of error in mobility measurements. The reader interested in further information on thermal diffusion can find it in the standard monographs by Chapman and Cowling (1970) and Hirschfelder et al. (1964); specialized reviews on thermal diffusion are also available (Grew and Ibbs, 1952; Mason et al., 1966).

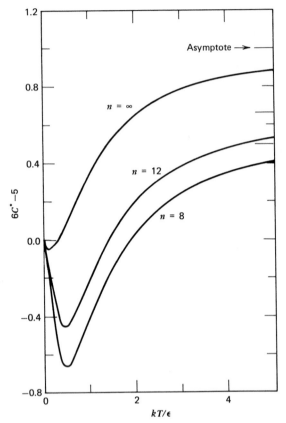

FIG. 5-3-7. The function $6C^* - 5$ for a series of $(n\text{-}4)$ potentials. This function is used to describe higher Chapman-Enskog approximations, deviations from Blanc's law, and thermal diffusion.

For an estimate of the effect of thermal diffusion on drift-tube measurements we can compare the magnitude of the thermal diffusive flux and the forced flux terms in (5-3-62),

$$\frac{\text{thermal flux}}{\text{forced flux}} = \frac{nD\alpha_T \, \mathbf{V} \ln T}{nKE},$$

$$= 8.62 \times 10^{-5} \alpha_T \frac{\mathbf{V} T(^\circ K)}{\mathbf{E}(\text{volts})}, \qquad (5\text{-}3\text{-}63)$$

for singly charged ions. For ordinary neutral gases α_T is of the order of magnitude unity. If we take rather extreme values of $100^\circ K/\text{cm}$ for the temperature gradient and 1 V/cm for the electric field, we find that the ratio of thermal flux to forced flux is only of order 10^{-2}, a rather small contribution.

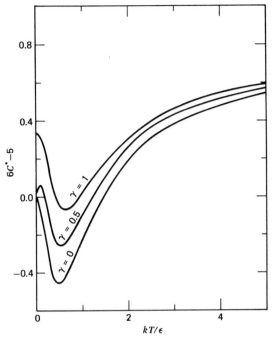

FIG. 5-3-8. The function $6C^* - 5$ for a series of (12-6-4) potentials. This function is used to describe higher Chapman-Enskog approximations, deviations from Blanc's law, and thermal diffusion.

Thus thermal diffusion is probably not an important source of error in mobility measurements unless α_T is very large. When we examine the kinetic-theory expression for α_T, we shall see that it is possible for α_T to be large if the ions are large.

The Chapman-Enskog first approximation for α_T for ions in a neutral gas is

$$[\alpha_T]_1 = (6C^* - 5)\left(\frac{S_\alpha}{Q_\alpha}\right), \qquad (5\text{-}3\text{-}64)$$

where

$$S_\alpha = \frac{15}{2}\frac{m(m - M)}{(m + M)^2} + 4\frac{mM}{(m + M)^2}A^* - \frac{M}{m}\left(\frac{2m}{m + M}\right)^{1/2}\frac{\overline{\Omega}_{nn}^{(2,\,2)}}{\overline{\Omega}_{in}^{(1,\,1)}}, \qquad (5\text{-}3\text{-}65)$$

$$Q_\alpha = \frac{2}{m(m + M)}(3m^2 + M^2 + \tfrac{8}{5}mMA^*)\left(\frac{2m}{m + M}\right)^{1/2}\frac{\overline{\Omega}_{nn}^{(2,\,2)}}{\overline{\Omega}_{in}^{(1,\,1)}}, \qquad (5\text{-}3\text{-}66)$$

where A^* and C^* are as already defined and the subscripts nn and in on the collision integrals refer to the neutral-neutral and ion-neutral interactions, respectively. This expression is strictly valid only for elastic collisions. The collision integrals can be eliminated in favor of experimental quantities; the integral $\overline{\Omega}_{in}^{(1,\,1)}$ is related to the ion-neutral diffusion coefficient (or mobility), and the integral $\overline{\Omega}_{nn}^{(2,\,2)}$ to the viscosity η of the pure neutral gas:

$$\left(\frac{2m}{m+M}\right)^{1/2}\frac{\overline{\Omega}_{nn}^{(2,\,2)}}{\overline{\Omega}_{in}^{(1,\,1)}} = \frac{5}{3}\frac{mM}{m+M}\frac{ND}{\eta}. \tag{5-3-67}$$

This relation is strictly correct only in the first Chapman-Enskog aproximation.

The sign convention used here for thermal diffusion is as follows. If α_T is positive, the ions move "down" the temperature gradient toward the lower temperature region. If $(6C^* - 5)$ is positive, then α_T is positive for $m \gg M$ but negative for $m \ll M$. The limiting form for $m \gg M$ is

$$[\alpha_T]_1 = \frac{5\sqrt{2}}{8}(6C^* - 5)\frac{\overline{\Omega}_{in}^{(1,\,1)}}{\overline{\Omega}_{nn}^{(2,\,2)}} = \frac{3}{4}(6C^* - 5)\frac{\eta}{ND}, \tag{5-3-68}$$

and for $m \ll M$ is

$$[\alpha_T]_1 = -\tfrac{1}{2}(6C^* - 5). \tag{5-3-69}$$

However, $(6C^* - 5)$ depends sensitively on the ion-neutral potential and can be positive, negative, or zero. For an r^{-n} potential it is equal to $(1 - 4/n)$ and so is positive for $n > 4$ and zero for $n = 4$. Thus thermal diffusion always vanishes in the polarization limit. For more complicated potentials like the $(n\text{-}4)$ and $(12\text{-}6\text{-}4)$ models $(6C^* - 5)$ can have a fairly complicated temperature dependence, as shown in Figs. 5-3-7 and 5-3-8. The curves in these figures can be used for making quick estimates of α_T.

If values of $(6C^* - 5)$ are not readily available but the mobility is known as a function of temperature, an estimate can be obtained from the formula

$$6C^* - 5 = -2\frac{d\ln[K_0]_1}{d\ln T}, \tag{5-3-70}$$

which follows directly from the recursion relation given in (5-3-26).

Examination of (5-3-69) shows that α_T can never be very large for light ions, but (5-3-68) reveals that α_T can be large for heavy ions and indeed enormous for ions as large as charged dust or aerosol particles. The reason is that D becomes very small for large particles, whereas η does not change. An order-of-magnitude estimate can be readily obtained by remembering that $\overline{\Omega} \approx \pi d^2$, where d is an equivalent rigid-sphere diameter.

So for ions of micron size moving through molecules of angstrom size

$$\alpha_T \sim (10^{-4}\text{cm}/10^{-8}\text{cm})^2 = 10^8. \qquad (5\text{-}3\text{-}71)$$

Temperature gradients therefore can be important for large ions.

5-4. MEDIUM-FIELD THEORY

When the field is not weak, we cannot make the linearization approximations used in the Chapman-Enskog treatment of the low-field mobility. In particular, the deviation term ϕ_i of the ion distribution function should not necessarily be assumed small, and it would certainly be incorrect to assume that ϕ_i is linear in E, as in (5-3-10). No extension of the Chapman-Enskog procedure to general nonlinear cases is known, but progress is possible for the special case of ion mobilities in medium fields. The reason is that the distribution function of the neutral gas can be taken as Maxwellian if the ion concentration is low $(n \ll N)$. Under these circumstances the Boltzmann equation becomes linear in the ion distribution function and for medium fields can be solved by expansion in orthogonal functions and formation of moments, much as in the Chapman-Enskog procedure.

Even though the deviation term ϕ_i cannot be linear in the field, it should still be a good approximation to take ϕ_i as linear in the concentration gradient; that is, the ion diffusion superposed on the drift velocity should still be adequately described by the usual diffusion equations of the Fickian type, although the longitudinal and transverse diffusion coefficients would not be equal. The problem could thus be tackled in two stages. In the first stage the diffusion is ignored and the Boltzmann equation solved for general electric fields not assumed to be small perturbations. In the second stage the gradient terms giving rise to diffusion are included in the Boltzmann equation but are treated as small perturbations as in the Chapman-Enskog approach.

In this section we give an account of the first stage of the procedure, as developed principally by Kihara (1953). The second stage has unfortunately never been carried through but should involve no new difficulties of principle.*

A. KIHARA EXPANSION. In order to ignore diffusion and treat only the mobility problem, we imagine a spatially uniform distribution of ions moving with a steady-state average velocity under the influence of an electric field.

---

* Note added in proof: Such work has now been reported by Whealton and Mason (1973) using Kihara's method, and by Kumar and Robson (1973) using somewhat different procedures.

Then the Boltzmann equation (5-3-1) becomes

$$\frac{e}{m} \mathbf{E} \cdot \mathbf{V}_v f_i = \sum_j \int \cdots \int (f_i' f_j' - f_i f_j) v_r I_s(\theta) \, d\Omega_{CM} \, d^3 V_j, \qquad (5\text{-}4\text{-}1)$$

where the subscript i refers to ions and j, to neutrals. We take the distribution functions of the neutrals to be Maxwellian:

$$f_j = f_j^{(0)} = N_j \left(\frac{M_j}{2\pi kT}\right)^{3/2} \exp\left(-\frac{M_j V_j^2}{2kT}\right), \qquad (5\text{-}4\text{-}2)$$

and, as before, we write the ion distribution function as

$$f_i = f_i^{(0)}(1 + \phi_i). \qquad (5\text{-}4\text{-}3)$$

The Boltzmann equation can then be written as a linear operator equation,

$$\frac{e}{m} \mathbf{E} \cdot \mathbf{V}_v f_i = -f_i^{(0)} \sum_j N_j(L_j \phi_i), \qquad (5\text{-}4\text{-}4)$$

where the linear collision operator L_j can operate on any function of the ion velocity

$$L_j \phi_i \equiv \frac{1}{N_j} \int \cdots \int f_j^{(0)}[\phi_i(\mathbf{v}) - \phi_i(\mathbf{v}')] v_r I_s(\theta) \, d\Omega_{CM} \, d^3 V_j. \qquad (5\text{-}4\text{-}5)$$

This looks much like the linearized Boltzmann equation (5-3-8) used in the Chapman-Enskog approach, with the omission of the time and space derivatives on the left-hand side and the deviation functions ϕ_j of the neutrals on the right-hand side. The important difference here is that we have *not* assumed that f_i differs only slightly from $f_i^{(0)}$. Thus f_i rather than $f_i^{(0)}$ appears on the left of (5-4-4) and the ϕ_i are not assumed to be small.

The linear collision operator L has some useful properties. If we define the inner product of two functions $\phi(\mathbf{v})$ and $\psi(\mathbf{v})$ of the ion velocity as

$$[\phi, \psi] \equiv \frac{1}{n} \int f_i^{(0)} \phi(\mathbf{v}) \psi(\mathbf{v}) \, d^3 v, \qquad (5\text{-}4\text{-}6)$$

it can be shown that L is symmetric and positive definite (Kihara, 1953):

$$[\psi, L\phi] = [\phi, L\psi], \qquad (5\text{-}4\text{-}7)$$

$$[\phi, L\phi] \geq 0. \qquad (5\text{-}4\text{-}8)$$

For simplicity we now assume that only one neutral component is present. We also lose no generality by choosing the z-axis to be in the direction of the field and writing $E_x = E_y = 0$, $E_z = E$. Multiplying both sides of (5-4-4) by $v_z \, d^3 v$ and integrating, we obtain

$$\frac{eE}{m} \int \frac{\partial f_i}{\partial v_z} v_z \, d^3v = -N \int f_i^{(0)} v_z (L\phi_i) \, d^3v. \qquad (5\text{-}4\text{-}9)$$

The left-hand side can be integrated by parts. The right-hand side can be transformed by use of the relation $L(1 + \phi_i) = L\phi_i$ and the symmetry property of L. The result can be written as

$$\frac{eE}{m} = N\langle Lv_z \rangle, \qquad (5\text{-}4\text{-}10)$$

where the average-value symbol $\langle \cdots \rangle$ is defined as

$$\langle Lv_z \rangle \equiv \frac{1}{n} \int f_i (Lv_z) \, d^3v. \qquad (5\text{-}4\text{-}11)$$

If this were the low-field case and we were following a Chapman-Enskog procedure, we would now expand the function (Lv_z) as v_z times a series of Sonine polynomials corresponding to (5-3-10) and (5-3-14). Then (5-4-10) would be the first moment equation. We follow a similar procedure for the medium-field case but a more general expansion is needed.

In order to make plausible the particular expansion procedure used by Kihara, it is useful to consider the Maxwellian r^{-4} model. (This will also make the connection with the Chapman-Enskog procedure more transparent.) The important feature of this model is that the function v_z is an eigenfunction of the operator L; this occurs because the transport cross sections $Q^{(l)}$ are inversely proportional to the relative collision velocity v_r for this model. This property suggests that we might use the complete set of Maxwellian-model eigenfunctions of L as a basis for the expansion of functions like (Lv_z) for general models. For low fields the Maxwellian-model eigenfunctions are the Sonine polynomials; for medium fields a more general set of eigenfunctions that does not correspond only to small distortions of the ion distribution function is needed.

To demonstrate the foregoing remarks we transform to center-of-mass coordinates, after which (Lv_z) can be reduced to the following form (Kihara, 1953):

$$Lv_z = \frac{1}{N} \frac{M}{m + M} \int f_j^{(0)} (v_z - V_z) v_r Q^{(1)} \, d^3V. \qquad (5\text{-}4\text{-}12)$$

For the Maxwellian model $v_r Q^{(1)}$ is a constant and we obtain the eigenvalue equation

$$Lv_z = \left[\frac{M}{m + M} v_r Q^{(1)} \right] v_z. \qquad (5\text{-}4\text{-}13)$$

Substitution of this expression back into (5-4-10) yields

$$\langle v_z \rangle = \frac{eE}{N v_r Q^{(1)}} \left(\frac{m + M}{mM} \right). \tag{5-4-14}$$

Since $\langle v_z \rangle = v_d$, the drift velocity, and since for the Maxwellian model

$$\overline{\Omega}^{(1,\,1)} = \frac{3}{8} \left(\frac{\pi \mu}{2kT} \right)^{1/2} v_r Q^{(1)}, \tag{5-4-15}$$

which follows directly from the definition (5-3-16) of the collision integrals, we find the mobility to be

$$K = \frac{3e}{16N} \left(\frac{2\pi}{\mu kT} \right)^{1/2} \frac{1}{\overline{\Omega}^{(1,\,1)}}. \tag{5-4-16}$$

This is the same as the first Chapman-Enskog approximation given by (5-3-18), but for the Maxwellian model we see that the result is not only exact but also valid for all fields and not limited to low fields.

The appropriate Maxwellian-model eigenfunctions of L were found by Kihara to be

$$\psi_l^{(r)} = w^l P_l \left(\frac{v_z}{v} \right) S_{l+1/2}^{(r)}(w^2), \tag{5-4-17}$$

$$l, r = 0, 1, 2, \ldots,$$

where $w^2 = mv^2/2kT$. The preferred direction in space caused by the electric field is taken into account through the Legendre polynomials $P_l(v_z/v)$, which are functions of the velocity component of the ions in the direction of the field. The eigenvalues are given by

$$L\psi_l^{(r)} = \lambda_l^{(r)} \psi_l^{(r)}.$$

The first few $\psi_l^{(r)}$ and $\lambda_l^{(r)}$, listed in Table 5-4-1 (Kihara, 1953; Mason and Schamp, 1958; Hahn, 1972), are special cases of what are now sometimes called the Burnett functions (Ford, 1968). For our purposes the most important function is $\psi_1^{(0)} = w_z$, since its average value gives the drift velocity.

We generate a set of moment equations by multiplying both sides of (5-4-4) by $\psi_l^{(r)} d^3v$ and integrating. The left-hand side can be integrated by parts, and the right-hand side can be transformed by use of $L(1 + \phi_i) = L\phi_i$ and the symmetry property of L. The result is

$$\frac{eE}{m} \left\langle \frac{\partial \psi_l^{(r)}}{\partial v_z} \right\rangle = N \langle L \psi_l^{(r)} \rangle. \tag{5-4-18}$$

TABLE 5-4-1. Eigenfunctions and eigenvalues of the Maxwellian-model linear collision operator

l, r	$\psi_l^{(r)}$	$\left(\dfrac{\pi\mu}{2kT}\right)^{1/2}\lambda_l^{(r)}$
0, 0	1	0
0, 1	$\frac{3}{2} - w^2$	$\dfrac{16}{3}\dfrac{mM}{(m+M)^2}\overline{\Omega}^{(1,1)}$
0, 2	$\frac{1}{2}(\frac{15}{4} - 5w^2 + w^4)$	$\dfrac{32}{15}\dfrac{mM}{(m+M)^4}[5(m^2 + M^2)\overline{\Omega}^{(1,1)} + 4mM\overline{\Omega}^{(2,2)}]$
0, 3	$\frac{1}{2}(\frac{35}{8} - \frac{35}{4}w^2 + \frac{7}{2}w^4 - \frac{1}{3}w^6)$	$\dfrac{16}{35}\dfrac{mM}{(m+M)^6}[35(m^2 + M^2)^2\overline{\Omega}^{(1,1)} + 56mM(m^2 + M^2)\overline{\Omega}^{(2,2)}$ $+ 64m^2M^2\overline{\Omega}^{(3,3)}]$
1, 0	w_z	$\dfrac{8}{3}\dfrac{M}{m+M}\overline{\Omega}^{(1,1)}$
1, 1	$w_z(\frac{5}{2} - w^2)$	$\dfrac{M}{15(m+M)^3}[5(3m^2 + M^2)\overline{\Omega}^{(1,1)} + 8mM\overline{\Omega}^{(2,2)}]$
1, 2	$\frac{1}{2}w_z(\frac{35}{4} - 7w^2 + w^4)$	$\dfrac{128}{105}\dfrac{M}{(m+M)^5}\left[\dfrac{35}{16}(5m^4 + 6m^2M^2 + M^4)\overline{\Omega}^{(1,1)}\right.$ $\left. + 7mM(2m^2 + M^2)\overline{\Omega}^{(2,2)} + 12m^2M^2\overline{\Omega}^{(3,3)}\right]$
2, 0	$\frac{1}{2}(3w_z^2 - w^2)$	$\dfrac{16}{15}\dfrac{M}{(m+M)^2}[5m\overline{\Omega}^{(1,1)} + 3M\overline{\Omega}^{(2,2)}]$
2, 1	$\frac{1}{2}(3w_z^2 - w^2)(\frac{7}{2} - w^2)$	$\dfrac{16}{105}\dfrac{M}{(m+M)^4}\left[\dfrac{35}{2}m(4m^2 + M^2)\overline{\Omega}^{(1,1)}\right.$ $\left. + 7M(11m^2 + 3M^2)\overline{\Omega}^{(2,2)} + 72mM^2\overline{\Omega}^{(3,3)}\right]$
2, 2	$\frac{1}{4}(3w_z^2 - w^2)(\frac{63}{4} - 9w^2 + w^4)$	$\dfrac{16}{35}\dfrac{M}{(m+M)^6}\left[35m^3(m^2 + M^2)\overline{\Omega}^{(1,1)}\right.$ $+ 7M(3m^2 + M^2)^2\overline{\Omega}^{(2,2)} + 16mM^2(7m^2 + 3M^2)\overline{\Omega}^{(3,3)}$ $\left. + \dfrac{512}{9}m^2M^3\overline{\Omega}^{(4,4)}\right]$
3, 0	$\frac{1}{2}(5w_z^3 - 3w_z w^2)$	$\dfrac{16}{21}\dfrac{M}{(m+M)^3}\left[\dfrac{21}{4}(2m^2 - M^2)\overline{\Omega}^{(1,1)}\right.$ $\left. + \dfrac{63}{5}mM\overline{\Omega}^{(2,2)} + 12M^2\overline{\Omega}^{(3,3)}\right]$
3, 1	$\frac{1}{2}w_z(5w_z^2 - 3w^2)(\frac{9}{2} - w^2)$	$\dfrac{16}{21}\dfrac{M}{(m+M)^5}\left[\dfrac{7}{4}(10m^4 - 3m^2M^2 - 3M^4)\overline{\Omega}^{(1,1)}\right.$ $+ \dfrac{21}{5}mM(7m^2 + M^2)\overline{\Omega}^{(2,2)}$ $\left. + \dfrac{12}{5}M^2(23m^2 + 5M^2)\overline{\Omega}^{(3,3)} + \dfrac{64}{3}mM^3\overline{\Omega}^{(4,4)}\right]$
4, 0	$\frac{1}{8}(35w_z^4 - 30w_z^2 w^2 + 3w^4)$	$\dfrac{16}{105}\dfrac{M}{(m+M)^4}\left[35m(2m^2 - M^2)\overline{\Omega}^{(1,1)}\right.$ $+ \dfrac{21}{2}M(12m^2 - 5M^2)\overline{\Omega}^{(2,2)} + 240mM^2\overline{\Omega}^{(3,3)}$ $\left. + \dfrac{280}{3}M^3\overline{\Omega}^{(4,4)}\right]$

This is the generalization of (5-4-10). From the recursion relations for Sonine and Legendre polynomials we can prove

$$(l + \tfrac{1}{2}) \frac{\partial \psi_l^{(r)}}{\partial w_z} = l(l + \tfrac{1}{2} + r)\psi_{l-1}^{(r)} - (l + 1)\psi_{l+1}^{(r-1)}, \tag{5-4-19}$$

with $\psi_{l+1}^{(-1)} \equiv 0$. Substitution of this expression back into (5-4-18) yields the moment equations

$$(l + \tfrac{1}{2})\langle L\psi_l^{(r)} \rangle = \frac{eE}{Nm}\left(\frac{m}{2kT}\right)^{1/2}[l(l + \tfrac{1}{2} + r)\langle \psi_{l-1}^{(r)} \rangle - (l + 1)\langle \psi_{l+1}^{(r-1)} \rangle]. \tag{5-4-20}$$

We now expand the unknown function $L\psi_l^{(r)}$ in terms of the $\psi_l^{(r)}$:

$$L\psi_l^{(r)} = \sum_{s=0}^{\infty} a_{rs}(l)\psi_l^{(s)}. \tag{5-4-21}$$

Since the $\psi_l^{(r)}$ are orthogonal, the expansion coefficients are readily found to be

$$a_{rs}(l) = \frac{[\psi_l^{(s)}, L\psi_l^{(r)}]}{[\psi_l^{(s)}, \psi_l^{(s)}]}. \tag{5-4-22}$$

The form (5-4-22) is deceptively simple in appearance. The reduction of the integrations in the numerator requires the same tedious manipulations that occur in the Chapman-Enskog theory; most of the integrations can be carried out without knowledge of the ion-neutral interaction but the final two integrations cannot. It is only for the r^{-4} potential that the reductions simplify to an eigenvalue problem. The final result is that the $a_{rs}(l)$ are expressed as linear combinations of the irreducible collision integrals of (5-3-16), just as the expansion coefficients c_{is} of the Chapman-Enskog theory were. The calculations are systematic but tedious and involved (Chapman and Cowling, 1970, Chapter 9; Ford, 1968). Expressions for the first few $a_{rs}(l)$ are summarized in Table 5-4-2 (Kihara, 1953; Mason and Schamp, 1958; Hahn, 1972; Hahn and Mason, 1973).

Substitution of the expansion (5-4-21) back into the moment equations (5-4-20) gives the following doubly infinite set of coupled algebraic equations:

$$(l + \tfrac{1}{2}) \sum_{s=0}^{\infty} a_{rs}(l)\langle \psi_l^{(s)} \rangle$$

$$= \frac{eE}{Nm}\left(\frac{m}{2kT}\right)^{1/2}[l(l + \tfrac{1}{2} + r)\langle \psi_{l-1}^{(r)} \rangle - (l + 1)\langle \psi_{l+1}^{(r-1)} \rangle]. \tag{5-4-23}$$

To find the mobility this set of equations must be solved to give $\langle \psi_1^{(0)} \rangle$ in terms of the $a_{rs}(l)$; the mobility then follows directly from

$$KE = v_d = \langle v_z \rangle = \left(\frac{2kT}{m}\right)^{1/2}\langle \psi_1^{(0)} \rangle. \tag{5-4-24}$$

The equations cannot in general be solved exactly and some approximation scheme must be adopted. Here we can be guided by the results for the Maxwellian model: comparing (5-4-21) with (5-4-18), we see that for this model $a_{rr}(l) = \lambda_l^{(r)}$ and $a_{rs}(l) = 0$ for $r \neq s$. For a first approximation we therefore set all off-diagonal $a_{rs}(l)$ equal to zero and obtain

$$(l + \tfrac{1}{2})a_{rr}(l)\langle\psi_l^{(r)}\rangle_1$$

$$= \frac{eE}{Nm}\left(\frac{m}{2kT}\right)^{1/2}[l(l + \tfrac{1}{2} + r)\langle\psi_{l-1}^{(r)}\rangle_1 - (l + 1)\langle\psi_{l+1}^{(r-1)}\rangle_1], \quad (5\text{-}4\text{-}25)$$

where the notation $\langle\cdots\rangle_1$ means the first approximation. In particular, we find

$$a_{00}(1)\langle\psi_1^{(0)}\rangle_1 = \frac{eE}{Nm}\left(\frac{m}{2kT}\right)^{1/2} \quad (5\text{-}4\text{-}26)$$

or

$$[K]_1 = \frac{e}{mNa_{00}(1)} = \frac{3e}{16N}\left(\frac{2\pi}{\mu kT}\right)^{1/2}\frac{1}{\overline{\Omega}^{(1,1)}}. \quad (5\text{-}4\text{-}27)$$

This is the same as the Chapman-Enskog low-field result. It is exact for the Maxwellian model but only approximate for other ion-neutral interactions. In particular, to find the field dependence of K we must go to higher approximations.

Higher approximations for $\langle\psi_l^{(0)}\rangle$ can now be obtained by inserting lower approximations for the other $\langle\psi_l^{(r)}\rangle$ in the terms involving the off-diagonal $a_{rs}(l)$. There is some choice at this point in the particular systematic procedure adopted to generate higher and higher approximations; ultimately all such procedures must converge to the same result but some might be more effective than others at lower levels of approximation. Two procedures have been suggested. The nth approximation $\langle\psi_l^{(r)}\rangle_n$ is generated, according to Mason and Schamp (1958), from the equation

$$(l + \tfrac{1}{2})a_{rr}(l)\langle\psi_l^{(r)}\rangle_n$$

$$= \frac{eE}{Nm}\left(\frac{m}{2kT}\right)^{1/2}[l(l + \tfrac{1}{2} + r)\langle\psi_{l-1}^{(r)}\rangle_{n-1} - (l + 1)\langle\psi_{l+1}^{(r-1)}\rangle_{n-1}]$$

$$- (l + \tfrac{1}{2})\sum_{s=0}^{n+r-1}(1 - \delta_{rs})a_{rs}(l)\langle\psi_l^{(s)}\rangle_{n-1}. \quad (5\text{-}4\text{-}28)$$

The summation on the right-hand side is terminated at $s = (n + r - 1)$ to keep the scheme consistent in regard to powers of $(E/N)^2$, which is here regarded as the expansion parameter; that is, each higher approximation results in the addition of the next higher power of $(E/N)^2$. An estimate of the

TABLE 5-4-2. Expansion coefficients $a_{rs}(l)$ for the medium-field mobility

$$a_{sr}(l) = a_{rs}(l)\,\frac{r!\,\Gamma(l+s+\frac{3}{2})}{s!\,\Gamma(l+r+\frac{3}{2})}$$

l	r, s	$\left(\dfrac{\pi\mu}{2kT}\right)^{1/2} a_{rs}(l)$
0	0, s	0
	1, 1	$\dfrac{16}{3}\dfrac{mM}{(m+M)^2}\overline{\Omega}^{(1,\,1)}$
	1, 2	$\dfrac{32}{15}\dfrac{mM^2}{(m+M)^3}[5\overline{\Omega}^{(1,\,1)}-6\overline{\Omega}^{(1,\,2)}]$
	1, 3	$\dfrac{16}{35}\dfrac{mM^3}{(m+M)^4}[35\overline{\Omega}^{(1,\,1)}-84\overline{\Omega}^{(1,\,2)}+48\overline{\Omega}^{(1,\,3)}]$
	2, 2	$\dfrac{16}{15}\dfrac{mM}{(m+M)^4}[5(2m^2+5M^2)\overline{\Omega}^{(1,\,1)}-60M^2\overline{\Omega}^{(1,\,2)}+48M^2\overline{\Omega}^{(1,\,3)}$ $+\,8mM\overline{\Omega}^{(2,\,2)}]$
	3, 3	$\dfrac{2mM}{(m+M)^6}\Big[(35M^4+56m^2M^2+8m^4)\overline{\Omega}^{(1,\,1)}$ $-\dfrac{168}{5}M^2(4m^2+5M^2)\overline{\Omega}^{(1,\,2)}$ $+\dfrac{96}{35}M^2(36m^2+133M^2)\overline{\Omega}^{(1,\,3)}-384M^4\overline{\Omega}^{(1,\,4)}+\dfrac{1152}{7}M^4\overline{\Omega}^{(1,\,5)}$ $+\dfrac{32}{5}mM(2m^2+7M^2)\overline{\Omega}^{(2,\,2)}-\dfrac{512}{5}mM^3\overline{\Omega}^{(2,\,3)}+\dfrac{512}{7}mM^3\overline{\Omega}^{(2,\,4)}$ $+\dfrac{512}{35}m^2M^2\overline{\Omega}^{(3,\,3)}\Big]$
1	0, 0	$\dfrac{8}{3}\dfrac{M}{m+M}\overline{\Omega}^{(1,\,1)}$
	0, 1	$\dfrac{8}{15}\left(\dfrac{M}{m+M}\right)^2[5\overline{\Omega}^{(1,\,1)}-6\overline{\Omega}^{(1,\,2)}]$
	0, 2	$\dfrac{8}{105}\left(\dfrac{M}{m+M}\right)^3[35\overline{\Omega}^{(1,\,1)}-84\overline{\Omega}^{(1,\,2)}+48\overline{\Omega}^{(1,\,3)}]$
	0, 3	$\dfrac{8}{315}\left(\dfrac{M}{m+M}\right)^4[105\overline{\Omega}^{(1,\,1)}-378\overline{\Omega}^{(1,\,2)}+432\overline{\Omega}^{(1,\,3)}-160\overline{\Omega}^{(1,\,4)}]$
	0, s	$(-1)^s\dfrac{2\pi^{1/2}}{\Gamma(s+\frac{5}{2})}\left(\dfrac{M}{m+M}\right)^{s+1}T^{s-1/2}\dfrac{d^s}{dT^s}[T^{1/2}\overline{\Omega}^{(1,\,1)}]$

TABLE 5-4-2. Continued.

l	r, s	$\left(\dfrac{\pi\mu}{2kT}\right)^{1/2} a_{rs}(l)$
1	1, 1	$\dfrac{4}{15}\dfrac{M}{(m+M)^3}[5(6m^2+5M^2)\overline{\Omega}^{(1,\,1)}-60M^2\overline{\Omega}^{(1,\,2)}$ $+48M^2\overline{\Omega}^{(1,\,3)}+16mM\overline{\Omega}^{(2,\,2)}]$
	1, 2	$\dfrac{128}{105}\dfrac{M^2}{(m+M)^4}\left[\dfrac{35}{32}(12m^2+5M^2)\overline{\Omega}^{(1,\,1)}-\dfrac{63}{16}(4m^2+5M^2)\overline{\Omega}^{(1,\,2)}\right.$ $\left.+\dfrac{57}{2}M^2\overline{\Omega}^{(1,\,3)}-15M^2\overline{\Omega}^{(1,\,4)}+7mM\overline{\Omega}^{(2,\,2)}-8mM\overline{\Omega}^{(2,\,3)}\right]$
	2, 2	$\dfrac{128}{105}\dfrac{M}{(m+M)^5}\left[\dfrac{35}{128}(40m^4+168m^2M^2+35M^4)\overline{\Omega}^{(1,\,1)}\right.$ $-\dfrac{21}{16}M^2(84m^2+35M^2)\overline{\Omega}^{(1,\,2)}$ $+\dfrac{3}{4}M^2(108m^2+133M^2)\overline{\Omega}^{(1,\,3)}-105M^4\overline{\Omega}^{(1,\,4)}+45M^4\overline{\Omega}^{(1,\,5)}$ $+\dfrac{7}{2}mM(4m^2+7M^2)\overline{\Omega}^{(2,\,2)}-56mM^3\overline{\Omega}^{(2,\,3)}$ $\left.+40mM^3\overline{\Omega}^{(2,\,4)}+12m^2M^2\overline{\Omega}^{(3,\,3)}\right]$
2	0, 0	$\dfrac{16}{15}\dfrac{M}{(m+M)^2}[5m\overline{\Omega}^{(1,\,1)}+3M\overline{\Omega}^{(2,\,2)}]$
	0, 1	$\dfrac{16}{15}\dfrac{M^2}{(m+M)^3}\left[5m\overline{\Omega}^{(1,\,1)}-6m\overline{\Omega}^{(1,\,2)}+3M\overline{\Omega}^{(2,\,2)}-\dfrac{24}{7}M\overline{\Omega}^{(2,\,3)}\right]$
	0, 2	$\dfrac{16}{15}\dfrac{M^3}{(m+M)^4}\left[5m\overline{\Omega}^{(1,\,1)}-12m\overline{\Omega}^{(1,\,2)}+\dfrac{48}{7}m\overline{\Omega}^{(1,\,3)}+3M\overline{\Omega}^{(2,\,2)}\right.$ $\left.-\dfrac{48}{7}M\overline{\Omega}^{(2,\,3)}+\dfrac{80}{21}M\overline{\Omega}^{(2,\,4)}\right]$
	1, 1	$\dfrac{128}{105}\dfrac{M}{(m+M)^4}\left[\dfrac{35}{16}m(4m^2+7M^2)\overline{\Omega}^{(1,\,1)}-\dfrac{147}{4}mM^2\overline{\Omega}^{(1,\,2)}\right.$ $+24mM^2\overline{\Omega}^{(1,\,3)}+\dfrac{7}{16}M(22m^2+21M^2)\overline{\Omega}^{(2,\,2)}-21M^3\overline{\Omega}^{(2,\,3)}$ $\left.+15M^3\overline{\Omega}^{(2,\,4)}+9mM^2\overline{\Omega}^{(3,\,3)}\right]$
3	0, 0	$\dfrac{16}{21}\dfrac{M}{(m+M)^3}\left[\dfrac{21}{2}m^2\overline{\Omega}^{(1,\,1)}-\dfrac{36}{5}M^2\overline{\Omega}^{(1,\,3)}\right.$ $\left.+\dfrac{63}{5}mM\overline{\Omega}^{(2,\,2)}+12M^2\overline{\Omega}^{(3,\,3)}\right]$

omitted terms can be obtained from their first approximations by adding to the right-hand side of (5-4-28) the terms

$$-(l + \tfrac{1}{2}) \sum_{s=n+r}^{\infty} (1 - \delta_{rs}) a_{rs}(l) \langle \psi_l^{(s)} \rangle_1. \tag{5-4-29}$$

It can be shown from (5-4-25) that each $\langle \psi_l^{(r)} \rangle_1$ is proportional to the $(l + 2r)$th power of E/N.

Repeated application of (5-4-28), followed by some algebraic manipulation, leads to an expansion for the mobility in powers of $(E/N)^2$. In fact, as expected from the simple free-flight theory of Section 5-2-A, the ratio E/N always occurs in the combination $[eE/NkT\overline{\Omega}^{(1,\,1)}]$. Thus it is convenient to define a dimensionless field-strength function:

$$\mathcal{E} \equiv \frac{eE}{mN} \left(\frac{m}{2kT} \right)^{1/2} \frac{1}{a_{00}(1)} = \frac{3\pi^{1/2}}{16} \left(\frac{m+M}{M} \right)^{1/2} \frac{eE}{NkT\overline{\Omega}^{(1,\,1)}}.$$

The expansion can then be written as

$$K = [K]_1 (g_0 + g_1 \mathcal{E}^2 + g_2 \mathcal{E}^4 + \cdots), \tag{5-4-31}$$

where each g_n is a complicated function of the $a_{rs}(l)$. Aside from convergence, which is considered in the next section, the main limitation of this expansion is the increasing complexity of the algebra as further terms are added, together with the difficulty of evaluating the complicated new $a_{rs}(l)$ that appear. Only the third approximation has been worked out completely (Mason and Schamp, 1958), in which the dimensionless coefficients g_n are given by the expressions

$$g_0 = 1 + \frac{a_{01}(1)a_{10}(1)}{a_{00}(1)a_{11}(1)} + \frac{a_{02}(1)a_{20}(1)}{a_{00}(1)a_{22}(1)} + \cdots, \tag{5-4-32}$$

$$g_1 = \Psi_1^{(1)} \left[-\frac{a_{01}(1)}{a_{00}(1)} + \frac{a_{02}(1)a_{21}(1)}{a_{00}(1)a_{22}(1)} + \cdots \right], \tag{5-4-33}$$

$$g_2 = \Psi_1^{(2)} \left[-\frac{a_{02}(1)}{a_{00}(1)} + \frac{a_{01}(1)a_{12}(1)}{a_{00}(1)a_{11}(1)} + \cdots \right]. \tag{5-4-34}$$

First approximations for the higher g_n come from the remainder terms of (5-4-29) and are

$$g_n = \Psi_1^{(n)} \left[-\frac{a_{0n}(1)}{a_{00}(1)} + \cdots \right]. \tag{5-4-35}$$

Each higher approximation adds a further term in $\mathcal{E}^2$ to (5-4-31) and other terms to the expressions for the coefficients g_n. Each g_n is thus itself expressed as an infinite series, and (5-4-31) is actually a doubly infinite series. The dimensionless factors $\Psi_1^{(n)}$ are

$$\Psi_1^{(n)} \equiv \frac{\langle \psi_1^{(n)} \rangle_1}{\langle \psi_1^{(0)} \rangle_1} \frac{1}{\mathcal{E}^{2n}} \tag{5-4-36}$$

and are given as closed expressions in terms of the diagonal $a_{rr}(l)$. The first three are

$$\Psi_1^{(1)} = -\frac{2}{3} \frac{a_{00}(1)}{a_{11}(1)} \left[\frac{5a_{00}(1)}{a_{11}(0)} + \frac{4a_{00}(1)}{a_{00}(2)} \right], \tag{5-4-37}$$

$$\Psi_1^{(2)} = \frac{48}{5} \frac{a_{00}(1)}{a_{22}(1)} \frac{a_{00}(1)}{a_{00}(2)} \frac{a_{00}(1)}{a_{11}(2)} \frac{a_{00}(1)}{a_{00}(3)}$$
$$- \frac{14}{3} \Psi_1^{(1)} \frac{a_{00}(1)}{a_{22}(1)} \left[\frac{a_{00}(1)}{a_{22}(0)} + \frac{4a_{00}(1)}{5a_{11}(2)} \right], \tag{5-4-38}$$

$$\Psi_1^{(3)} = -\frac{96}{35} \frac{a_{00}(1)}{a_{33}(1)} \frac{a_{00}(1)}{a_{00}(2)} \frac{a_{00}(1)}{a_{22}(2)} \frac{a_{00}(1)}{a_{00}(3)} \frac{a_{00}(1)}{a_{11}(3)} \left[\frac{81a_{00}(1)}{5a_{11}(2)} + \frac{16a_{00}(1)}{a_{00}(4)} \right]$$
$$+ \frac{432}{25} \Psi_1^{(1)} \frac{a_{00}(1)}{a_{33}(1)} \frac{a_{00}(1)}{a_{11}(2)} \frac{a_{00}(1)}{a_{22}(2)} \frac{a_{00}(1)}{a_{11}(3)}$$
$$- \frac{6}{5} \Psi_1^{(2)} \frac{a_{00}(1)}{a_{33}(1)} \left[\frac{5a_{00}(1)}{a_{33}(0)} + \frac{4a_{00}(1)}{a_{22}(2)} \right]. \tag{5-4-39}$$

Even for $\Psi_1^{(3)}$ not all the necessary $a_{rs}(l)$ have been evaluated, but Kihara (1953) has pointed out that $(-1)^n \Psi_1^{(n)}$ is always positive.

Another approximation procedure has been suggested by Kihara (1953); instead of (5-4-28), the generalization for the nth approximation would be

$$(l + \tfrac{1}{2})\langle \psi_l^{(r)} \rangle_n \left[\sum_{s=0}^{\infty} a_{rs}(l) \frac{\langle \psi_l^{(s)} \rangle_{n-1}}{\langle \psi_l^{(r)} \rangle_{n-1}} \right]$$
$$= \frac{eE}{Nm} \left(\frac{m}{2kT} \right)^{1/2} [l(l + \tfrac{1}{2} + r)\langle \psi_{l-1}^{(r)} \rangle_n - (l + 1)\langle \psi_{l+1}^{(r-1)} \rangle_n]. \tag{5-4-40}$$

Application of this expression leads to the expansion

$$K = \frac{[K]_1}{h_0 + h_1 \mathcal{E}^2 + h_2 \mathcal{E}^4 + \cdots}, \tag{5-4-41}$$

where each h_n is a complicated function of the $a_{rs}(l)$. Only the second approximation is simple enough to be useful:

$$h_n = \Psi_1^{(n)} \frac{a_{0n}(1)}{a_{00}(1)} + \cdots. \tag{5-4-42}$$

Like (5-4-31), the expression (5-4-41) is in principle a doubly infinite series.

Expressions for the g_n and h_n in forms better suited to numerical computation are given in Section 5-4-B.

To summarize, the foregoing results are valid for a binary-collision mechanism, elastic collisions, and medium field strengths. The limitation to medium fields comes from the form of the expansion in powers of $(E/N)^2$. There is no assumption, however, that the deviation from the equilibrium distribution function of the ions is small or that the flux is a linear function of the field strength. The Chapman-Enskog low-field result is a special case of these results. The medium-field results, however, have never been extended to include inelastic collisions or multicomponent mixtures, as have the Chapman-Enskog results. There would be no difficulty of principle in making such extensions but the calculations would be complicated and lengthy.

B. CONVERGENCE OF APPROXIMATIONS.   Two sorts of convergence errors appear in (5-4-31) and (5-4-41) for the mobility: the convergence error of the expansion in powers of $\mathcal{E}^2$ and the convergence errors of the expressions for the expansion coefficients g_n and h_n. We consider first the convergence of the expressions for the expansion coefficients and then the convergence of the overall expansion.

It is worthwhile to comment that there are three cases known for which exact answers are available. The first is the Maxwellian model, which has already been mentioned. The second is heavy ions in a light gas $(m \gg M)$ for which the mobility is independent of field strength. Kihara (1953) has given a general proof of this result, subject to the condition that the drift velocity of the ions is much smaller than the thermal velocity of the neutrals. In terms of the formulas given in Section 5-4-A, the ratio $a_{0n}(1)/a_{00}(1)$ vanishes for $n \neq 0$ if $m \gg M$, and so do all the expansion coefficients h_n of (5-4-41). The third case is again the Lorentzian mixture $(m \ll M)$, for which a special expansion valid at all field strengths can be devised. This special case is discussed in Section 5-5.

Considerable simplification without loss of accuracy in the coefficients g_n and h_n can often be obtained by systematically discarding higher order derivatives of the quantities $[T^{1/2}\overline{\Omega}^{(l, l)}]$. Reduction of the formulas is accomplished by application of the recursion relation,

$$\overline{\Omega}^{(l, s+1)} = \overline{\Omega}^{(l, s)} + \frac{T}{s + 2}\frac{d\overline{\Omega}^{(l, s)}}{dT}. \qquad (5\text{-}4\text{-}43)$$

This relation can be proved easily by direct differentiation of the definition (5-3-16) of the collision integrals. It is consistent to the present order of approximation for g_n to neglect all third and higher order derivatives, as well as products of two second derivatives, a second derivative and two first derivatives, or of four first derivatives. It is also convenient to use the following dimensionless ratios of collision integrals, which are all of order of magnitude unity:

$$A^* = \frac{\overline{\Omega}^{(2,\,2)}}{\overline{\Omega}^{(1,\,1)}}, \tag{5-4-44}$$

$$B^* = \frac{5\overline{\Omega}^{(1,\,2)} - 4\overline{\Omega}^{(1,\,3)}}{\overline{\Omega}^{(1,\,1)}}, \tag{5-4-45}$$

$$C^* = \frac{\overline{\Omega}^{(1,\,2)}}{\overline{\Omega}^{(1,\,1)}}, \tag{5-4-46}$$

$$E^* = \frac{\overline{\Omega}^{(2,\,3)}}{\overline{\Omega}^{(2,\,2)}}, \tag{5-4-47}$$

$$F^* = \frac{\overline{\Omega}^{(3,\,3)}}{\overline{\Omega}^{(1,\,1)}}. \tag{5-4-48}$$

In terms of these ratios the expansion coefficients g_n are

$$g_0 = 1 + \frac{M^2(6C^* - 5)^2}{30m^2 + M^2(25 - 12B^*) + 16mMA^*} + \cdots, \tag{5-4-49}$$

$$\frac{g_1}{\Psi_1^{(1)}} = \frac{1}{5}\frac{M}{m + M}(6C^* - 5) + \frac{1}{20}\frac{M^3}{m + M}$$

$$\begin{aligned}&\quad [4(6C^* - 5) - 3(5 - 4B^*)]\\ &\times \frac{\times\,[\tfrac{7}{8}(3m^2 + M^2)(6C^* - 5) + mM(8E^* - 7)A^*]}{[\tfrac{35}{16}(5m^4 + 6m^2M^2 + M^4) + 7mM(2m^2 + M^2)A^* + 12m^2M^2F^*]}\\ &+ \cdots,\end{aligned} \tag{5-4-50}$$

$$\frac{g_2}{\Psi_1^{(2)}} = \frac{1}{35}\left(\frac{M}{m + M}\right)^2[4(6C^* - 5) - 3(5 - 4B^*)] + \frac{16}{35}\left(\frac{M}{m + M}\right)^2$$

$$\times \frac{(6C^* - 5)[\tfrac{7}{8}(3m^2 + M^2)(6C^* - 5) + mM(8E^* - 7)A^*]}{5(3m^2 + M^2) + 8mMA^*} + \cdots, \tag{5-4-51}$$

with

$$\Psi_1^{(1)} = -\frac{50}{3m}\left(\frac{9m + 3MA^*}{5m + 3MA^*}\right)\frac{(m + M)^3}{30m^2 + M^2(25 - 12B^*) + 16mMA^*}, \tag{5-4-52}$$

$$\Psi_1^{(2)} \approx \frac{48}{5} \frac{[\lambda_1^{(0)}]^4}{\lambda_1^{(2)}\lambda_2^{(0)}\lambda_1^{(1)}\lambda_3^{(0)}} + \frac{28}{45} \frac{[\lambda_1^{(0)}]^4[4\lambda_0^{(1)} + 5\lambda_2^{(0)}][4\lambda_0^{(2)} + 5\lambda_2^{(1)}]}{\lambda_0^{(1)}\lambda_0^{(2)}\lambda_1^{(1)}\lambda_1^{(2)}\lambda_2^{(0)}\lambda_2^{(1)}}, \qquad (5\text{-}4\text{-}53)$$

the $\lambda_l^{(r)}$ being the eigenvalues given in Table 5-4-1. The corresponding h_n are given to one lower order of approximation (except for h_0, for which it is easy to include the next correction):

$$h_0 = 1 - \frac{M^2(6C^* - 5)^2}{30m^2 + 10M^2 + 16mMA^*} + \cdots, \qquad (5\text{-}4\text{-}54)$$

$$h_1 = \frac{10}{3} \frac{M}{m} \left(\frac{9m + 3MA^*}{5m + 3MA^*}\right) \frac{(6C^* - 5)(m + M)^2}{30m^2 + 10M^2 + 16mMA^*} + \cdots, \qquad (5\text{-}4\text{-}55)$$

$$h_2 = \frac{1}{35} \left(\frac{M}{m + M}\right)^2 [3(5 - 4B^*) - 4(6C^* - 5)]\Psi_1^{(2)} + \cdots. \qquad (5\text{-}4\text{-}56)$$

The convergence of the g_n can now be checked. The case of g_0 has already been considered in Section 5-3; the expression (5-4-49) for g_0 is essentially the same as the formula (5-3-19) for $K/[K]_1$ at zero field, as it should be. The only difference is that B^* was set equal to $\frac{3}{4}$ in (5-3-19). Setting $B^* = \frac{5}{4}$ is equivalent to replacing $a_{11}(1)$ by the corresponding eigenvalue $\lambda_1^{(1)}$ or, what is the same thing, neglecting all derivatives of the quantity $[T^*\bar{\Omega}^{(1,\ 1)}]$ in $a_{11}(1)$. The difference is minor.

We next examine the coefficient g_1, which determines the initial slope of a K versus $(E/N)^2$ isotherm. The expression (5-4-33) or (5-4-50) is second approximation for g_1 as written, the implication being that the first term of g_1 is usually considerably larger than the second. For $m \gg M$ the second term is negligible compared with the first; poor convergence is thus expected for $m \ll M$, for which case (5-4-50) reduces to

$$\frac{[g_1]_2}{[g_1]_1} = 1 + \frac{1}{10} [4(6C^* - 5) - 3(5 - 4B^*)], \qquad (5\text{-}4\text{-}57)$$

where the notation $[g_1]_2$ means the second approximation to g_1. We test for convergence by numerical computation for specific models of ion-neutral interactions, just as we did for the zero-field case in Section 5-3-B. The results are shown in Table 5-4-3 for inverse-power potentials, in Table 5-4-4 for $(n\text{-}4)$ potentials, and in Table 5-4-5 for $(12\text{-}6\text{-}4)$ potentials. The pattern of converence appears to be rather similar to that for the coefficient g_0 but perhaps slightly poorer. Except for the long-range r^{-2} potential, the maximum deviation of $[g_1]_2$ from $[g_1]_1$ is within 20%. If

TABLE 5-4-3. Convergence of the first expansion coefficient g_1 for K in powers of $(E/N)^2$ as indicated by the ratio $[g_1]_2/[g_1]_1$: results for the potential $V(r) = \pm C/r^n$ for the unfavorable case of $m \ll M$

n	$\dfrac{[g_1]_2}{[g_1]_1}$
2	0.700
3	0.922
4	1.000
6	1.056
8	1.075
10	1.084
25	1.097
50	1.099
∞	1.100

TABLE 5-4-4. Convergence of the first expansion coefficient g_1 for K in powers of $(E/N)^2$ as indicated by the ratio $[g_1]_2/[g_1]_1$: results for $(n\text{-}4)$ potentials for the unfavorable case of $m \ll M$

$\dfrac{kT}{\varepsilon}$	$\dfrac{[g_1]_2}{[g_1]_1}$		
	$n = 8$	$n = 12$	$n = \infty$
0	1.000	1.000	1.000
0.5	0.822	0.881	1.013
1	0.899	0.840	1.058
2	1.005	0.926	1.088
4	1.058	1.015	1.098
9	1.075	1.068	1.100
∞	1.075	1.089	1.100

TABLE 5-4-5. Convergence of the first expansion coefficient g_1 for K in powers of $(E/N)^2$ as indicated by the ratio $[g_1]_2/[g_1]_1$: results for (12-6-4) potentials for the unfavorable case of $m \ll M$

$\dfrac{kT}{\varepsilon}$	$\dfrac{[g_1]_2}{[g_1]_1}$		
	$\gamma = 0$	$\gamma = 0.5$	$\gamma = 1$
0	1.000	1.000	1.056
0.5	0.881	0.953	1.021
1	0.840	0.891	0.938
2	0.926	0.950	0.965
4	1.015	1.025	1.032
9	1.068	1.071	1.073
∞	1.089	1.089	1.089

the rate of convergence of g_1 were about the same as that for g_0, the maximum error involved in using $[g_1]_2$ in place of the exact value of g_1 would be about 5 to 10%, occurring for $m \ll M$. Usually the error would be much less, so that it would seem safe to take the second approximation to g_1 as sufficiently accurate in almost all cases.

The final coefficient that can be examined is g_2, given by (5-4-34) or (5-4-51). These expressions are only first approximations as written, the two terms being usually of comparable magnitude, and the convergence error in g_2 can therefore be estimated only by analogy with the convergence errors in g_0 and g_1. From these we would estimate that the worst possible error in g_2 is of the order of 25%. Such a relatively large error is actually not too serious because the whole term involving g_2 must be small if the expansion in powers of $(E/N)^2$ is to be usable at all.

The convergence errors in the expansion coefficients h_n are more difficult to estimate, since the expressions (5-4-55) and (5-4-56) are only first approximations. As a rough guess, we could suppose the errors to be about the same as those for the corresponding g_n.

It is more difficult to generalize about the convergence of the expansion in powers of $\mathcal{E}^2$ than about the convergence of the individual expansion coefficients. As a rough rule of thumb, we suggest that the expansion should not be trusted when the field dependence of K or v_d amounts to more than about 10% of the zero-field value. The particular value of E/N at which this occurs depends strongly on the masses, the temperature, and the ion-neutral force law; for example, the convergence is very good for $m \gg M$ at any temperature. It is also very good for any mass ratio at temperatures near the polarization limit, or, as is explained in Section 5-4-E, at temperatures for which the zero-field mobility has a minimum or maximum. In such favorable cases the three terms given explicitly in (5-4-31) or (5-4-41) may be adequate out to values of E/N of 10^2 to 10^3 Td. At the other extreme we find poor convergence for $m \ll M$ at temperatures far from the polarization limit or from a minimum or maximum in the zero-field value. In such an unfavorable case the limit of usefulness of (5-4-31) or (5-4-41) may be as low as 3 Td. For the more typical case of $m \approx M$ the expressions (5-4-31) and (5-4-41) are useful to at least 10 Td for the unfavorable case of rigid-sphere collisions and should be much better for most real cases.

As an illustration, Fig. 5-4-1 shows v_d as a function of E/N (in the form of dimensionless ratios) for $m \ll M$ and rigid-sphere collisions. The main curve is the free-flight interpolation formula (5-2-17) and the dashed straight lines are the asymptotes for low fields (slope $= 1$) and for high fields (slope $= \frac{1}{2}$). The expansions (5-4-31) and (5-4-41) are seen to be accurate up to at least $\mathcal{E}(M/m)^{1/2} \approx 0.6$, which is indeed the value at

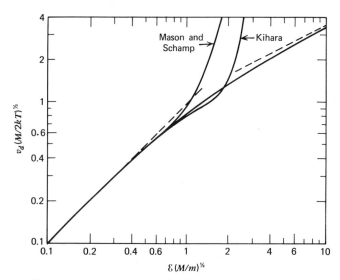

FIG. 5-4-1. Drift velocity as a function of electric field for rigid spheres with $m \ll M$. The main solid curve is the interpolation formula (5-2-17) with $\xi = 3(6\pi)^{1/2}/16$. The other solid curves are the $(E/N)^2$ expansions of Kihara (5-4-41) and Mason and Schamp (5-4-31) with $g_0 = h_0 = 1$. The dashed lines are the asymptotes for low fields (slope $= 1$) and for high fields (slope $= \frac{1}{2}$). The dimensionless field-strength function $\mathcal{E}$ is defined by (5-4-30).

which the drift velocity is about 10% different from the extrapolated low-field drift velocity. Neither of the expansions (5-4-31) or (5-4-41) seems to be markedly superior to the other, appreciable deviations starting for each at about the same field strength.

Finally, it is useful to have the dimensionless field-strength function in practical units, as follows:

$$\mathcal{E} = 0.020837 \left(\frac{m}{T}\right)^{1/2} [K_0]_1 \left(\frac{E}{N}\right), \tag{5-4-58}$$

where m is the ion mass in molar units (g/mole), T is in degrees Kelvin, $[K_0]_1$ is the first approximation to the mobility in square centimeters per volt per second at standard gas density, and E/N is in Townsends. This form, together with (5-4-49) through (5-4-56), is the recommended one for numerical computation.

C. CONNECTION WITH ELEMENTARY THEORIES.   The general structure of the connection between the simple free-flight theory and the Kihara expansion would seem to be clearly indicated by analogy with the corresponding connection for the Chapman-Enskog theory; that is, we would expect that a suitable iterative solution of the Boltzmann equation (5-4-4) would yield the simple free-flight results as a first approximation. No such formal connection, however, has ever been worked out. The most we can claim at present is that the simple free-flight theory can indeed give an expansion for the mobility or drift velocity in powers of $(eE/NkTQ)^2$, just as the Kihara expansion does. This was shown explicitly by (5-2-18) and (5-2-19). Presumably further iterations, or following the ion collision history back through more than the last collision, would yield improved values of the free-flight coefficients of such an expansion but this has not been proved.

D. QUANTUM EFFECTS.   All the results of the Kihara expansion are valid in quantum mechanics as well as in classical mechanics. Moreover, exactly the same collision integrals occur in the Kihara expansion as in the Chapman-Enskog theory. Nothing therefore needs to be added to the discussion of Section 5-3-E.

E. RELATION BETWEEN FIELD DEPENDENCE AND TEMPERATURE DEPENDENCE OF MOBILITY.   The simple arguments of the free-flight theory in Section 5-2-A suggest that a relation should exist between the behavior of the mobility with temperature and with field strength. Qualitatively, the ion energy is expected to be proportional to kT and to v_d^2 additively, as shown in (5-2-15); since v_d is approximately proportional to E/N, we expect the mobility to have similar dependences on T and on $(E/N)^2$. This relation can now be made precise by the results already obtained in this section.

To demonstrate, it is sufficient to consider just the first approximations to g_n or h_n. The general term in the expansion can then be written as

$$h_n \, \mathcal{E}^{2n} = b_n (\mathcal{E}^2 T)^n \, \frac{1}{[T^{1/2}\overline{\Omega}^{(1,\,1)}]} \, \frac{d^n}{dT^n} \, [T^{1/2}\overline{\Omega}^{(1,\,1)}],$$

$$= b_n (\mathcal{E}^2 T)^n \, \frac{1}{[K_0]_1} \, \frac{d^n [K_0]_1}{dT^n}, \qquad (5\text{-}4\text{-}59)$$

where

$$b_n = (-1)^n \, \frac{3}{4} \, \frac{\pi^{1/2}}{\Gamma(n+\tfrac{5}{2})} \left(\frac{M}{m+M} \right)^n \Psi_1^{(n)}. \qquad (5\text{-}4\text{-}60)$$

If the temperature dependence of $[K_0]_1$ is fairly weak, the dominant effect occurs through the terms in $(\mathcal{E}^2 T)^n$. Thus $\mathcal{E}^2$ and T have similar effects on the mobility.

Equation 5-4-59 illustrates another interesting effect. If $[K_0]_1$ has a maximum or minimum at a particular temperature, the expansion coefficient $a_{01}(1)$ is zero. This has two results: first, the leading correction term to the zero-field mobility vanishes and $[K_0]_1$ is exact to first order at zero field strength; second, the leading expansion coefficients g_1 and h_1 also vanish to first order, hence mobility is approximately independent of field strength. Such a temperature is analogous to the Boyle temperature of a gas, at which the gas is ideal to a good approximation, and g_1 and h_1 are analogous to the second virial coefficient in the equation of state. Moreover, for $T < T_{\max}$ the mobility increases initially with increasing field strength, and for $T > T_{\max}$ the mobility initially decreases. The reverse is true for $T_{\min}$; that is, for $T < T_{\min}$ the mobility decreases initially, and vice versa.

The effect is illustrated in Fig. 5-4-2 for a (12-4) potential, in which the mobility as a function of field strength is shown for two temperatures, one below $T_{\max}$ and one above. The ions and neutrals have the same mass for simplicity, and the shapes of the curves are similar to the shape of the zero-field mobility as a function of temperature. In fact, comparison with Fig, 5-3-1 shows that a good qualitative indication of the K_0 versus $\mathcal{E}^2$ curve at a given temperature can be obtained simply by taking that portion of the $K_0(0)$ versus T curve corresponding to higher temperatures and relabelling the T axis as the $\mathcal{E}^2$ axis.

Information on the ion-neutral force law thus can be obtained from study of the variation of K_0 with E/N as well as from the variation of the zero-field K_0 with T. Lack of accurate data has so far impeded such study.

The values of kT/ε for which maxima and minima occur are given in Table 5-4-6 for several potentials. These results are given only for the classical-mechanical approximation; quantum effects can introduce additional maxima and minima (see Fig. 5-3-6 for an example).

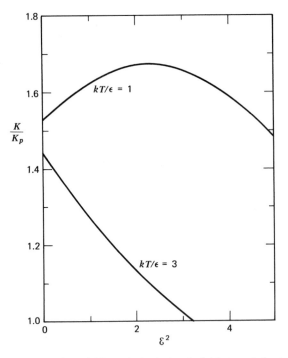

FIG. 5-4-2. Mobility as a function of (dimensionless) electric field strength for temperatures below and above the temperature at which the zero-field mobility has a maximum (see Fig. 5-3-1). The curves are calculated for a (12-4) potential with $m = M$.

F. PARTITION OF ION ENERGY. The present theory also enables us to determine how the ion energy is divided between drift energy and random field energy or between the longitudinal and transverse components of the random field energy. In the simple free-flight theory of Section 5-2-A we took these to be

$$\frac{\text{random field energy}}{\text{drift energy}} \equiv \frac{\frac{1}{2}m(\overline{v^2} - v_d^2) - \frac{3}{2}kT}{\frac{1}{2}mv_d^2} \approx \frac{M}{m},$$

$$\frac{\text{longitudinal random field energy}}{\text{transverse random field energy}} \equiv \frac{\frac{1}{2}m(\overline{v_L^2} - v_d^2) - \frac{1}{2}kT}{\frac{1}{2}m\overline{v_T^2} - \frac{1}{2}kT} \approx \frac{\zeta_L}{\zeta_T},$$

according to (5-2-9), (5-2-26), and (5-2-27), respectively. In terms of the functions $\psi_l^{(r)}$, given in Table 5-4-1, these ratios are

$$\frac{\text{random field energy}}{\text{drift energy}} = \frac{-\langle\psi_0^{(1)}\rangle - \langle\psi_1^{(0)}\rangle^2}{\langle\psi_1^{(0)}\rangle^2}, \tag{5-4-61}$$

$$\frac{\text{longitudinal random field energy}}{\text{transverse random field energy}} = \frac{2\langle\psi_2^{(0)}\rangle - \langle\psi_0^{(1)}\rangle - 3\langle\psi_1^{(0)}\rangle^2}{-\langle\psi_2^{(0)}\rangle - \langle\psi_0^{(1)}\rangle}.$$

$$\tag{5-4-62}$$

Thus we need only solve the moment equations (5-4-23) for the required $\langle\psi_l^{(r)}\rangle$.

For orientation we treat the Maxwell model first, for which exact solutions valid at all field strengths can be found. The necessary average eigenfunctions (Kihara, 1953) are readily found to be

$$\langle\psi_0^{(1)}\rangle = -2\varepsilon^2 \frac{\lambda_1^{(0)}}{\lambda_0^{(1)}}, \tag{5-4-63}$$

$$\langle\psi_1^{(0)}\rangle = \varepsilon, \tag{5-4-64}$$

$$\langle\psi_2^{(0)}\rangle = 2\varepsilon^2 \frac{\lambda_1^{(0)}}{\lambda_2^{(0)}}. \tag{5-4-65}$$

TABLE 5-4-6. Values of temperatures at which $[K_0]_1$ is a minimum or maximum for several ion-neutral potential models. At these temperatures mobility is approximately independent of field strength

Potential		$\dfrac{kT_{\min}}{\varepsilon}$	$\dfrac{kT_{\max}}{\varepsilon}$
$(n\text{-}4)$:	$n = 8$	...	1.90
	$n = 12$	...	1.44
	$n = \infty$	...	0.36
$(12\text{-}6\text{-}4)$:	$\gamma = 0$	...	1.44
	$\gamma = 0.25$	...	1.33
	$\gamma = 0.50$	0.19	1.22
	$\gamma = 0.75$	0.29	1.09
	$\gamma = 1$	0.40	0.94
$(12\text{-}4)$ core:	$a^* = 0$	...	1.44
	$a^* = 0.1$	0.05	1.25
	$a^* = 0.2$	0.21	1.03
	$a^* = 0.3$	0.40	0.71
	$a^* \approx 0.32$	0.5	0.5
	$a^* > 0.32$	...	...

From these we obtain, after a little algebra,

$$\frac{\text{random field energy}}{\text{drift energy}} = 2\frac{\lambda_1^{(0)}}{\lambda_0^{(1)}} - 1 = \frac{M}{m}, \tag{5-4-66}$$

$$\frac{\text{longitudinal random field energy}}{\text{transverse random field energy}} = \frac{5m - 2mA^* + MA^*}{(m+M)A^*}. \tag{5-4-67}$$

These equations are in accord with the free-flight results. They also agree with detailed calculations by Wannier (1953) for a model with constant free-flight time. In particular, for constant free-flight time and an isotropic scattering pattern we find $A^* = \frac{5}{6}$, which causes (5-4-67) to reduce to the value of ζ_L/ζ_T given by (5-2-32) and (5-2-33).

For the general case we solve the moment equations (5-4-23) by successive approximations. The first gives the same result as the Maxwellian model. Higher approximations show that the partitioning of energy depends on the field strength. Making a second approximation according to the scheme given by (5-4-40), we obtain the following results after some manipulation:

$$\frac{\text{random field energy}}{\text{drift energy}} = \frac{M}{m} + \left(\frac{m+M}{m}\right)\left(\frac{\Delta_1 - \Delta_2}{1 + \Delta_2}\right), \tag{5-4-68}$$

longitudinal random field energy
transverse random field energy

$$= \frac{(5m - 2mA^* + MA^*) + \frac{1}{3}\left(\frac{m+M}{M}\right)\left[(5m + 3MA^*)\left(\frac{\Delta_1 - \Delta_2}{1 + \Delta_2}\right) + 10m\left(\frac{\Delta_1 - \Delta_3}{1 + \Delta_3}\right)\right]}{(m+M)A^* + \frac{1}{3}\left(\frac{m+M}{M}\right)\left[(5m + 3MA^*)\left(\frac{\Delta_1 - \Delta_2}{1 + \Delta_2}\right) - 5m\left(\frac{\Delta_1 - \Delta_3}{1 + \Delta_3}\right)\right]}$$

$$\tag{5-4}$$

where

$$\Delta_1 = \frac{a_{01}(1)}{a_{00}(1)}\Psi_1^{(1)}\varepsilon^2 + \frac{a_{02}(1)}{a_{00}(1)}\Psi_1^{(2)}\varepsilon^4 + \cdots, \tag{5-4-70}$$

$$\Delta_2 = \frac{a_{12}(0)}{a_{22}(0)}\Psi_1^{(1)}\varepsilon^2 + \frac{a_{13}(0)}{a_{33}(0)}\Psi_1^{(2)}\varepsilon^4 + \cdots, \tag{5-4-71}$$

$$\Delta_3 = \frac{a_{01}(2)}{a_{00}(2)}\Psi_2^{(1)}\varepsilon^2 + \frac{a_{02}(2)}{a_{00}(2)}\Psi_2^{(2)}\varepsilon^4 + \cdots, \tag{5-4-72}$$

$$\Psi_l^{(n)} = \frac{\langle \psi_l^{(n)} \rangle_1}{\langle \psi_l^{(0)} \rangle_1}\frac{1}{\varepsilon^{2n}}. \tag{5-4-73}$$

No numerical calculations with these formulas have yet been made to see how the magnitudes of the correction terms vary with field strength, temperature, masses, or ion-neutral forces. In fact, many of the needed expansion coefficients have not even been evaluated in terms of the collision integrals.*

5-5. HIGH-FIELD THEORY

No comprehensive theory valid for arbitrary masses and ion-neutral forces exists for ion mobility and diffusion in high electric fields. At present only partial results and special cases are available. Judiciously used, they are often adequate to give a clear qualitative picture, even when a quantitatively accurate description may not be accessible. They can also serve as guides for devising connection formulas that give reasonable representations of results for arbitrary fields.

In this section we discuss three soluble models and two general computational techniques for which a few results are available. Exact solutions can be found if the ion mass is either very much greater or very much smaller than the neutral mass; these solutions are not limited to any particular form of ion-neutral interaction. A third exact solution can be found if the interaction is such that the mean free time between collisions is independent of energy; this solution is valid for any mass ratio. An r^{-4} potential gives a constant mean free time. One general computational method has been suggested by Wannier (1953) in which moment relations are used to improve indefinitely an initial trial function for the distribution function. This method has been little used; its most notable success has been the application to rigid-sphere ions and neutrals of equal mass. Another general computational procedure is computer simulation, which amounts to performing the experiment on a computer instead of in the laboratory and may become more popular as computer speeds and capacities increase.

A. RAYLEIGH MODEL $(m \gg M)$. A mixture consisting of a heavy test particle in a sea of light particles was first discussed by Lord Rayleigh (1891) in connection with gas theory and it seems appropriate to call this model after him. It has also been called a quasi-Lorentzian gas (Mason, 1957b; Chapman and Cowling, 1970, Section 10.53).

The fact that the model of heavy ions in a light gas can be treated exactly at nonzero field strengths was first pointed out by Kihara (1952). The results are remarkably simple and can be understood without going into

---

* *Note added in proof*: Some of these calculations have now been reported by Hahn and Mason (1973).

details of the mathematical proof. When $m \gg M$ and the ion drift velocity is much smaller than the thermal velocity of the gas molecules,

$$v_d^2 \ll \frac{kT}{M}, \tag{5-5-1}$$

the ion motion is decoupled from the neutral motion. The condition (5-5-1) means that the field strength cannot be made indefinitely large. The ion velocity is then an eigenfunction of the collision operator L:

$$L\mathbf{v} = \lambda_1^{(0)}\mathbf{v},$$

where $\lambda_1^{(0)}$ is of the form given in Table 5-4-1 but with no restriction on the ion-neutral interaction. Moreover, the ion distribution function is Maxwellian about the drift velocity:

$$f_i = n\left(\frac{m}{2\pi kT}\right)^{3/2} \exp\left\{-\left(\frac{m}{2kT}\right)[v_x^2 + v_y^2 + (v_z - v_d)^2]\right\}.$$

The first Chapman-Enskog result is thus exact to all orders of approximation at all field strengths as long as the condition (5-5-1) holds.

In a sense the strict Rayleigh model is uninteresting because nothing new happens at high field strengths, but at least it furnishes a known limit to which any connection formula or more general theory must reduce. Wannier (1953) and Smirnov (1966), however, have considered the case in which M/m is not vanishingly small and the electric field is so high that the drift velocity is much larger than the thermal velocities; the ion distribution function can then be taken as a delta function at v_d. The relation (5-4-10) can be integrated directly to yield (Smirnov, 1966)

$$\frac{eE}{\mu N} = \frac{e^{-Mv_d^2/2kT}}{v_d^2}\left(\frac{2kT}{\pi M}\right)^{1/2} \int_0^\infty e^{-Mv_r^2/2kT}v_r^2 Q^{(1)}(v_r)$$

$$\times \left[\frac{Mv_d v_r}{kT}\cosh\left(\frac{Mv_d v_r}{kT}\right) - \sinh\left(\frac{Mv_d v_r}{kT}\right)\right] dv_r. \tag{5-5-2}$$

This relation reduces to the first Chapman-Enskog result if the drift velocity is small compared with the thermal velocities and (5-5-2) is expanded in the small parameter $(Mv_d v_r/kT)$. Thus (5-5-2) holds both at low and at high fields and is presumably at least a good interpolation formula for intermediate fields. Moreover, for very large drift velocities we can take $v_r = v_d$ and obtain

$$\frac{eE}{\mu N} = v_d^2 Q^{(1)}(v_d), \tag{5-5-3}$$

which is essentially the free-flight result (5-2-10). This result was also obtained by Wannier (1953), who showed that the ion distribution was no longer Maxwellian about the drift velocity but had elliptic distortion as well.

B. LORENTZ MODEL $(m \ll M)$. A mixture consisting of a light test particle in a sea of heavy particles (essentially fixed scattering centers) was first studied by H. A. Lorentz and is usually called a Lorentzian gas (Chapman and Cowling, 1970, Section 10.5).

The Lorentz and Rayleigh models have a number of features in common— the decoupling of the equations for light and heavy particles, in particular. The motion of a heavy particle is hardly perturbed by a collision with a light particle in both models. The difference in the two models is that we follow the heavy particle in the Rayleigh model but the light particle in the Lorentz model. The collisions strongly perturb the motion of the light particles, and so their distribution function is far from Maxwellian. The fact that the perturbation is strong, however, can be used as the basis of a different approximation, since the perturbation acts on the direction of the ion velocity and not on its magnitude (since $m \ll M$). Thus the ion distribution function in a Lorentzian gas can be assumed to be nearly isotropic, although it may be far from Maxwellian.

Even at high fields, therefore, we can treat the effect of the field as a perturbation on an isotropic (but not Maxwellian) distribution. The isotropic part of the distribution, if it is far from Maxwellian, must also depend on the magnitude of the field but not its direction. Thus we write

$$f_i = f_i^{(0)}(v, E)\left[1 + \frac{e}{m}(\mathbf{v} \cdot \mathbf{E})\phi_i(v, E) + \cdots\right]. \qquad (5\text{-}5\text{-}4)$$

The distribution function of the neutrals is still Maxwellian, since $n \ll N$. These distributions are substituted back into the Boltzmann equation, and terms that involve only scalars, that involve the vector $\mathbf{E}$, and so on, are equated. If further terms in the expansion (5-5-4) are ignored and terms smaller than the magnitude of m/M are neglected, two equations that result for $f_i^{(0)}$ and ϕ_i can be solved exactly. The mathematical details are given in Section 19.61 of Chapman and Cowling (1970), who also have provided a short historical account of the problem; an improved derivation appears in Wannier (1971). The results are

$$\ln f_i^{(0)} = \ln A - \int \frac{mv^3\, dv}{kTv^2 + \frac{1}{3}M[eE/mNQ^{(1)}]^2}, \qquad (5\text{-}5\text{-}5)$$

$$\phi_i = \frac{[mv/NQ^{(1)}]}{kTv^2 + \frac{1}{3}M[eE/mNQ^{(1)}]^2}, \qquad (5\text{-}5\text{-}6)$$

where A is a normalization constant and $Q^{(1)}$ is the usual velocity-dependent diffusion cross section.

If $Q^{(1)}$ is given, (5-5-5) and (5-5-6) can be evaluated and the drift velocity and mean energy calculated by integration of the expressions

$$\mathbf{v}_d = \frac{1}{n} \int \mathbf{v} f_i^{(0)} \frac{e}{m} (\mathbf{v} \cdot \mathbf{E}) \phi_i \, d^3 v, \tag{5-5-7}$$

$$\tfrac{1}{2} m \overline{v^2} = \frac{1}{n} \int \left(\tfrac{1}{2} m v^2 \right) f_i^{(0)} \, d^3 v. \tag{5-5-8}$$

The transverse and longitudinal diffusion coefficients can be calculated also by adding a small perturbation term proportional to the concentration gradient to the distribution function (5-5-4) (Wannier, 1953). The solution has been obtained by Skullerud (1969b). The result for the transverse diffusion coefficient is quite simple:

$$D_T = \frac{1}{3nN} \int \frac{v}{Q^{(1)}} f_i^{(0)} \, d^3 v. \tag{5-5-9}$$

The reasons for this simplicity are that for the Lorentz model the magnitude of the ion velocity is hardly changed by collisions and the persistence of velocity is small (being proportional to m/M) and can be taken as zero for the transverse motion. Thus the free-flight results of Section 5-2-A should be exact for D_T; in fact (5-5-9) is equivalent to $\tfrac{1}{3} \overline{v^2 \tau}$, which is the same as the free-flight result (5-2-22) or (5-2-23) if τ is constant. The result for D_L is more complicated, however:

$$D_L = D_T + \frac{M}{m} \int_0^\infty \left[\frac{nU(v)}{4\pi v^2 f_i^{(0)}} - \frac{2}{3} \left(\frac{eE}{mNQ^{(1)}} \right) \right] U(v) \phi_i \, d^3 v, \tag{5-5-10}$$

where

$$U(v) = \frac{4\pi}{n} \int_0^v [u(v) - \bar{u}] f_i^{(0)} v^2 \, dv, \tag{5-5-11}$$

and $u(v)$ is an instantaneous drift velocity

$$u(v) = \frac{1}{3} \frac{eE}{mN} \frac{1}{v^2} \frac{d}{dv} \left[\frac{v^2}{Q^{(1)}} \right], \tag{5-5-12}$$

$$\bar{u} = \frac{1}{n} \int u(v) f_i^{(0)} \, d^3 v. \tag{5-5-13}$$

Similar results have been obtained, but in a less explicit form, by Parker and Lowke (1969), who have also shown how the theory can be extended to include inelastic collisions (Lowke and Parker, 1969).

In general, numerical integration is required to evaluate the above expressions, but a few special cases can be calculated analytically. The simplest case occurs for low fields, for which

$$\frac{1}{3}\frac{M}{v^2}\left[\frac{eE}{mNQ^{(1)}}\right]^2 \ll kT. \tag{5-5-14}$$

The second terms in the denominators of (5-5-5) and (5-5-6) can then be neglected: $f_i^{(0)}$ becomes Maxwellian and the drift velocity and diffusion coefficients reduce to

$$v_d = \frac{1}{3nN}\frac{eE}{kT}\int \frac{v}{Q^{(1)}}f_i^{(0)}\,d^3v, \tag{5-5-15}$$

$$D_L = D_T = \left(\frac{kT}{eE}\right)v_d. \tag{5-5-16}$$

These are exactly equivalent to the result (5-3-22) for the diffusion coefficient of a Lorentzian mixture at zero field. The mean energy, of course, is entirely thermal.

For high fields the inequality of (5-5-14) is reversed:

$$\frac{1}{3}\frac{M}{v^2}\left[\frac{eE}{mNQ^{(1)}}\right]^2 \gg kT \tag{5-5-17}$$

and (5-5-5) and (5-5-6) reduce to

$$\ln f_i^{(0)} = \ln A - \frac{3m^3N^2}{M(eE)^2}\int [Q^{(1)}]^2 v^3\,dv, \tag{5-5-18}$$

$$\phi_i = \frac{3m^3N}{M(eE)^2}Q^{(1)}. \tag{5-5-19}$$

Numerical integration is still required unless the velocity dependence of $Q^{(1)}$ is simple; for rigid spheres $Q^{(1)}$ is constant and (5-5-18) becomes

$$f_i^{(0)} = A \exp\left\{-\frac{3m^3N^2v^4[Q^{(1)}]^2}{4M(eE)^2}\right\}, \tag{5-5-20}$$

$$A = \frac{n}{\pi\Gamma(\frac{3}{4})}\left\{\frac{3m^3N^2[Q^{(1)}]^2}{4M(eE)^2}\right\}^{3/4}. \tag{5-5-21}$$

The drift velocity is then found to be

$$v_d = \frac{1}{2}\left(\frac{4}{3}\right)^{3/4}\frac{1}{\Gamma(\frac{3}{4})}\left[\frac{\pi}{(mM)^{1/2}}\frac{eE}{NQ^{(1)}}\right]^{1/2} = \frac{0.897}{(mM)^{1/4}}\left[\frac{eE}{NQ^{(1)}}\right]^{1/2}, \tag{5-5-22}$$

which was the expression compared with the free-flight results in (5-2-11). The mean energy is

$$\tfrac{1}{2}m\overline{v^2} = \left(\frac{M}{3m}\right)^{1/2} \frac{\Gamma(\tfrac{5}{4})}{\Gamma(\tfrac{3}{4})} \frac{eE}{NQ^{(1)}},\tag{5-5-23}$$

from which we find

$$\frac{\text{random field energy}}{\text{drift energy}} = \frac{3}{\pi} \Gamma(\tfrac{5}{4})\Gamma(\tfrac{3}{4}) \frac{M}{m} = 1.061 \frac{M}{m}.\tag{5-5-24}$$

This is close to the free-flight result (1.061 instead of 1). It is easy to show that because of the isotropy of $f_i^{(0)}$

$$\frac{\text{longitudinal random field energy}}{\text{transverse random field energy}} = 1.\tag{5-5-25}$$

The last two results mean that almost all the field energy is random and isotropically distributed because rigid spheres are isotropic scatterers and almost no energy is exchanged between ions and neutrals.

The value of D_T at high fields is readily found from (5-5-9) to be

$$D_T = \frac{1}{3} \left(\frac{4}{3}\right)^{3/4} \frac{1}{\Gamma(\tfrac{3}{4})} \frac{1}{NQ^{(1)}} \left(\frac{M}{m}\right)^{1/4} \left[\frac{eE}{mNQ^{(1)}}\right]^{1/2}$$

$$= \frac{1}{3} \frac{0.877}{NQ^{(1)}} \left(\frac{M}{m}\right)^{1/4} \left[\frac{eE}{mNQ^{(1)}}\right]^{1/2} = \frac{M}{3}(1.214)\frac{v_d^3}{eE}.\tag{5-5-26}$$

This is in fair agreement (1.214 instead of 1) with the free-flight expression (5-2-30) plus the expression (5-2-32) for ζ_T based on a constant mean free time. The value of D_L has been found numerically to be (Skullerud, 1969b)

$$\frac{D_L}{D_T} = 0.4910,\tag{5-5-27}$$

which is quite different from the value of unity expected for a constant mean-free-time model (r^{-4} potential). The ratio D_L/D_T is therefore sensitive to the ion-neutral potential as well as to the mass ratio. The free-flight arguments in Section 5-2-A suggested $D_L/D_T \geq 1$ on the basis of persistence-of-velocity effects; these are largely mass effects, as shown by (5-2-32) and (5-2-33). Persistence effects can also be attributed to the "softness" of the potential or of the diffusion cross section; the more rapidly $Q^{(1)}$ decreases with increasing energy, the larger we would expect the ratio D_L/D_T to become.

The behavior of the ratio D_L/D_T has been calculated numerically by Skullerud (1969b) for a model in which $Q^{(1)}$ varies as a power of v; this is equivalent to a potential varying as r^{-n}, for which $Q^{(1)}$ varies as $v^{-4/n}$, given in (5-2-38). The results are shown in Fig. 5-5-1; the ratio diverges for $n \leq 2$. It is apparent that D_L/D_T depends strongly on the potential and is equal to or greater than unity only for $n \leq 4$.

The transition between the low-field and high-field values of D_T and D_L is shown in Fig. 5-5-2 for rigid spheres, as calculated numerically by Skullerud (1969b). The transition is smooth, the only surprise being a shallow minimum in D_L.

As a final special case we note that all the integrations can be performed if $Q^{(1)}$ varies as $1/v$. In particular, $f_i^{(0)}$ is Maxwellian in form but with an effective "temperature" given by

$$kT_{\text{eff}} = kT + \tfrac{1}{3}M\left[\frac{eE}{mNvQ^{(1)}}\right]^2. \qquad (5\text{-}5\text{-}28)$$

However, $Q^{(1)} \propto 1/v$ is the Maxwellian model, which can be treated exactly without any conditions on the mass ratio and is discussed next.

C. MAXWELL MODEL (CONSTANT MEAN FREE TIME).   As first noticed by Maxwell (1867), a great simplification occurs in kinetic theory if the cross sections are proportional to $1/v_r$. All desired quantities can then be found

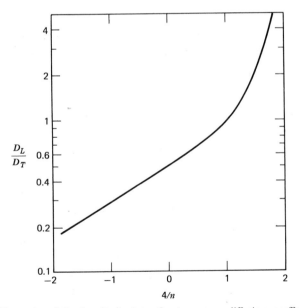

FIG. 5-5-1. The ratio of the longitudinal to the transverse diffusion coefficient at high fields for r^{-n} potentials with $m \ll M$.

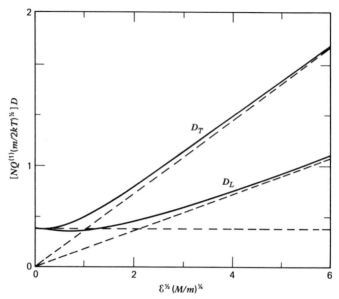

FIG. 5-5-2. Transition between low- and high-field transverse and longitudinal diffusion coefficients for rigid spheres with $m \ll M$.

without having to determine the distribution function itself. The reason is that the product $v_r Q^{(l)}$ can be removed from the integral in the collision operator of the Boltzmann equation; what remains in the integrand can be used to form one of the desired moments directly. Almost all present general treatments of kinetic theory are based in some sense on the Maxwell model as a first-order approximation. Application of the model to ion mobility and diffusion is due primarily to Kihara (1952, 1953) and to Wannier (1953).

Several workers have independently noted that the solution of the Boltzmann equation can be treated as an eigenvalue problem for the Maxwell model (Chapman and Cowling, 1970, Section 10.331). This is the point of view adopted by Kihara. Wannier used somewhat different methods, based on the fact that the mean free time between collisions is a constant for the Maxwell model, independent of temperature and field intensity. This property was implicitly invoked in the simple free-flight theory of Section 5-2-A.

The eigenvalue procedure for the Maxwell model was used as the starting point for the medium-field theory discussed in Section 5-4-A and does not need to be repeated. Indeed, for the Maxwell model, the results obtained there are not limited to medium fields; all terms in the expansion in powers of $(E/N)^2$ are zero except the first, and the results are exact at all field intensities. In particular, the drift velocity is exactly

$$v_d = \frac{eE}{\mu N v_r Q^{(1)}} = \frac{3}{16}\left(\frac{2\pi}{\mu kT}\right)^{1/2}\frac{eE}{N\overline{\Omega}^{(1,\,1)}}. \tag{5-5-29}$$

The mobility is thus seen to be independent of field. Similarly, the partition of ion energy, discussed for medium fields in Section 5-4-F, is found to be field-independent for the Maxwell model, and the thermal and field energies are additive:

$$\tfrac{1}{2}m\overline{v^2} = \tfrac{3}{2}kT + \tfrac{1}{2}mv_d^2 + \tfrac{1}{2}Mv_d^2, \tag{5-5-30}$$

$$\frac{\text{random field energy}}{\text{drift energy}} = \frac{M}{m}, \tag{5-5-31}$$

$$\frac{\text{longitudinal random field energy}}{\text{transverse random field energy}} = \frac{5m - 2mA^* + MA^*}{(m + M)A^*}, \tag{5-5-32}$$

where $A^* = \tfrac{5}{6}Q^{(2)}/Q^{(1)}$, a constant for the Maxwell model.

The precise numerical value of A^* depends on the nature of the ion-neutral interaction. Even though a cross section varying as $1/v_r$ implies a potential varying as $1/r^4$, there are at least four reasonable possibilities for the latter. First, the potential may be repulsive or attractive. Then, if attractive, there may be different conditions at $r = 0$: a thin repulsive rigid core, a random core that absorbs the colliding particles and then re-emits them at random angles, or a transparent core for which two colliding particles pass through each other. The values must be found by numerical integration and are listed in Table 5-5-1 (Kihara et al., 1960; Higgins and Smith, 1968). The most physically realistic model is probably the attractive potential with a rigid core, which corresponds to the polarization potential, but the others are useful mathematical models.

TABLE 5-5-1. Transport cross sections for various forms of the Maxwell model $V(r) = \pm C/r^4$, as discussed in the text

Model	$A^* = \dfrac{5}{6}\dfrac{Q^{(2)}}{Q^{(1)}}$	$\left(\dfrac{E}{C}\right)^{1/2}Q^{(1)}$
Repulsion	1.292	3.75
Attraction, rigid core	0.871	6.95
Attraction, random core	0.960	7.59
Attraction, transparent core	0.735	8.23
Isotropic scattering	5/6	2π

Another useful mathematical model combines the Maxwell model with an isotropic scattering pattern (Wannier, 1953); no known potential corresponds to such scattering, but the combination has simple mathematical properties and the corresponding A^* is also listed in Table 5-5-1.

Wannier (1953) has indicated how ion diffusion at high fields can be handled as a perturbation problem and has carried the calculations through in detail for the Maxwell model, using Maxwell's original method. The results have the simple form

$$D_T = \frac{m + M}{M} \frac{\overline{v_T^2}}{N v_r Q^{(1)}}, \tag{5-5-33}$$

$$D_L = \frac{m + M}{M} \frac{(\overline{v_L^2} - v_d^2)}{N v_r Q^{(1)}}, \tag{5-5-34}$$

where $v_r Q^{(1)}$ is a constant for the Maxwell model. As with A^*, the precise numerical value of $v_r Q^{(1)}$ depends on the model assumed for the ion-neutral interaction. For a potential of the form $V(r) = \pm C/r^4$ the quantity $[(E/C)^{1/2} Q^{(1)}]$ is a pure number; values are listed in Table 5-5-1 for the same models discussed in connection with A^*. These values can be converted to $v_r Q^{(1)}$ through the relation $E = \frac{1}{2}\mu v_r^2$. Inserting the appropriate energy averages in (5-5-33) and (5-5-34), we obtain expressions for the diffusion coefficients explicit in the field:

$$D_T[N v_r Q^{(1)}] = \frac{kT}{\mu} + \frac{(m + M)^4 A^*}{mM^2(5m + 3MA^*)} \left[\frac{eE}{mN v_r Q^{(1)}}\right]^2, \tag{5-5-35}$$

$$D_L[N v_r Q^{(1)}] = \frac{kT}{\mu} + \frac{(m + M)^3(5m - 2mA^* + MA^*)}{mM^2(5m + 3MA^*)} \left[\frac{eE}{mN v_r Q^{(1)}}\right]^2. \tag{5-5-36}$$

The first term on the right is the low-field approximation, which can be verified by substitution for the cross section in terms of the more usual collision integral,

$$v_r Q^{(1)} = \frac{8}{3} \left(\frac{2kT}{\pi\mu}\right)^{1/2} \overline{\Omega}^{(1, 1)}, \tag{5-5-37}$$

a relation valid only for the Maxwell model. The diffusion coefficients are seen to be given simply as the sum of the low-field and high-field limits; again, this result is valid only for the Maxwell model.

It is interesting to combine some of the foregoing expressions to eliminate the explicit dependence on cross sections. From (5-5-29), (5-5-33), and (5-5-34) we obtain a sort of generalized Einstein relation (Wannier, 1963),

$$\frac{eD_{T, L}}{K} = 2(\text{random energy along } T, L). \tag{5-5-38}$$

From (5-5-29), (5-5-35), and (5-5-36) we obtain D_T and D_L in terms of v_d:

$$D_T = kT\left(\frac{v_d}{eE}\right) + \frac{(m + M)MA^*}{5m + 3MA^*}\left(\frac{v_d^3}{eE}\right), \qquad (5\text{-}5\text{-}39)$$

$$D_L = kT\left(\frac{v_d}{eE}\right) + \frac{M(5m - 2mA^* + MA^*)}{5m + 3MA}\left(\frac{v_d^3}{eE}\right). \qquad (5\text{-}5\text{-}40)$$

These make explicit the general forms (5-2-30) and (5-2-31) obtained by the simple free-flight theory. Finally, the Lorentz model results suggest that it might be useful to write D_L in terms of D_T:

$$D_L = D_T + \frac{mM(5 - 3A^*)}{5m + 3MA^*}\left(\frac{v_d^3}{eE}\right). \qquad (5\text{-}5\text{-}41)$$

In some ways the Maxwell model is disappointing in that results valid at low fields are merely found to be valid also at high fields. Relations such as those of the foregoing paragraph, however, may well have a wider validity than the Maxwell model, at least to a useful degree of approximation. In this regard a convolution theorem proved by Wannier (1953) for the constant mean-free-time model is interesting in that the effects of field and temperature are separated. The ion distribution function for arbitrary field and temperature is reduced to two components, one containing only the field, the other only the temperature. We give here only the precise statement of the theorem, not the proof. It is necessary to make a temporary additional restriction on the general Maxwell model in order to prove the theorem without divergence difficulties; the restriction consists of some cutoff procedure to prevent the small-angle scattering from contributing to the cross sections. In any transport problem this restriction can be dropped ultimately, since weighting factors $(1 - \cos^l \theta)$ eventually appear and suppress the small-angle divergence naturally.

The differential cross section for the Maxwell model can be written as the product of two factors,

$$I_s(v_r, \theta) = q(v_r)i_s(\theta). \qquad (5\text{-}5\text{-}42)$$

Since $v_r q(v_r)$ is constant, it can be removed from the collision operator of the Boltzmann equation and a constant mean free time τ defined:

$$\frac{1}{\tau} \equiv N v_r q(v_r). \qquad (5\text{-}5\text{-}43)$$

As in Section 5-4-A, we assume a uniform steady-state distribution of ions, take the distribution function of the neutrals to be Maxwellian, and let the field be uniform and in the z-direction; the Boltzmann equation then becomes

$$\frac{eE}{m}\frac{\partial f_i(\mathbf{v})}{\partial v_z} = \frac{1}{N\tau}\int \cdots \int [f_i(\mathbf{v}')f_j^{(0)}(\mathbf{V}') - f_i(\mathbf{v})f_j^{(0)}(\mathbf{V})]i_s(\theta)\, d\Omega_{CM}\, d^3V. \qquad (5\text{-}5\text{-}44)$$

The cutoff trick enables the two terms on the right-hand side to be integrated separately; without a cutoff the two integrations diverge individually, although their difference remains finite. The result is the equation to which the convolution theorem applies:

$$\frac{eE\tau}{m}\frac{\partial f_i(\mathbf{v})}{\partial v_z} + f_i(\mathbf{v}) = \frac{1}{N}\int \cdots \int f_i(\mathbf{v}')f_j^{(0)}(\mathbf{V}')i_s(\theta)\, d\Omega_{CM}\, d^3V, \qquad (5\text{-}5\text{-}45)$$

where $\int i_s(\theta)\, d\Omega_{CM} = 1$. Two limiting cases of (5-5-45) can be distinguished: a high-field equation in which $f_j^{(0)}$ is replaced by a delta function and an equilibrium equation in which the field term is dropped:

$$\frac{eE\tau}{m}\frac{\partial h_i(\mathbf{v})}{\partial v_z} + h_i(\mathbf{v}) = \frac{1}{N}\int \cdots \int h_i(\mathbf{v}')\,\delta(\mathbf{V}')i_s(\theta)\, d\Omega_{CM}\, d^3V, \qquad (5\text{-}5\text{-}46)$$

$$f_i^{(0)}(\mathbf{v}) = \frac{1}{N}\int \cdots \int f_i^{(0)}(\mathbf{v}')f_j^{(0)}(\mathbf{V}')i_s(\theta)\, d\Omega_{CM}\, d^3V, \qquad (5\text{-}5\text{-}47)$$

where we have written h_i for the high-field form of f_i. The theorem then states that the general solution $f_i(\mathbf{v})$ of (5-5-45) is the convolution of the solution $h_i(\mathbf{v})$ of (5-5-46) and the solution $f_i^{(0)}(\mathbf{v})$ of (5-5-47):

$$f_i(\mathbf{v}) = \int h_i(\mathbf{u})f_i^{(0)}(\mathbf{v} - \mathbf{u})\, d^3u. \qquad (5\text{-}5\text{-}48)$$

In other words, it is necessary to solve only the high-field equation in order to obtain the general solution.

D. IMPROVEMENT OF TRIAL FUNCTIONS BY MOMENT EQUATIONS. This method of computation was devised by Wannier (1953) to permit the determination of desired averages for arbitrary mass ratios and arbitrary ion-neutral interactions but has in practice been applied only for mass ratio unity with constant mean free time (as a test case) and with constant mean free path (rigid spheres). The procedure is to expand the distribution function into spherical harmonics and form moments from the Boltzmann equation. These moment equations can be thought of as conditions that a distribution function must satisfy. A trial distribution function with undetermined constants can be chosen and the constants found by requiring that the first few moment equations be satisfied. More constants and more moment equations can be included to secure improvement. The procedure is thus somewhat like the Rayleigh-Ritz method, except that no extremum principle is involved. Success, of course, depends on the degree of shrewdness shown in the choice of trial functions.

The details are as follows. At high fields the relative velocity $\mathbf{v}_r$ is equal to the ion velocity $\mathbf{v}$, and the distribution function of the neutrals can be replaced with a delta function. Making the cutoff assumption and splitting $I_s(v, \theta)$ into two parts, as in the convolution theorem of the preceding section, we can write the Boltzmann equation for high fields as

$$\frac{eE}{m} \frac{\partial h(\mathbf{v})}{\partial v_z} + \frac{h(\mathbf{v})}{\tau(v)} = \frac{1}{N} \int \cdots \int h(\mathbf{v}') \, \delta(\mathbf{V}') \frac{i_s(\theta)}{\tau(v')} \, d\Omega_{CM} \, d^3 V. \quad (5\text{-}5\text{-}49)$$

There is no loss of generality in expansion of $h(\mathbf{v})$ into spherical harmonics:

$$h(\mathbf{v}) = \sum_{v=0}^{\infty} h_v(v) P_v\left(\frac{v_z}{v}\right). \quad (5\text{-}5\text{-}50)$$

Substitution of (5-5-50) into (5-5-49) and formation of moments leads to the following set of algebraic moment equations

$$(2v + 1)\left\langle \frac{1 - \Pi_{s,\,v}(\theta)}{(eE/m)\tau(v)} \, v^s P_v\left(\frac{v_z}{v}\right) \right\rangle$$

$$= v(v + s + 1)\left\langle v^{s-1}P_{v-1}\left(\frac{v_z}{v}\right) \right\rangle + (v + 1)(s - v)\left\langle v^{s-1}P_{v+1}\left(\frac{v_z}{v}\right) \right\rangle, \quad (5\text{-}5\text{-}51)$$

where

$$\Pi_{s,\,v}(\theta) \equiv \left(\frac{v}{v'}\right)^s P_v(\cos \vartheta), \quad (5\text{-}5\text{-}52)$$

in which ϑ is the angle between $\mathbf{v}$ and $\mathbf{v}'$. The function $\Pi_{s,\,v}$ depends also on the mass ratio; for $m = M$ it has the simple form

$$\Pi_{s,\,v}(\theta) = \cos^s \frac{\theta}{2} P_v\left(\cos \frac{\theta}{2}\right). \quad (5\text{-}5\text{-}53)$$

As usual the pointed brackets represent averages over the distribution function. The remainder of the procedure is dictated by the velocity dependence of $\tau(v)$. For rigid spheres of diameter d the mean free time is

$$\frac{1}{\tau(v)} = Nv(\pi d^2), \quad (5\text{-}5\text{-}54)$$

so that a convenient dimensionless variable to use in (5-5-51) is

$$\mathbf{w} = \mathbf{v}\left(\frac{mN\pi d^2}{eE}\right)^{1/2}. \quad (5\text{-}5\text{-}55)$$

The moment equations (5-5-51) then take the form

$$(2v + 1)[1 - \langle \Pi_{s,v}(\theta) \rangle]\langle s + 1, v \rangle$$
$$= v(v + s + 1)\langle s - 1, v - 1 \rangle + (v + 1)(s - v)\langle s - 1, v + 1 \rangle, \quad (5\text{-}5\text{-}56)$$

in which $\langle s, v \rangle$ denotes $\langle w^s P_v(v_z/v) \rangle$. Wannier used three relations derived from this set to determine three constants in a trial function for $h_0(w)$ which has four terms consisting of an exponential integral, a Gaussian function, and modified Hankel functions of order 0 and 1. (The fourth constant was determined by normalization.) Once $h_0(w)$ is found, the desired averages can be obtained by integration. The drift velocity is

$$v_d = 1.1467 \left(\frac{eE}{mN\pi d^2} \right)^{1/2}. \quad (5\text{-}5\text{-}57)$$

The partition of ion energy is given by

$$\frac{\text{random field energy}}{\text{drift energy}} = 0.789, \quad (5\text{-}5\text{-}58)$$

$$\frac{\text{longitudinal random field energy}}{\text{transverse random field energy}} = 1.54. \quad (5\text{-}5\text{-}59)$$

It is interesting to compare these numbers with the corresponding ones for the Maxwell model with isotropic scattering. From (5-5-29), (5-5-31), and (5-5-32) we find $\sqrt{2} = 1.414$ in place of 1.1467, 1.000 in place of 0.789, and 2.50 in place of 1.54.

Wannier has also shown how this computation method can be extended to include diffusion and has calculated the longitudinal diffusion coefficient for rigid spheres of equal mass. The numerical accuracy is lower than for the drift velocity and energy partition, since less was known about the perturbation part of the distribution function that describes diffusion and consequently the choice of trial functions could not be so shrewd. The result obtained is

$$D_L = \frac{0.22}{N\pi d^2} \left(\frac{eE}{mN\pi d^2} \right)^{1/2}, \quad (5\text{-}5\text{-}60)$$

which can be written as

$$D_L = 0.15 \frac{mv_d^3}{eE}. \quad (5\text{-}5\text{-}61)$$

This last formula is in poor agreement with the analogous result (5-5-40) for the isotropic Maxwell model, for which the corresponding numerical coefficient is 0.56 instead of 0.15. However, (5-5-60) is in reasonable agreement with the generalized Einstein relation (5-5-38), provided we interpret K as equal to (dv_d/dE) rather than as v_d/E (the distinction does not matter for the Maxwell model); this procedure yields

$$D_L = \frac{m}{e}\,(\overline{v_L^2} - v_d^2)\,\frac{dv_d}{dE} = \frac{0.26}{N\pi d^2}\left(\frac{eE}{mN\pi d^2}\right)^{1/2} \qquad (5\text{-}5\text{-}62)$$

on inserting numerical values from (5-5-57), (5-5-58), and (5-5-59).

This method should be capable of further exploitation, especially in view of the great advances in high-speed computers since 1953.

E. COMPUTER SIMULATION (MONTE CARLO). A Monte Carlo calculation in effect follows one ion through a large number of collisions, choosing the scattering angle and the time between collisions according to some random scheme. A pioneering calculation of this sort was described by Wannier (1953), who used it to determine the ion velocity distribution function at high fields for $m = M$. On the average, the collisions corresponded to isotropic scattering and constant mean-free-flight time. These results were used to help in the choice of trial functions used in the calculations described in the preceding Section 5-5-D.

More recently Skullerud (1968) published a brief description of a Monte Carlo procedure in which the computational simplicity of constant τ is extended to include an arbitrary dependence of τ on v. The procedure was tested on the case of rigid spheres with $m = M$ and yielded a numerical coefficient for the drift velocity of 1.143 ± 0.004 in place of the value 1.1467 given in (5-5-57) (private communication of H. R. Skullerud to E. W. McDaniel, 1969). The results have since been greatly extended to include diffusion coefficients, velocity averages, and speed distributions (Skullerud, 1972).

F. PARTITION OF ION ENERGY IN MULTICOMPONENT MIXTURES. The moment equations of Section 5-5-D can be used to calculate the partitioning of ion energy between drift energy and random field energy, at least in an approximate way. These results are needed for mixtures to find the composition dependence of the drift velocity at high fields, as discussed in Section 5-2-C. We consider first a single gas and then multicomponent mixtures.

We start with the moment equations (5-5-51). The functions $\Pi_{s,\,v}(\theta)$ (Wannier, 1953) are given by

$$\Pi_{s,\,v}(\theta) = \left(\frac{v}{v'}\right)^s P_v(\cos\vartheta), \qquad (5\text{-}5\text{-}63)$$

$$\frac{v}{v'} = \frac{(m^2 + M^2 + 2mM\cos\theta)^{1/2}}{m + M}, \qquad (5\text{-}5\text{-}64)$$

$$\cos \vartheta = \frac{(m + M \cos \theta)}{(m^2 + M^2 + 2mM \cos \theta)^{1/2}}.$$
(5-5-65)

The moment equations we need correspond to $s = 1$, $v = 1$ and $s = 2$, $v = 0$:

$$\frac{M}{m + M} \left\langle \frac{(1 - \cos \theta)}{(eE/m)\tau(v)} \, v \left(\frac{v_z}{v} \right) \right\rangle = 1,$$
(5-5-66)

$$\frac{mM}{(m + M)^2} \left\langle \frac{(1 - \cos \theta)}{(eE/m)\tau(v)} \, v^2 \right\rangle = \left\langle v \left(\frac{v_z}{z} \right) \right\rangle.$$
(5-5-67)

Now we invoke an approximation, or the special properties of a model, to evaluate the above averages. We assume that the averages of products can be split into products of averages; the moment equations can then be combined to give

$$\frac{mM}{(m + M)^2} \, \overline{v^2} = \frac{M}{m + M} v_d^2,$$
(5-5-68)

since $\overline{v^2} = \langle v^2 \rangle$ and $v_d = \langle v_z \rangle$. This reduces to the result we have used several times before:

$$m\overline{v^2} = mv_d^2 + Mv_d^2.$$

This splitting of averages is exact for a model in which $(1 - \cos \theta)$ and τ are both independent of v but can be regarded as approximately correct for any model.

On the same basis the model shows, according to the convolution theorem discussed in Section 5-5-C, that thermal energies and field energies are additive (Wannier, 1953).

The extension to mixtures is straightforward (Mason and Hahn, 1972). If we trace the derivation of the moment equations (5-5-51) from the Boltzmann equation for a mixture, we find the only change is that the left-hand side of (5-5-51) becomes a summation:

$$(2v + 1) \sum_j x_j \left\langle \frac{1 - \Pi_{s,v}(\theta_j)}{(eE/m)\tau_j} \, v^s P_v \left(\frac{v_z}{v} \right) \right\rangle,$$
(5-5-69)

where τ_j is the mean time between ion collisions in pure gas j at the same total density as the mixture. For ion collisions with species j the factors of $\Pi_{s,v}(\theta_j)$ are

$$\frac{v}{v'} = \frac{(m^2 + M_j^2 + 2mM_j \cos \theta_j)^{1/2}}{m + M_j},$$
(5-5-70)

$$\cos \vartheta = \frac{m + M_j \cos \theta_j}{(m^2 + M_j^2 + 2mM_j \cos \theta_j)^{1/2}}.$$
(5-5-71)

The moment equations for $s = 1$, $v = 1$ and $s = 2$, $v = 0$ are

$$\sum_j \frac{x_j M_j}{m + M_j} \left\langle \frac{(1 - \cos \theta_j)}{(eE/m)\tau_j} v_z \right\rangle = 1, \tag{5-5-72}$$

$$\sum_j \frac{x_j m M_j}{(m + M_j)^2} \left\langle \frac{(1 - \cos \theta_j)}{(eE/m)\tau_j} v^2 \right\rangle = \langle v_z \rangle. \tag{5-5-73}$$

We again split the average of products into products of averages; the average of $(1 - \cos \theta_j)\tau_j$ is proportional to $\langle v_{r(j)} Q_{ij}^{(1)} \rangle$, and the two moment equations can be combined into the form

$$\sum_j \frac{x_j m M_j}{(m + M_j)^2} \langle v_{r(j)} Q_{ij}^{(1)} \rangle \overline{v^2} = \sum_j \frac{x_j M_j}{m + M_j} \langle v_{r(j)} Q_{ij}^{(1)} \rangle v_d^2. \tag{5-5-74}$$

This reduces to (5-5-68) for a single gas. If, by analogy with a pure gas, we write

$$m\overline{v^2} = m v_d^2 + \overline{M} v_d^2, \tag{5-5-75}$$

where $\overline{M}$ is some average mass of the gas mixture, then comparison of (5-5-74) and (5-5-75) shows that $\overline{M}$ is to be identified as

$$\overline{M} = \sum_j \frac{\omega_j M_j}{\sum \omega_j}, \tag{5-5-76}$$

where

$$\omega_j = \frac{x_j M_j \langle v_{r(j)} Q_{ij}^{(1)} \rangle}{(m + M_j)^2}. \tag{5-5-77}$$

For high fields $v_{r(j)}$ is proportional to v_d and cancels out of (5-5-76). These results were used in Section 5-2-C without proof.

We can also determine how the random field energy is divided between longitudinal and transverse components in a mixture

$$\frac{\text{longitudinal random field energy}}{\text{transverse random field energy}} \equiv \frac{\overline{v_z^2} - v_d^2}{\overline{v_x^2}}. \tag{5-5-78}$$

We need find only $\overline{v_z^2}$, since $\overline{v_x^2} = \overline{v_y^2} = \overline{v^2} - \overline{v_z^2}$, and for this purpose we use the moment equation for $s = 2$, $v = 2$:

$$\sum_j \frac{x_j M_j}{(m + M_j)^2} \left\langle \frac{4m(1 - \cos \theta_j) + 3M_j(1 - \cos^2 \theta_j)}{(eE/m)\tau_j} (3v_z^2 - v^2) \right\rangle = 8\langle v_z \rangle. \tag{5-5-79}$$

Again we split the averages of products into products of averages and note that

$$\frac{\langle(1-\cos^2\theta_j)/\tau_j\rangle}{\langle(1-\cos\theta_j)/\tau_j\rangle} = \frac{2\langle v_{r(j)} Q_{ij}^{(2)}\rangle}{3\langle v_{r(j)} Q_{ij}^{(1)}\rangle} = \frac{4}{5} A_{ij}^*. \tag{5-5-80}$$

Combining (5-5-72), (5-5-73), and (5-5-79), we find after considerable algebra that

$$\frac{\text{longitudinal random field energy}}{\text{transverse random field energy}} = \frac{5m - 2m\overline{A}^* + \overline{M}\overline{A}^*}{(m+\overline{M})\overline{A}^*}. \tag{5-5-81}$$

where $\overline{M}$ is defined by (5-5-76) and (5-5-77) and

$$A^* = \sum_j \frac{\omega_j M_j A_{ij}^*}{\overline{M} \sum_j \omega_j}. \tag{5-5-82}$$

This reduces to (5-5-32) for the case of a single gas.

5-6. CONNECTION FORMULAS

We now have enough accurate results set down to make it worthwhile to try to connect them up with semiempirical interpolation formulas. We are guided in this attempt by the free-flight theory of Section 5-2-A, which in fact yielded several approximate connection formulas, such as (5-2-17), (5-2-30), and (5-2-31).

A. FIELD DEPENDENCE. There are two parts to the problem. The first is to devise formulas giving an explicit dependence on field strength. The second is to average the transport cross sections $Q^{(l)}$ in an appropriate way—at low fields they are averaged over a velocity distribution (5-3-16) to yield the temperature-dependent collision integrals $\Omega^{(l,s)}$, but at high fields they correspond to the drift energy. It is easiest to dispose of the second part of the problem first. Referring to (5-3-16), we see that $Q^{(l)}(E)$ is averaged over relative energies $E = \frac{1}{2}\mu v_r^2$ with a weight factor $(E^{s+1}e^{-E/kT})$ corresponding to the spread in thermal energies. As a simple solution we imagine that the same weight factor is appropriate at all fields but is centered about the drift energy so that the energy E in (5-3-16) is replaced by an energy E', given by

$$E' = E(\text{thermal}) + E(\text{field}) = E + \frac{1}{2}\mu[v_r(\text{field})]^2, \tag{5-6-1}$$

where $v_r(\text{field})$ is the relative speed between ions and neutrals due to the electric field. There are both drift and random components to $v_r(\text{field})$ for which we can use the Maxwellian-model approximation

$$\tfrac{1}{2}m[v_r(\text{field})]^2 = \tfrac{1}{2}mv_d^2 + \tfrac{1}{2}Mv_d^2,$$

which leads to

$$\tfrac{1}{2}\mu[v_r(\text{field})]^2 = \tfrac{1}{2}Mv_d^2. \tag{5-6-2}$$

Thus (5-3-16) is replaced by

$$\overline{\Omega}^{(l,\,s)}(T, v_d) = \frac{\displaystyle\int_0^\infty e^{-(E+\frac{1}{2}Mv_d^2)/kT}(E + \tfrac{1}{2}Mv_d^2)^{s+1}Q^{(l)}(E + \tfrac{1}{2}Mv_d^2)\,dE}{\displaystyle\int_0^\infty e^{-(E+\frac{1}{2}Mv_d^2)/kT}(E + \tfrac{1}{2}Mv_d^2)^{s+1}\,dE}$$

$$= \frac{\displaystyle\int_{\frac{1}{2}Mv_d^2}^\infty e^{-E'/kT}(E')^{s+1}Q^{(l)}(E')\,dE'}{\displaystyle\int_{\frac{1}{2}Mv_d^2}^\infty e^{-E'/kT}(E')^{s+1}\,dE'}, \tag{5-6-3}$$

the integral in the denominator being a normalization factor. For $\tfrac{1}{2}Mv_d^2 \ll kT$, (5-6-3) reduces to (5-3-16), and for $\tfrac{1}{2}Mv_d^2 \gg kT$ it reduces to $Q^{(l)}(\tfrac{1}{2}Mv_d^2)$. However, (5-6-3) has no theoretical status other than a reasonable interpolation formula.

To fit the drift velocity we use the general form of the free-flight equation (5-2-5), which then allows us two choices: the value of the parameter ξ and the division of the field energy into random and drift components. In the simplest procedure, and one that works quite well, we choose ξ to reproduce the low-field Chapman-Enskog result and to partition the field energy according to the Maxwell model. This leads to the expression

$$v_d = \frac{3}{16h_0} \frac{(2\pi/\mu kT)^{1/2}}{[1 + (Mv_d^2/3kT)]^{1/2}} \frac{eE}{N\overline{\Omega}^{(1,1)}(T, v_d)}. \tag{5-6-4}$$

where h_0 is given by (5-4-54). This is apparently a quadratic in v_d^2 and is just a refinement of the free-flight result (5-2-17). However, there is extra dependence on v_d hidden in $\overline{\Omega}^{(1,1)}(T, v_d)$, except for rigid-sphere interactions. Now we can test (5-6-4) to see how well it reproduces known high-field results. These results are the Maxwell r^{-4} model for all mass ratios, the Rayleigh $m \gg M$ model for all force laws, given by (5-5-3), the Lorentz $m \ll M$ model for all force laws, given by (5-5-18) and (5-5-19), and the special result (5-5-57) valid for rigid spheres with $m = M$. It is easy to get explicit answers for only the inverse-power potential $V(r) = \pm C/r^n$; the numerical comparison in given in Table 5-6-1 (Hahn and Mason, 1972). The agreement at high fields is seen to be quite reasonable, the largest deviation being less than 20%. We can also test (5-6-4) at intermediate fields but only for the Lorentz model given by (5-5-5) and (5-5-6). These equations have been integrated numerically for rigid spheres (Hahn and

Mason, 1972), and the results are compared with (5-6-4) in Fig. 5-6-1, where the exact value of $h_0 = 9\pi/32$ has been used to secure exact agreement at low fields. The agreement is quite good over the whole range.

To improve the agreement at intermediate fields we can choose the energy partitioning to require that the coefficient of the first $(E/N)^2$ Kihara expansion term be given correctly. This leads to the expression

$$v_d = \frac{3}{16h_0} \frac{(2\pi/\mu kT)^{1/2}}{[1 + h_0 h_1 (mv_d^2/kT)]^{1/2}} \frac{eE}{N\bar{\Omega}^{(1,1)}(T, v_d)}, \qquad (5\text{-}6\text{-}5)$$

where h_0 and h_1 are given by (5-4-54) and (5-4-55), respectively. Although

TABLE 5-6-1. Test of the connection formula (5-6-4) at high fields for the potential $V(r) = \pm C/r^n$

n	v_d(approximate)/v_d(accurate)		
	$m \gg M$	$m = M$	$m \ll M$
4	0.814	0.814	0.814
6	0.857		0.902
8	0.872		0.943
10	0.879		0.966
12	0.884		0.982
25	0.894		1.022
50	0.898		1.041
∞	0.902	0.944	1.060

this result produces improved agreement at medium fields, the price exacted comes as poorer agreement at high fields, and the result is not worthwhile unless the ion-neutral short-range repulsion is rather steep.

Connection formulas for D_T and D_L are harder to devise, since no accurate results are known for medium fields, and the few results available for high fields suggest that the ratio D_L/D_T depends rather strongly on the ion-neutral force law (see Fig. 5-5-1). The safest procedure would seem to be to express D_T and D_L in terms of v_d. Examination of the meager available results suggest that both D_T and D_L follow the generalized Einstein relation of (5-5-38) but that the "mobility" should be taken as v_d/E for D_T and as (dv_d/dE) for D_L. Partitioning the random energy according to the Maxwell model for lack of more accurate information, we find

$$D_T = \frac{1}{e}\left[kT + \frac{(m+M)MA^*}{5m+3MA^*}v_d^2\right]\frac{v_d}{E}. \tag{5-6-6}$$

which is the same as (5-5-39) for the Maxwell model, but is proposed here as general and

$$D_L \approx \frac{1}{e}\left[kT + \frac{M(5m-2mA^*+MA^*)}{5m+3MA^*}v_d^2\right]\frac{dv_d}{dE}. \tag{5-6-7}$$

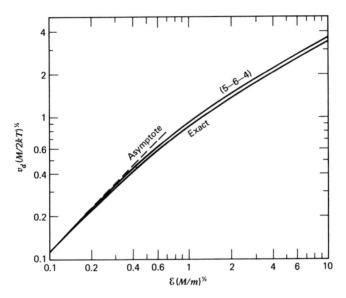

FIG. 5-6-1. Test of the connection formula (5-6-4) for intermediate fields: drift velocity as a function of field strength for rigid spheres with $m \ll M$. The dimensionless field-strength function $\mathscr{E}$ is defined by (5-4-30).

Both of these results are exact at low fields for all interactions and at all fields for the Maxwellian-model interaction. At very high fields they give the asymptotic relation

$$\frac{D_L(\infty)}{D_T(\infty)} = \frac{5m-2mA^*+MA^*}{(m+M)A^*}\frac{d\ln v_d}{d\ln E}. \tag{5-6-8}$$

As a check on the foregoing three relations we have available the calculations by Skullerud (1969b) for $m \ll M$ and r^{-n} potentials and the calculation of

D_L by Wannier (1953) for rigid spheres with $m = M$. Dimensional considerations show that for r^{-n} potentials,

$$\frac{d \ln v_d}{d \ln E} = \frac{n}{2(n-2)}. \tag{5-6-9}$$

Thus for $m \ll M$ and $n = 4, 8, \infty$ we find that (5-6-8) predicts values of D_L/D_T of 1, 2/3, and 1/2, respectively. The respective accurate values are 1, 0.66 (as read from a graph), and 0.4910, which is rather good agreement. The absolute values of D_T and D_L, as predicted by (5-6-6) and (5-6-7), are not so satisfactory, however; for rigid spheres with $m \ll M$ the ratios of the predicted to the accurate values are 0.824 for D_T and 0.839 for D_L. The agreement is poorer for rigid spheres with $m = M$, for which the ratio of the predicted D_L to the value calculated by Wannier is 1.7.

In short, the connection formulas for v_d are quite good, but much more work is needed before formulas of comparable accuracy can be found for D_T and D_L.*

B. COMPOSITION DEPENDENCE.   Here we combine the momentum-transfer results of Section 5-2-C with the results on the partition of ion energy in Section 5-5-F to produce a rule that will give the drift velocity in a gas mixture in terms of the drift velocities in the pure component gases, valid at all field strengths. The previously obtained formulas, with $Q_{D(j)}$ assumed constant, are

$$\frac{1}{v_d^2(E)} = \sum_j \frac{x_j}{v_{d(j)}^2(E)} \left[\left(\frac{m + \overline{M}}{m + M_j} \right) + \frac{3kT}{M_j v_d^2(E)} \right]^{1/2} \left[1 + \frac{3kT}{M_j v_{d(j)}^2(E)} \right]^{-1/2},$$

$$\tag{5-6-10}$$

with the mean mass $\overline{M}$ given by (5-5-76) and (5-5-77). Although no experimental data exist at present for testing these results, it is interesting to examine the effects predicted (Mason and Hahn, 1972). The results for K^+ ions in $H_2 + N_2$ mixtures at about $300°K$ are shown in Fig. 5-6-2, based on measurements of K^+ in H_2 (Fleming et al., 1969a) and

---

* Note added in proof: The best formulas to date have been found by adding corrections obtained by Kihara's method to the Maxwell model results (Whealton and Mason, 1973):

$$D_T \approx D_T^{(0)} \left(1 + \frac{1}{2} \frac{d \ln K}{d \ln E} \right),$$

$$D_L \approx D_L^{(0)} \left(1 + \frac{3}{2} \frac{d \ln K}{d \ln E} \right),$$

where $D_T^{(0)}$ and $D_L^{(0)}$ are given by (5-5-39) and (5-5-40), respectively.

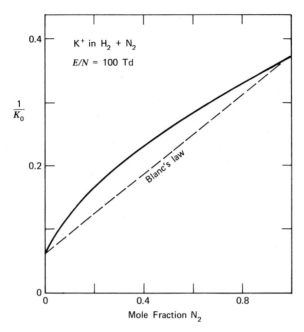

FIG. 5-6-2. Predicted deviations from Blanc's law for K^+ ions in $H_2 + N_2$ mixtures at $E/N = 100$ Td and 300°K. The solid curve is that given by (5-6-10).

of K^+ in N_2 (Fleming et al., 1969b). Although Blanc's law is followed at low fields, deviations up to about 20% are expected for $E/N = 100$ Td. The deviations for a mixture of gases of closer mass would be expected to be smaller; for instance, the maximum deviation from Blanc's law for K^+ ions in $N_2 + O_2$ mixtures is predicted to be only about 0.05% at 300°K even at $E/N = 300$ Td (Mason and Hahn, 1972). Similar results hold for the diffusion coefficients (Whealton and Mason, 1972).

5-7. RESONANT CHARGE TRANSFER

When ions move in their parent gas, an ion and a neutral can interchange roles by the resonant transfer of an electron. This resonant charge transfer is usually so probable that it dominates all other elastic scattering processes except at very low energies at which the long-range polarization scattering eventually dominates. The result is that the collision dynamics is very different, and the temperature and field dependence of the mobility are greatly changed from what they would have been in the absence of charge transfer. No changes are necessary, however, in the kinetic-theory portion of the discussion; all the changes occur in the cross sections and collision integrals only.

First we give a simple but substantially correct semiclassical description of mobility and diffusion with charge transfer, in which the results appear in a physically transparent form. Then we give a discussion of the special nature of the ion-neutral interaction when resonant charge transfer is possible; this is a necessary preliminary for the full quantum-mechanical treatment given in the two sections following. Finally, some methods are given for obtaining quick approximate estimates of the magnitudes and energy dependences of the relevant cross sections.

A. SEMICLASSICAL DESCRIPTION.   The basis of this description was given by Holstein (1952) and elaborated by Mason, Vanderslice, and Yos (1959). The nuclei are assumed to follow classical trajectories, which are not affected by an electron exchange that takes place with a probability P_{ex}. The only effect of exchange is to convert the apparent classical deflection angle of the ion from θ to $\pi - \theta$, so that a glancing collision is converted into an apparent head-on collision. The diffusion cross section thus appears abnormally large. In this semiclassical picture nothing can be said about P_{ex} except that it may depend on energy and impact parameter; actual values would have to come from experiment or from some quantum-mechanical calculation.

If the probability of conversion of θ to $\pi - \theta$ is P_{ex}, the probability of θ being unaffected is $1 - P_{ex}$ and the transport cross sections become

$$\left[1 - \frac{1 + (-1)^l}{2(1 + l)}\right]Q^{(l)}(E) = 2\pi \int_0^\infty (1 - P_{ex})(1 - \cos^l \theta)b \, db$$
$$+ 2\pi \int_0^\infty P_{ex}[1 - \cos^l(\pi - \theta)]b \, db, \quad (5\text{-}7\text{-}1)$$

the quantity in brackets on the left being just the normalization factor. From this expression we see immediately the distinct difference of character for even and odd l, since

$$\cos^l(\pi - \theta) = (-1)^l \cos \theta. \quad (5\text{-}7\text{-}2)$$

Thus for even l the value of P_{ex} cancels out in (5-7-1) and the value of $Q^{(l)}$ is unchanged; but for odd l we obtain, after a little rearrangement,

$$Q^{(l)}(E) = 4\pi \int_0^\infty P_{ex} b \, db + 2\pi \int_0^\infty (1 - 2P_{ex})(1 - \cos^l \theta)b \, db. \quad (5\text{-}7\text{-}3)$$

If exchange is dominant, the second integral in (5-7-3) will be small because $P_{ex} \approx \frac{1}{2}$ for small impact parameters and $\cos \theta \approx 1$ for large impact parameters. If exchange is minor, the first integral will be small and the usual expression for $Q^{(l)}$ without exchange will be recovered. The total cross section for charge transfer is

$$Q_T(E) = 2\pi \int_0^\infty P_{ex} b \, db, \qquad (5\text{-}7\text{-}4)$$

a quantity that is experimentally measurable (say in a beam experiment). Thus, when exchange is dominant, (5-7-3) becomes approximately

$$Q^{(l)} \approx 2Q_T, \qquad l \text{ odd}, \qquad (5\text{-}7\text{-}5)$$

a result we have already quoted without proof in Section 5-2-B. Since $Q^{(1)}$ is the diffusion cross section, it is apparent that the charge transfer cross section controls ion mobility and diffusion.

As discussed in Section 5-7-B, the possibility of resonant charge transfer is always associated with a pair of ion-neutral potentials, one corresponding to a total wavefunction that is symmetric under exchange of nuclei and the other to the antisymmetric wavefunction. It is easy to show that the transport cross sections in such cases are simple averages of the cross sections for scattering by the individual potentials (Mason et al., 1959),

$$Q^{(l)} = \tfrac{1}{2}[Q_+^{(l)} + Q_-^{(l)}], \qquad (5\text{-}7\text{-}6)$$

where $+$ refers to the symmetric potential and $-$ to the antisymmetric. The semiclassical description then proceeds as above to yield

$$Q^{(l)}(l \text{ even}) = 2\pi\left(\frac{l+1}{l}\right)\int_0^\infty [1 - \tfrac{1}{2}(\cos^l \theta^+ + \cos^l \theta^-)]b \, db, \quad (5\text{-}7\text{-}7)$$

$$Q^{(l)}(l \text{ odd}) = 2Q_T + 2\pi \int_0^\infty (1 - 2P_{ex})[1 - \tfrac{1}{2}(\cos^l \theta^+ + \cos^l \theta^-)]b \, db.$$
$$(5\text{-}7\text{-}8)$$

Thus for odd l we still obtain $Q^{(l)} \approx 2Q_T$ even when the multiple potentials are taken into account.

The foregoing arguments suffice for low and medium fields, in which it is necessary only to modify the cross sections and collision integrals in the Chapman-Enskog or Kihara expansions. A slightly more drastic approximation makes the problem simple enough to be solved for all field strengths. This approximation states that in a collision the trajectories of the nuclei are negligibly distorted from straight lines and that the only effect is for the ion and atom to interchange roles. Thus, if the velocities of ion and atom before collision are $\mathbf{v}$ and $\mathbf{V}$, respectively, they become $\mathbf{v}' = \mathbf{V}$ and $\mathbf{V}' = \mathbf{v}$ after collision. The idea of undistorted trajectories must obviously be untrue for close collisions, although it is good for glancing collisions. Perhaps a better statement of the approximation is that it is assumed that only undistorted trajectories contribute appreciably to the cross section. The result is that the Boltzmann equation (5-3-1) takes on a simplified form in which the collision kernel changes as follows:

$$[f_i(\mathbf{v}')f_j(\mathbf{V}') - f_i(\mathbf{v})f_j(\mathbf{V})] \to [f_i(\mathbf{V})f_j(\mathbf{v}) - f_i(\mathbf{v})f_j(\mathbf{V})]. \qquad (5\text{-}7\text{-}9)$$

The Boltzmann equation can then be solved by suitable expansion techniques at low fields (Smirnov, 1967a; Fahr and Müller, 1967). Expansion techniques (Smirnov, 1967a) or free-path arguments (Fahr and Müller, 1967; Skullerud, 1969a) can also be used at high fields. The latter are based on the fact that each collision converts a fast ion and a nearly stationary neutral to a fast neutral and a nearly stationary ion. Thus the ion always comes essentially to rest after each collision; there is no problem with persistence of velocities and simple free-path arguments give exact answers.

We quote the results obtained for rigid spheres; the original papers must be consulted for results applicable to other velocity dependences of the cross section. At low fields the results are of the form

$$v_d(0) = \frac{A}{(mkT)^{1/2}} \frac{eE}{NQ_T} \qquad (5\text{-}7\text{-}10)$$

Smirnov (1968a) finds $A = 0.341$; Fahr and Müller (1967) find $A = 0.330$; the Chapman-Enskog result, with $Q^{(1)} = 2Q_T$, gives $A = 3\pi^{1/2}/16 = 0.332$. At high fields both Smirnov and Fahr and Müller obtain

$$v_d(\infty) = \left(\frac{2}{\pi}\frac{eE}{mNQ_T}\right)^{1/2}. \qquad (5\text{-}7\text{-}11)$$

It is interesting to compare this with Wannier's high-field result (5-5-57) for rigid spheres. Taking $\pi d^2 = 2Q_T$, we find a numerical coefficient of $1.1467/2^{1/2} = 0.811$ instead of the value $(2/\pi)^{1/2} = 0.798$ from (5-7-8). Thus the simple replacement of $Q^{(1)}$ by $2Q_T$ gives remarkably good agreement for the drift velocity at both low and high fields.

The agreement is somewhat poorer, however, for the longitudinal diffusion coefficient for which Skullerud (1969a) finds

$$D_L(\infty) = 3\left(1 - \frac{\pi^{1/2}}{2}\right)\frac{(2/\pi)^{1/2}}{N\pi d^2}\left(\frac{eE}{mN\pi d^2}\right)^{1/2},$$

$$= \frac{0.272}{N\pi d^2}\left(\frac{eE}{mN\pi d^2}\right)^{1/2} = 0.190\frac{mv_d^3}{eE}, \qquad (5\text{-}7\text{-}12)$$

where the last result follows by substitution from (5-7-11). The numerical coefficient found by Wannier is given in (5-5-60) as 0.22 instead of 0.272.

Smirnov and Fahr and Müller also obtain v_d at intermediate fields. Smirnov (1967b) approximates his result with the formula

$$v_d(E) = v_d(0)\left[1 + 0.13\left(\frac{eE}{2kTNQ_T}\right)^2\right]^{-1/4}, \qquad (5\text{-}7\text{-}13)$$

which gives both $v_d(0)$ and $v_d(\infty)$ correctly. Fahr and Müller obtain

$$v_d(E) = v_d(0)\left[1 + \frac{9\pi^2}{512}\left(\frac{eE}{kTNQ_T}\right)\right]^{-1/2}, \qquad (5\text{-}7\text{-}14)$$

where we have chosen the numerical constant to give $v_d(\infty)$ correctly. The form of (5-7-14) can be obtained from the free-flight result (5-2-16) by replacing the value of v_d^2 in the denominator of the right-hand side of (5-2-16) with its high-field value.

B. ION-NEUTRAL POTENTIAL. The potential energy curve for the inter-action between an ion A^+ and its parent neutral atom A can be considered to be the energy of the molecule $A^+ + A$ as a function of internuclear separation; that is, $V(r)$ is the eigenvalue for the molecular wavefunction, but if Ψ_A is the wave function for the configuration $A^+ + A$, there is another wavefunction Ψ_B corresponding to the configuration $A + A^+$. If the two nuclei are identical, the correct total wavefunctions must be either symmetric or antisymmetric with respect to nuclear interchange:

$$\Psi^+ = 2^{-1/2}(\Psi_A + \Psi_B), \qquad (5\text{-}7\text{-}15a)$$

$$\Psi^- = 2^{-1/2}(\Psi_A - \Psi_B). \qquad (5\text{-}7\text{-}15b)$$

These two wavefunctions give rise to two different potential curves, $V^+(r)$ and $V^-(r)$. In briefer language the exact resonance splits the degenerate potentials corresponding to Ψ_A and Ψ_B into a pair of potentials. Both must be taken into account in a full quantum-mechanical calculation of the collisions between A^+ and A.

As an example, the two curves for $He^+ + He$ are shown in Fig. 5-7-1, based on the *ab initio* calculations of Gupta and Matsen (1967) with a 26-term wavefunction. The weak minimum at large r in the $V^-(r)$ curve, due to the polarization potential, does not show on the scale of this figure. For comparison the potential for the similar but nonresonant system $Li^+ + He$, as obtained from mobility measurements (Mason and Schamp, 1958), is also shown. The profound effect of exact resonance is apparent.

The foregoing two-state description holds only if the wavefunction Ψ_A is itself nondegenerate. It thus applies only to the case in which both ion and atom are in S states, giving rise to Σ_g and Σ_u molecular states. Examples would be H^+ in H, H^- in H, He^+ in He, Li^+ in Li,

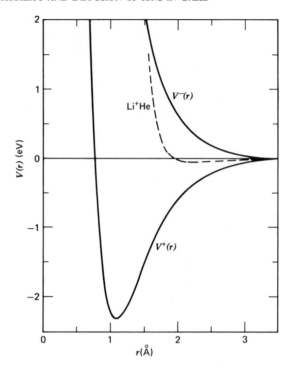

FIG. 5-7-1. Interaction of He^+ with He, showing effect of resonance. The similar but nonresonant interaction of Li^+ with He is shown for comparison.

Na^+ in Na, and so on, but if either one is in a P state or higher more molecular states are possible and more potential energy curves result; for instance, in the case of Ne^+ in Ne, Ar^+ in Ar, Kr^+ in Kr, and Xe^+ in Xe the ion is in a ^2P state and the atom in a ^1S state. Four molecular states then arise: $^2\Sigma_g$, $^2\Sigma_u$, $^2\Pi_g$, and $^2\Pi_u$, with four associated potential energy curves. The determination of the molecular states arising from a given pair of atomic states is a straightforward exercise in the addition of electronic angular momentum and is summarized by the Wigner-Witmer rules (Herzberg, 1950, Section VI,1).

Since Σ states have zero electronic angular momentum along the internuclear axis and Π, Δ, etc., states have nonzero angular momentum, the Σ states have only half the statistical weight of the other states for which the angular momentum vector can point in either direction along the axis.

C. QUANTUM-MECHANICAL DESCRIPTION. Since charge exchange involves at least two states, there should be wave-interference affects in the cross sections from scattering by the two states analogous to Newton's rings in optics. There might also be resonances due to the existence of virtual bound states (e.g., H_2^+ and He_2^+ are stable molecules), analogous to the orbiting resonances discussed in Section 5-3-E. These effects, as well as the effects of nuclear spin and Bose-Einstein (BE) or Fermi-Dirac (FD) statistics, should all appear in a proper quantum-mechanical description. In addition, we should expect to find the physical simplicities of the semiclassical description contained in the full quantum-mechanical description, hopefully in a transparent way.

If we assume that the total Hamiltonian for the system of electrons and nuclei contains only electrostatic terms and that the Born-Oppenheimer separation of nuclear and electronic motion is valid, a general result can be proved fairly straightforwardly (Smith, 1967; Heiche, 1967). This result states that there is no coupling between molecular states of different electronic orbital angular momentum, so that the cross sections are sums of terms, each of which depends only on one g-u pair of molecular states, multiplied by a suitable statistical weight factor. In the case of Ne^+ in Ne, for example, there will be one cross section for the Σ_g, Σ_u pair and another for the Π_g, Π_u pair, and the complete cross section will be

$$Q^{(l)} = \tfrac{1}{3}Q_\Sigma^{(l)} + \tfrac{2}{3}Q_\Pi^{(l)}. \tag{5-7-16}$$

We therefore lose no generality by limiting the following discussion to a single g-u pair.

Since there is no way in principle to tell whether a scattered ion has undergone exchange, the correct scattering amplitude must be a linear combination of $f(\theta)$ and $f(\pi - \theta)$, the sign depending on whether the nuclei are bosons or fermions. For the nuclear-symmetric potential the correct combination is

$$f^+\begin{pmatrix} BE \\ FD \end{pmatrix} = 2^{-1/2}[f^+(\theta) \pm f^+(\pi - \theta)], \tag{5-7-17}$$

and for the nuclear-antisymmetric potential it is

$$f^-\begin{pmatrix} BE \\ FD \end{pmatrix} = 2^{-1/2}[f^-(\theta) \mp f^-(\pi - \theta)], \tag{5-7-18}$$

where we neglect nuclear spin degeneracy for the moment. The total scattering amplitude for the pair of potentials is therefore

$$f\begin{pmatrix} BE \\ FD \end{pmatrix} = 2^{-1/2}\left[f^+\begin{pmatrix} BE \\ FD \end{pmatrix} + f^-\begin{pmatrix} BE \\ FD \end{pmatrix}\right], \tag{5-7-19}$$

which can be arranged into the form

$$f\binom{BE}{FD} = \tfrac{1}{2}\{[f^+(\theta) + f^-(\theta)] \pm [f^+(\pi - \theta) - f^-(\pi - \theta)]\}. \quad (5\text{-}7\text{-}20)$$

If the scattering is strongly peaked around 0 and π, so that interference between the two peaks is negligible, it is reasonable to attribute the forward peak to direct scattering and the backward peak to exchange scattering and to write a differential cross section for charge transfer as the second part of (5-7-20),

$$I_T(\theta) = \tfrac{1}{4}|f^+(\pi - \theta) - f^-(\pi - \theta)|^2. \quad (5\text{-}7\text{-}21)$$

The scattering amplitudes in terms of the phase shifts are

$$f^\pm(\theta) = \frac{1}{2i\kappa} \sum_{l=0}^{\infty} (2l + 1)(e^{2i\delta_l^\pm} - 1)P_l(\cos\theta), \quad (5\text{-}7\text{-}22)$$

where, as usual, $\kappa = \mu v/\hbar$, δ_l^+ is the lth partial wave for scattering by $V^+(r)$, and δ_l^- is the lth partial wave for scattering by $V^-(r)$. On substituting (5-7-22) into (5-7-21) and integrating over all angles to obtain Q_T, we can manipulate the result into the form (Massey and Smith, 1933),

$$Q_T = \frac{\pi}{\kappa^2} \sum_{l=0}^{\infty} (2l + 1) \sin^2(\delta_l^+ - \delta_l^-). \quad (5\text{-}7\text{-}23)$$

The results for the transport cross sections are a little more complicated. Because of the odd parity of the Legendre polynomials,

$$P_l[\cos(\pi - \theta)] = (-1)^l P_l(\cos\theta), \quad (5\text{-}7\text{-}24)$$

the expression (5-7-20) looks like a similar expression for scattering without charge exchange, but the even and odd phases must be calculated alternately for the two different potential curves (Massey and Mohr, 1934). The diffusion cross sections for the hypothetical spinless particles under discussion then become

$$Q_0^{(1)}(BE) = \frac{4\pi}{\kappa^2} \sum_{l=0}^{\infty} (l + 1) \sin^2(\beta_l - \beta_{l+1}), \quad (5\text{-}7\text{-}25)$$

$$Q_0^{(1)}(FD) = \frac{4\pi}{\kappa^2} \sum_{l=0}^{\infty} (l + 1) \sin^2(\gamma_l - \gamma_{l+1}), \quad (5\text{-}7\text{-}26)$$

where

$$\begin{aligned} \beta_l &= \delta_l^- \quad \text{and} \quad \gamma_l = \delta_l^+ \quad \text{for } l \text{ even,} \\ \beta_l &= \delta_l^+ \quad \text{and} \quad \gamma_l = \delta_l^- \quad \text{for } l \text{ odd.} \end{aligned} \quad (5\text{-}7\text{-}27)$$

The subscript zero on $Q^{(1)}$ emphasizes that the nuclei are imagined to be spinless for the moment. The expressions for $Q^{(2)}$ are

$$Q_0^{(2)}(BE) = \frac{4\pi}{\kappa^2} \left(\frac{3}{2}\right) \sum_{l=0}^{\infty} \frac{(l+1)(l+2)}{(2l+3)} \sin^2 (\beta_l - \beta_{l+2}), \quad (5\text{-}7\text{-}28)$$

$$Q_0^{(2)}(FD) = \frac{4\pi}{\kappa^2} \left(\frac{3}{2}\right) \sum_{l=0}^{\infty} \frac{(l+1)(l+2)}{(2l+3)} \sin^2 (\gamma_l - \gamma_{l+2}). \quad (5\text{-}7\text{-}29)$$

The character of $Q^{(1)}$ and $Q^{(2)}$ can be seen to be quite different. The sum for $Q^{(2)}$ is made up of two types of term, one depending only on $V^+(r)$ through $(\delta_l^+ - \delta_{l+2}^+)$, and the other only on $V^-(r)$ through $(\delta_l^- - \delta_{l+2}^-)$. Nothing resembling interference occurs. The sum for $Q^{(1)}$, however, involves $V^+(r)$ and $V^-(r)$ in the same terms, through $(\delta_l^+ - \delta_{l+1}^-)$ and $(\delta_l^- - \delta_{l+1}^+)$, so that interference effects are important. The physical significance of this will become more apparent when we make the connection with the semi-classical results.

Similar expressions that hold for $Q^{(3)}$ and $Q^{(4)}$ are obtained from the expressions (5-3-35) and (5-3-36) for no exchange by replacement of δ_l with β_l for BE and γ_l for FD. The difference in character of the transport cross sections depends on whether the index of the cross section is odd or even.

Only minor modification is needed to include the effects of nuclear spin. A nucleus of spin s has $(2s+1)$ states. A system containing two such identical nuclei therefore has $(2s+1)^2$ degenerate states, of which $(s+1)(2s+1)$ are symmetric and $s(2s+1)$ are antisymmetric. The transport cross sections therefore become

$$Q^{(l)}\left(\begin{array}{c} BE \\ FD \end{array}\right) = \frac{s+1}{2s+1} Q_0^{(l)}\left(\begin{array}{c} BE \\ FD \end{array}\right) + \frac{s}{2s+1} Q_0^{(l)}\left(\begin{array}{c} FD \\ BE \end{array}\right). \quad (5\text{-}7\text{-}30)$$

This expression is physically transparent, but it gives the impression that the final results are strongly spin-dependent, which is not usually true. A better way to exhibit the effects of spin and statistics is to rewrite (5-7-30) as

$$Q^{(l)}\left(\begin{array}{c} BE \\ FD \end{array}\right) = \tfrac{1}{2}[Q_0^{(l)}(BE) + Q_0^{(l)}(FD)] + \frac{1}{2(2s+1)} \left[Q_0^{(l)}\left(\begin{array}{c} BE \\ FD \end{array}\right) - Q_0^{(l)}\left(\begin{array}{c} FD \\ BE \end{array}\right)\right].$$

$$(5\text{-}7\text{-}31)$$

The second term is usually much smaller than the first.

All of the foregoing expressions hold only for Σ, Δ, etc., states of the ion-atom pair. The expressions for Π, Φ, etc., states are obtained by interchange of β_l and γ_l or, equivalently, of the labels BE and FD (Smith, 1967; Heiche, 1967).

The foregoing formulas for the various cross sections are not in a form that readily suggests relations among them, but the semiclassical description of Section 5-7-A indicates how to rearrange the exact quantum-mechanical results to obtain such relations (Heiche and Mason, 1970). Consider first the even-index transport cross sections $Q^{(2)}$, $Q^{(4)}$, etc. According to the semiclassical description, these cross sections are unaffected by exchange, so we expect them to be simple averages of the corresponding cross sections for $V^+(r)$ and $V^-(r)$ individually. Moreover, recalling the semiclassical relations $\kappa b = l + \frac{1}{2}$ and $\theta = 2(d\delta_l/dl)$, we define "angles" as

$$\theta_{ln}^{\pm} \equiv \left(\frac{2}{n}\right)(\delta_{l+n}^{\pm} - \delta_l^{\pm}). \qquad (5\text{-}7\text{-}32)$$

Then from (5-7-31) we see that we can regroup terms in the summations over phase shifts and obtain

$$Q^{(2)} = \tfrac{1}{2}[Q_+^{(2)} + Q_-^{(2)}] + \frac{(-1)^{2s+\Lambda}}{2s+1}\left(\frac{2\pi}{\kappa^2}\right)\sum_{l=0}^{\infty}(-1)^l C_l^{(2)}, \qquad (5\text{-}7\text{-}33)$$

where $\Lambda = 0, 1, 2, 3, \ldots$ for $\Sigma, \Pi, \Delta, \Phi, \ldots$ states, respectively. The statistics is now taken into account automatically by the $(-1)^{2s}$ factor, since particles with integral s obey BE statistics and those with half-integral s obey FD statistics. The $Q_{\pm}^{(2)}$ in the first term are

$$Q_{\pm}^{(2)} = \frac{4\pi}{\kappa^2}\sum_{l=0}^{\infty}\frac{(l+1)(l+2)}{(2l+3)}\sin^2\theta_{l2}^{\pm}. \qquad (5\text{-}7\text{-}34)$$

Thus we obtain the sort of average expected from the semiclassical description. The spin-dependent second term of (5-7-33) has no simple semiclassical analogue, since it involves spin and since the factor $(-1)^l$ causes the individual terms to alternate in sign. The expression for $C_l^{(2)}$, however, can be written in a form having a semiclassical appearance:

$$C_l^{(2)} = \frac{(l+1)(l+2)}{(2l+3)}(\sin^2\theta_{l2}^{-} - \sin^2\theta_{l2}^{+}). \qquad (5\text{-}7\text{-}35)$$

Analogous expressions hold for $Q_{\pm}^{(4)}$ and $C_l^{(4)}$.

A simple result also holds for P_{ex} and Q_T. Comparing the semiclassical (5-7-4) and the quantum-mechanical (5-7-23) expressions for Q_T, we see that

$$Q_T = \frac{2\pi}{\kappa^2}\sum_{l=0}^{\infty}(l+\tfrac{1}{2})P_{ex}, \qquad (5\text{-}7\text{-}36)$$

$$P_{ex} \equiv \sin^2\zeta_l \equiv \sin^2(\delta_l^+ - \delta_l^-). \qquad (5\text{-}7\text{-}37)$$

The results are less obvious for the odd-index transport cross sections. Substituting for $\theta_{ln}^{\pm}$ and ζ_l, we obtain from (5-7-31) the result

$$Q^{(1)} = \frac{2\pi}{\kappa^2} \sum_{l=0}^{\infty} (l+1)[\sin^2 (\zeta_l - \tfrac{1}{2}\theta_{l1}^+) + \sin^2 (\zeta_l + \tfrac{1}{2}\theta_{l1}^-)]$$

$$+ \frac{(-1)^{2s+\Lambda}}{2s+1} \left(\frac{2\pi}{\kappa^2}\right) \sum_{l=0}^{\infty} (-1)^l (l+1)[\sin^2 (\zeta_l - \tfrac{1}{2}\theta_{l1}^+) - \sin^2 (\zeta_l + \tfrac{1}{2}\theta_{l1}^-)].$$

$$(5\text{-}7\text{-}38)$$

With the help of the trigonometric identity,

$$\sin^2 (\zeta \mp \tfrac{1}{2}\theta^{\pm}) = \sin^2 \zeta + \tfrac{1}{2}(1 - 2\sin^2 \zeta)(1 - \cos \theta^{\pm}) \mp \tfrac{1}{2} \sin 2\zeta \sin \theta^{\pm},$$

$$(5\text{-}7\text{-}39)$$

this can be written in the form

$$Q^{(1)} = 2Q_T + \frac{2\pi}{\kappa^2} \sum_{l=0}^{\infty} (l+1)(1 - 2\sin^2 \zeta_l)[1 - \tfrac{1}{2}(\cos \theta_{l1}^+ + \cos \theta_{l1}^-)]$$

$$- \frac{\pi}{\kappa^2} \sum_{l=0}^{\infty} (l+1) \sin 2\zeta_l(\sin \theta_{l1}^+ - \sin \theta_{l1}^-) + \frac{2\pi}{\kappa^2} \sum_{l=0}^{\infty} \sin^2 \zeta_l$$

$$+ \frac{(-1)^{2s+\Lambda}}{2s+1} \left(\frac{2\pi}{\kappa^2}\right) \sum_{l=0}^{\infty} (-1)^l C_l^{(1)}, \qquad (5\text{-}7\text{-}40)$$

where

$$C_l^{(1)} = -\tfrac{1}{2}(l+1)[(1 - 2\sin^2 \zeta_l)(\cos \theta_{l1}^+ - \cos \theta_{l1}^-)$$

$$+ \sin 2\zeta_l(\sin \theta_{l1}^+ + \sin \theta_{l1}^-)]. \quad (5\text{-}7\text{-}41)$$

If we interpret $\sin^2 \zeta_l$ as P_{ex} and replace summations with integrations, then the first two terms of $Q^{(1)}$ will have an obvious analogy to the semiclassical (5-7-8). The third term has no obvious physical interpretation, but it has a resemblance to Landau-Zener transition effects coming from the term (Landau and Lifshitz, 1958, p. 311; Mott and Massey, 1965, pp. 352–353),

$$\sin 2\zeta_l = 2[\sin^2 \zeta_l(1 - \sin^2 \zeta_l)]^{1/2} \to 2[P_{ex}(1 - P_{ex})]^{1/2}. \quad (5\text{-}7\text{-}42)$$

The fourth term is a small piece left over from Q_T, caused by the difference between $(2l+1)$ and $2(l+1)$. Since it lacks the factor $(l+1)$, it is probably negligible except at very low energies at which only a few phase shifts contribute. The last term, involving $C_l^{(1)}$, involves spin and has no semiclassical analogue. Its effect tends to wash out because of the oscillation caused by the $(-1)^l$ factor.

Similar results hold for $Q^{(3)}$, but this cross section appears only in kinetic-theory correction terms.

The behavior of the quantum-mechanical cross sections is illustrated for He^+ in He in Fig. 5-7-2. The effect of charge transfer is shown by comparison of $Q^{(1)}$ calculated correctly according to (5-7-25) (since $s = 0$ for ^4He) with values of $Q^{(1)}$ calculated separately for $V^+(r)$ and $V^-(r)$ without charge transfer. The scale factors in the figure are r_e and D_e, the position and depth of the minimum in $V^+(r)$; the potential actually used in the calculations is not so accurate as that shown in Fig. 5-7-1, but it is similar enough to exhibit the features of interest here (Heiche and Mason, 1970). The most striking feature at first sight is the oscillatory nature of the curves. This is due to the presence of orbiting resonances, corresponding to the possibility of unstable classical orbits for integral values of orbital angular momentum, as shown in Fig. 5-3-5. The resonances can be seen to coincide approximately with the integral values of l marked at the top of the figure until for large l they become too close together and tend to wash out. The resonances disappear for the $V^-(r)$ cross section above an energy of about $E/D_e = 10^{-5}$ but for the other curves persist to much higher energies because of the deep well of the $V^+(r)$ potential.

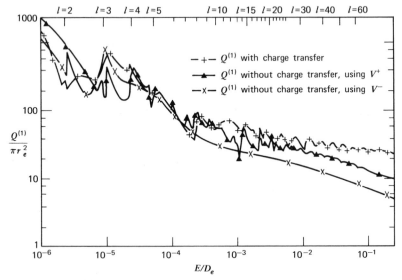

FIG. 5-7-2. Increase of diffusion cross section $Q^{(1)}$ by charge transfer, illustrated by calculations for He^+ in He (Heiche and Mason, 1970). The symbols merely identify the curves and are not the calculated points. The structures in the curves are orbiting resonances, due mostly to the deep well of the $V^+(r)$ potential.

Disregarding the resonance structure, we see that all three curves are about the same, on the average, up to an energy slightly greater than 10^{-4} because the scattering is dominated by the polarization tail of the potential, which is the same for all three. At higher energies charge transfer begins to dominate, and $Q^{(1)}$ with charge transfer becomes more than twice as big as *either* cross section without charge transfer.

The different characteristics of Q_T, $Q^{(1)}$, $Q^{(2)}$, and $Q^{(3)}$ are illustrated in Fig. 5-7-3 for He^+ in He (Heiche and Mason, 1970). A number of features stand out. First, the curves for $Q^{(1)}$ and $Q^{(3)}$ are similar but that for $Q^{(2)}$ is quite different. In particular, $Q^{(2)}$ is slightly greater than $Q^{(1)}$ and $Q^{(3)}$ in the polarization region (this is the normal situation in which charge transfer does not occur) but is much smaller and has a markedly different energy dependence in the charge-transfer region. The reason is that charge transfer affects only the odd-index transport cross sections. Second, the resonances are more numerous in Q_T than in $Q^{(1)}$ or $Q^{(3)}$. This phenomenon has been attributed by Dickinson (1968a) to the fact that only resonances involving odd l appear in $Q^{(1)}$, since $s = 0$ for ^{4}He. Third, the approximation $Q^{(1)} \approx 2Q_T$ is remarkably good in an average sense, even into the region in which the polarization potential dominates the scattering. The agreement in the polarization region happens to be fortuitous in this case, but it is, of course, a general phenomenon in the charge-transfer region.

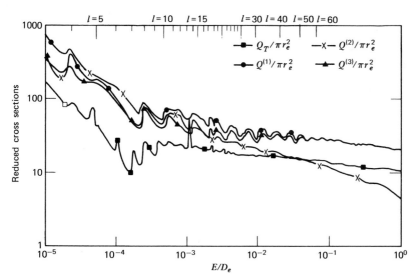

FIG. 5-7-3. Charge-transfer cross section Q_T and transport cross sections $Q^{(1)}$, $Q^{(2)}$, and $Q^{(3)}$ calculated for He^+ in He (Heiche and Mason, 1970). The symbols merely identify the curves.

Figure 5-7-4 is the analog of Fig. 5-7-3 for Cs^+ in Cs, as calculated by Heiche and Mason (1970). The oscillatory behavior of the curves is much less pronounced. This is partly due to the fact that Cs is heavier than He, and thus has a shorter de Broglie wavelength, and partly due to the nuclear spin of $\frac{7}{2}$ for Cs, which allows the resonances for the BE and FD contributions to combine; hence they tend to cancel one another out.

The influence of charge transfer on the temperature and field dependence of the mobility is considered in the next section.

D. TEMPERATURE AND FIELD DEPENDENCE OF MOBILITY. Resonant charge transfer has a profound effect on the behavior of the mobility as a function of temperature and field strength, though not in the way that might be expected from a cursory inspection of the results of the preceding section. The most prominent features of the cross sections, namely the orbiting resonances, are mostly washed out by the integration over energy to form the collision integrals, leaving at most a few gradual undulations. What persists is the general magnitude and energy dependence of the cross sections. Since charge transfer effectively converts a glancing collision into a head-on collision, the cross section is very large and the mobility correspondingly low. The result is that the mobility as a function of temperature drops from its polarization limit to a much lower value controlled by charge transfer.

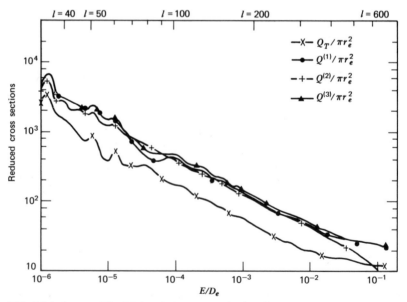

FIG. 5-7-4. Same as Fig. 5-7-3, calculated for Cs^+ in Cs (Heiche and Mason, 1970).

To illustrate this behavior the calculated and experimental values of the zero-field mobility of He^+ in He are shown in Fig. 5-7-5. The two calculated curves (Dickinson, 1968a; Heiche and Mason, 1970) are based on somewhat different potentials, hence differ. The only remnant of the many orbiting resonances is the undulation around 20°K; above that temperature range the mobility falls smoothly with increasing temperature. For contrast the mobility of Li^+ in He is also shown (these are the same results given in Fig. 5-3-3). Except for the possibility of resonant charge transfer, the systems Li^+ in He and He^+ in He are rather similar, and the comparison shows the great influence of charge transfer on mobility. Thus at 0°K the polarization limit of Li^+ in He is slightly less than that of He^+ in He because of the mass difference, but by 500°K the Li^+ mobility is more than three times larger than the He^+ mobility.

A similar comparison is made in Fig. 5-7-6 for the more massive systems Cs^+ in Cs and Cs^+ in Xe. Here no low-temperature quantum-mechanical undulations are visible. The calculated curve for Cs^+ in Cs shows a smooth transition from polarization behavior to charge-transfer behavior (Heiche and Mason, 1970). The Cs^+ in Xe calculations were obtained by fitting a (12-6-4) potential to the experimental measurements (Mason and Schamp,

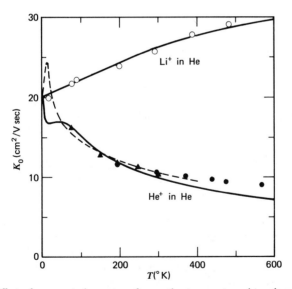

FIG. 5-7-5. Effect of resonant charge transfer on the temperature dependence of mobility. The results for Li^+ in He are the same as in Fig. 5-3-3. The dashed curve for He^+ in He was calculated by Dickinson (1968a) and the solid curve, by Heiche and Mason (1970). Experimental results are: ● Orient (1967); ▲ Patterson (1970b).

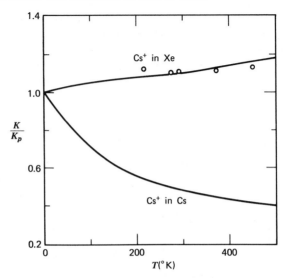

FIG. 5-7-6. Effect of resonant charge transfer on the temperature dependence of mobility. Experimental results for Cs^+ in Xe are from Hoselitz (1941).

1958). Because the polarizability of Cs is so much larger than that of Xe, the results are plotted as K/K_p to make the curves meet at $0°K$. The contrast in temperature dependence is striking and again results in about a factor of 3 difference at $500°K$.

The parallel behavior of mobility with temperature and with field strength has been discussed at some length in Section 5-4-E. Charge transfer requires no modification of these arguments, which are independent of the detailed behavior of the cross sections. We may conclude therefore that when charge transfer is important the mobility will normally decrease with increasing field strength (unless some quantum-mechanical undulations are present, as in Fig. 5-7-5). Such behavior is amply confirmed by experiment (see, for example, Patterson, 1970b).

E. APPROXIMATE CROSS SECTIONS.   In Section 5-2-B we discussed how quick estimates of the magnitude and energy dependence of diffusion cross sections could be made. In Section 5-3-E we also discussed how quantum-mechanical cross sections went over to their semiclassical and classical limits. Here we give the corresponding discussion for resonant charge transfer.

We consider four commonly used approximations: the semiclassical, impact-parameter, random-phase, and polarization approximations, which are all closely related to approximations discussed for no charge exchange.

In the semiclassical approximation the sums over l are replaced by integrations over dl, and the phase shifts are calculated by the JWKB approximation, given by (5-3-38). The integration over dl completely eliminates the effects of nuclear spin and the difference between Σ, Π, etc., states. The results, however, do not reduce to a completely classical form as they did in no charge transfer; in particular, (5-7-40) does not reduce to (5-7-8), for the third term in (5-7-40) involving $[\sin 2\zeta(\sin \theta^+ - \sin \theta^-)]$ survives the passage to the semiclassical limit. Except for light particles at low energies, this is expected to be an accurate approximation.

In the impact-parameter approximation the additional step is made of expanding the JWKB expression for the phase shift to obtain (Massey and Mohr, 1934)

$$\delta_l{}^\pm \approx -\frac{\kappa}{2E}\int_b^\infty \frac{V^\pm(r)}{(r^2 - b^2)^{1/2}}\, r\, dr, \tag{5-7-43}$$

where $\kappa b = l + \frac{1}{2}$. This expression is closely related to the Kennard small-angle approximation (5-2-42) for the classical deflection angle; they are connected by the semiclassical formula $\theta = 2(d\delta_l/dl)$. The approximation corresponds physically to replacing the actual collision trajectory with a straight line and is thus expected to be accurate only at high energies. It is often usable, however, at lower energies if the large phase shifts corresponding to small impact parameters happen to contribute to the cross section in an essentially random manner. The advantage of this approximation over the semiclassical approximation is that (5-7-43) is much easier to evaluate than the full JWKB expression and can often be integrated in terms of known functions.

The random-phase approximation is a further simplification of the impact-parameter approximation. It is the quantum-mechanical analogue of the "random-angle" approximation of Section 5-2-B and approximates the integrations over dl (or db), which are all of the form

$$\int_0^\infty F(b) \sin^2 G(b) b\, db. \tag{5-7-44}$$

If $G(b)$ is large and varies rapidly with b, then $\sin^2 G(b)$ oscillates rapidly between 0 and 1 and is replaced by its mean value of $\frac{1}{2}$ out to some value b^*, after which it is taken as zero. With the Firsov (1951) choice of b^*

$$|G(b^*)| = \frac{1}{\pi} \tag{5-7-45}$$

(5-7-44) is replaced by

$$\frac{1}{2}\int_0^{b^*} F(b) b\, db, \tag{5-7-46}$$

and the integration can then usually be done easily. For the transport cross sections, which involve differences of phase shifts with different l values, the further approximation

$$\delta_{l+n} = \delta_l + n \frac{d\delta_l}{dl} \qquad (5\text{-}7\text{-}47)$$

is made, which is consistent with taking l continuous, provided the higher terms in the Taylor expansion can be neglected.

The polarization approximation assumes that the polarization tail of the potential dominates the scattering. Then Q_T vanishes and the transport cross sections take the form

$$Q_p^{(l)} = 4\pi A^{(l)}(4) \left(\frac{\alpha e^2}{2E}\right)^{1/2}, \qquad (5\text{-}7\text{-}48)$$

where α is the polarizability of the neutral. Values of the constants $A^{(l)}(4)$ have been tabulated (Kihara et al., 1960; Higgins and Smith, 1968; Heiche and Mason, 1970); $4\pi A^{(1)}(4)$ appears in Table 5-5-1. The expression (5-7-48) yields the polarization limit of the mobility given by (5-3-29).

Comparisons of these approximations with accurate quantum-mechanical calculations are made in Figs. 5-7-7 and 5-7-8 for He^+ in He and Cs^+ in Cs, respectively. The integration over dl in the semiclassical approximation washes out the structure due to the orbiting resonances, as shown in the curves for Q_T. The advantage of the integration is that in practice the integrand needs to be evaluated at many fewer values of l than are needed in the summation. When the resonances are small, as for Cs^+ in Cs, the semiclassical approximation is almost identical to the full quantum-mechanical results. Although the figures show only Q_T and $Q^{(1)}$, it should be understood that the effect of the semiclassical approximation for other cross sections is simply to put a smooth average curve through the orbiting resonances.

In the impact-parameter approximation the expressions for the phase shifts can usually be integrated analytically, but the integrations for the cross sections must still be carried out numerically. The agreement shown in Figs. 5-7-7 and 5-7-8 with the accurate cross sections is very good, apart from the expected loss of the orbiting resonances. The only failure occurs for Q_T in the region in which polarization is dominant. This failure is to be expected because Q_T depends only on the difference $(\delta^+ - \delta^-)$, which in the impact-parameter approximation is proportional to $(V^+ - V^-)$, so that the polarization potential cancels out.

The random-phase approximation permits the integrations for the cross sections to be carried out analytically. In Figs. 5-7-7 and 5-7-8 it is

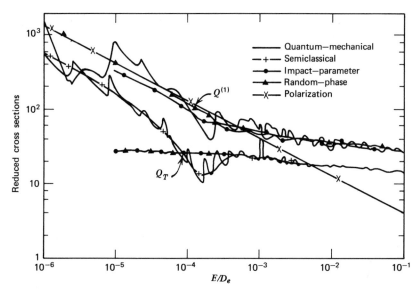

FIG. 5-7-7. Results of various approximations for the charge-transfer and diffusion cross sections (reduced by division by πr_e^2), for He$^+$ in He (Heiche and Mason, 1970).

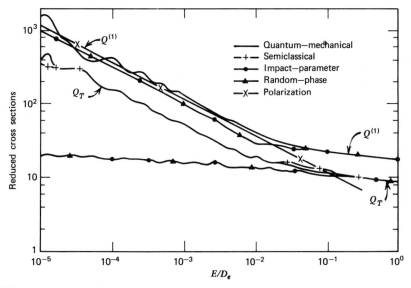

FIG. 5-7-8. Results of various approximations for the charge-transfer and diffusion cross section (reduced by division by πr_e^2) for Cs$^+$ in Cs (Heiche and Mason, 1970).

remarkable how close the much simpler random-phase approximation is to the impact-parameter approximation.

The polarization limit for $Q^{(1)}$ is also given in Figs. 5-7-7 and 5-7-8. As expected, it is accurate at low energies in an average sense. A similar result holds for $Q^{(2)}$ (Heiche and Mason, 1970).

Detailed examination of Figs. 5-7-7 and 5-7-8 shows that the approximation $Q^{(1)} \approx 2Q_T$ is very good indeed above the polarization region. Moreover, the location of the transition between dominant polarization and dominant charge transfer is easily located at the intersection of $Q_p^{(1)}$ and the random-phase approximation for $2Q_T$. Nothing peculiar appears to happen in this transition region. These observations can be used as the basis for a simple approximation for the mobility over the entire temperature range corresponding to elastic scattering.

Finally, some indication of the general energy dependence of Q_T, hence of $Q^{(1)}$, can be made, based on the random-phase approximation. Since the g-u splitting is an exchange effect, it is perhaps not surprising that it falls off approximately exponentially:

$$V^-(r) - V^+(r) \approx Ae^{-r/a}, \tag{5-7-49}$$

where A and a are positive constants. From (5-7-43) the phase-shift difference then becomes

$$\delta^+ - \delta^- \approx \frac{\kappa}{2E} AbK_1\left(\frac{b}{a}\right), \tag{5-7-50}$$

where $K_1(x)$ is the modified Bessel function of the second kind. Using the random-phase approximation with the asymptotic expression $K_1(x) \approx (\pi/2x)^{1/2}e^{-x}$, we obtain

$$Q_T = \frac{1}{2}\pi b^{*2}, \tag{5-7-51}$$

$$\frac{1}{\pi} = \frac{\kappa b^*}{2}\frac{A}{E}\left(\frac{\pi a}{2b^*}\right)^{1/2}e^{-b^*/a}. \tag{5-7-52}$$

Assuming that the major dependence of (5-7-52) on b^* comes from the exponential term, we can eliminate b^* between these two equations to obtain the energy dependence of Q_T (Dalgarno, 1958):

$$Q_T = (a_1 - a_2 \ln E)^2, \tag{5-7-53}$$

where a_1 is a weak (logarithmic) function of b^*. In practice, Q_T can usually be fitted over a wide range of energy by an expression like (5-7-53) with constant a_1 and a_2.

The formula (5-7-53) was given in Section 5-2-B, where it was noted that the energy dependence was the same as that of the diffusion cross section for an exponential repulsion potential. We can see now that the basic reason for the similar energy dependence is the exponential form of the interaction and that it will follow from almost any approximation of a dimensional character, whether it be a random-angle, random-phase, or effective collision diameter approximation.

The polarization and random-phase results can be combined to give a formula for the mobility whose error is at most 25% over the entire temperature range. If a_1 and a_2 in (5-7-53) are found from experimental results, then from $Q^{(1)} \approx 2Q_T$ and the polarization limit we obtain

$$K_0(T) \approx \text{minimum} \left\{ \frac{13.876}{(\mu\alpha)^{1/2}}; \right.$$

$$\left. 9254(\mu T)^{-1/2}[(a_1 + 3.664a_2 - a_2 \log_{10} T)^2 + 0.745a_2^2]^{-1} \right\}, \quad (5\text{-}7\text{-}54)$$

where $K_0(T)$ is in square centimeters per volt per second at standard gas density, μ is in g/mole, α is in cubic angstroms, T is in degrees Kelvin, and a_1 and a_2 are from (5-7-53) with Q_T in square angstroms, E in electron volts, and the logarithm to base 10. A more accurate result can be obtained by arranging a smoother transition between $Q_p^{(1)}$ and $2Q_T$ rather than merely switching from one to the other at their intersection, as (5-7-54) does.

5-8. ION TRANSFER

Having seen the important effects of resonant charge transfer on ion mobility, we may expect that any process by which charge passes easily from an ion to a neutral will qualitatively and quantitatively affect the mobility. In particular, the transfer of an ionic fragment rather than an electron should be such a process. In this section we present two examples of ion transfer that seem to be well documented. Many others are suspected. Unfortunately, unlike resonant charge transfer, no real theory exists in usable form for the description of these collisions. The reason is that ion transfer is a rearrangement collision, indeed a chemical reaction, and constitutes a difficult theoretical problem.

The first example concerns the mobility of H_3^+ ions in H_2, where a proton transfer is possible:

$$H_3^+ + H_2 = H_2 + H_3^+. \quad (5\text{-}8\text{-}1)$$

This is perhaps not exactly resonant because of some mismatch in equilibrium internuclear distances, but it is nevertheless highly probable. Mason and Vanderslice (1959) determined the potential energy between H_3^+ and H_2, partly from theory and partly from analysis of measurements of the elastic scattering of beams of H_3^+ by H_2 gas, and used this potential to calculate the mobility according to the methods discussed in Section 5-3. They obtained a mobility of 22 cm²/V-sec at room temperature. Varney (1960) suggested that the mobility should be much lower because of proton transfer. Subsequent experimental work established that a hydrogen ion of mobility 11 to 12 cm²/V-sec usually observed in mobility experiments was indeed H_3^+ (Barnes et al., 1961; Albritton et al., 1968; Miller et al., 1968). These results are shown in Fig. 5-8-1. Not only is the mobility lower by a factor of 2 at room temperature by virtue of the proton transfer but the whole temperature dependence of the mobility curve is altered. Without proton transfer the mobility would increase with increasing temperature from its polarization limit of 14.2 cm²/V-sec; instead, the proton transfer causes the mobility to decrease.

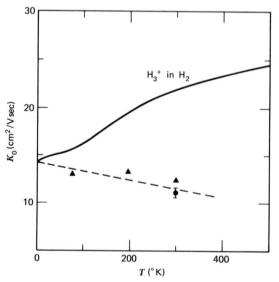

FIG. 5-8-1. Mobility of H_3^+ in H_2 as a function of temperature, showing the effect of proton transfer. The solid curve is a calculated result based on the correct potential between H_3^+ and H_2 but neglecting proton transfer; the measured value is lower by a factor of about 2 at room temperature. The dashed curve has no theoretical status other than to connect the measured points (from Tables 7-1-G-1 and 7-3-D-1) with the polarization limit.

The second example concerns the mobility of He_2^+ ions in He, where He^+ ion transfer is possible:

$$He_2^+ + He = He + He_2^+, \qquad (5\text{-}8\text{-}2)$$

the process being resonant if He_2^+ is in its ground state $(^2\Sigma_u)$. It was long thought that an ion of mobility 20 cm^2/V-sec at room temperature referred to He_2^+ in the ground state and theoretical calculations neglecting ion transfer were consistent with this interpretation (Geltman, 1953; Mason and Schamp, 1958). Later identification of another mass-8 ion of mobility 16 cm^2/V-sec led Beaty, Browne, and Dalgarno (1966) to suggest that the faster ion was the metastable $^4\Sigma_u$ state of He_2^+ arising from the interaction of $He^+(^2S)$ and an excited $He(^3S)$ atom, for which ion transfer with ground-state $He(^1S)$ would be nonresonant. This interpretation now seems generally accepted. The results are shown in Fig. 5-8-2. The calculated curve is based on the potential calculated quantum-mechanically by Geltman, with the collision-integral calculation of Mason and Schamp. The measurements are distinctly lower and in fact lie below the polarization limit, so that the whole temperature dependence of the mobility curve is altered.

The effects of ion transfer are thus seen to be analogous to charge transfer, even though a quantitative theory is lacking.

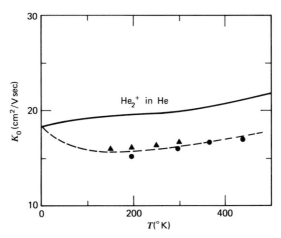

FIG. 5-8-2. Mobility of He_2^+ in He as a function of temperature, showing the effect of He^+ transfer. The solid curve is the calculated result with no ion transfer; the measured values are appreciably lower. The dashed curve has no theoretical status other than to connect the measured points with the polarization limit. Experimental points are: ● Orient (1967); ▲ Patterson (1970b).

REFERENCES

Albritton, D. L., T. M. Miller, D. W. Martin, and E. W. McDaniel (1968), *Phys. Rev.* **171**, 94.

Alievskiĭ, M. Ya., and V. M. Zhdanov (1969), *Soviet Phys.-JEPT* (*English Transl.*) **28**, 116 [*Zh. Eksperim. Teor. Fiz.* **55**, 221 (1968)].

Arthurs, A. M., and A. Dalgarno (1960), *Proc. Roy. Soc.* (*London*) **A256**, 552.

Barnes, W. S., D. W. Martin, and E. W. McDaniel (1961), *Phys. Rev. Letters* **6**, 110.

Bates, D. R., and A. H. Boyd (1962), *Proc. Phys. Soc.* (*London*) **80**, 1301.

Beaty, E. C., J. C. Browne, and A. Dalgarno (1966), *Phys. Rev. Letters* **16**, 723.

Biondi, M. A., and L. M. Chanin (1961), *Phys. Rev.* **122**, 843.

Blanc, A. (1908), *J. Phys.* **7**, 825.

Chapman, S. (1967), "The Kinetic Theory of Gases Fifty Years Ago," in *Lectures in Theoretical Physics, Vol. 9C: Kinetic Theory* (W. E. Brittin, Ғd), Gordon and Breach, New York, pp. 1–13.

———, and T. G. Cowling (1970), *The Mathematical Theory of Non-Uniform Gases*, 3rd ed., Cambridge University Press, London.

Cowling, T. G. (1960), *Molecules in Motion*, Harper, New York.

Dalgarno, A. (1958), *Phil. Trans. Roy. Soc.* (*London*) **A250**, 426.

———, M. R. C. McDowell, and A. Williams (1958), *Phil. Trans. Roy. Soc.* (*London*) **A250**, 411.

Dalgarno, A., and A. Williams (1958), *Proc. Phys. Soc.* (*London*) **72**, 274.

Dickinson, A. S. (1968a), *J. Phys.* **B1**, 387.

——— (1968b), *J. Phys.* **B1**, 395.

Fahr, H., and K. G. Müller (1967), *Z. Phys.* **200**, 343.

Firsov, O. B. (1951), *Zh. Eksperim. Teor. Fiz.* **21**, 1001.

Fleming, I. A., R. J. Tunnicliffe, and J. A. Rees (1969a), *J. Phys.* **B2**, 780.

——— (1969b), *J. Phys.* **D2**, 551.

Ford, G. W. (1968), *Phys. Fluids* **11**, 515.

Frenkel, S. P. (1940), *Phys. Rev.* **57**, 661.

Geltman, S. (1953), *Phys. Rev.* **90**, 808.

Grew, K. E., and T. L. Ibbs (1952), *Thermal Diffusion in Gases*, Cambridge University Press, London.

Gupta, B. K., and F. A. Matsen (1967), *J. Chem. Phys.* **47**, 4860.

Hahn, H. (1972), thesis, Brown University.

———, and E. A. Mason (1972), *Phys. Rev.* **A5**, 1573.

——— (1973), *Phys. Rev.* (to be published).

Hassé, H. R. (1926), *Phil. Mag.* **1**, 139.

———, and W. R. Cook (1931), *Phil. Mag.* **12**, 554.

Heiche, G. (1967), U. S. Naval Ordnance Lab. Rept. **NOL TR** 67–150.

———, and E. A. Mason (1970), *J. Chem. Phys.* **53**, 4687.

Herzberg, G. (1950), *Spectra of Diatomic Molecules*, 2nd ed., Van Nostrand, New York.

Higgins, L. D., and F. J. Smith (1968), *Mol. Phys.* **14**, 399.

Hirschfelder, J. O., C. F. Curtiss, and R. B. Bird (1964), *Molecular Theory of Gases and Liquids*, Wiley, New York.

Hirschfelder, J. O., and M. A. Eliason (1957), *Ann. N. Y. Acad. Sci.* **67**, 451.

Hoffman, D. K., and J. S. Dahler (1969), *J. Stat. Phys.* **1**, 521.

Holstein, T. (1952), *J. Phys. Chem.* **56**, 832.

——— (1955), *Phys. Rev.* **100**, 1230 (A).

Hoselitz, K. (1941), *Proc. Roy. Soc. (London)* **A177**, 200.

Jeans, J. H. (1925), *The Dynamical Theory of Gases*, 4th ed., Cambridge University Press, London. Reprinted by Dover, New York, 1954.

Kagan, Yu., and A. M. Afanas'ev (1962), *Soviet Phys.-JETP (English Transl.)* **14**, 1096 [*Zh. Eksperim. Teor. Fiz.* **41**, 1536 (1961)].

Kennard, E. H. (1938), *Kinetic Theory of Gases*, McGraw-Hill, New York.

Kihara, T. (1952), *Rev. Mod. Phys.* **24**, 45.

——— (1953), *Rev. Mod. Phys.* **25**, 844.

———, M. H. Taylor, and J. O. Hirschfelder (1960), *Phys. Fluids* **3**, 715.

Kramers, H. A. (1949), *Nuovo Cimento Suppl.* **6**, 297.

Kumar, K., and R. E. Robson (1973), *Australian J. Phys.* (to be published).

Landau, L. D., and E. M. Lifshitz (1958), *Quantum Mechanics, Non-Relativistic Theory*, Pergamon, London.

Langevin, P. (1905), *Ann. Chim. Phys.* **5**, 245. A translation is given by E. W. McDaniel, *Collision Phenomena in Ionized Gases*, Wiley, New York, 1964, Appendix II.

Lowke, J. J., and J. H. Parker, Jr. (1969), *Phys. Rev.* **181**, 302.

Margenau, H. (1941), *Phil. Sci.* **8**, 603.

Mason, E. A. (1957a), *J. Chem. Phys.* **27**, 75.

——— (1957b), *ibid.*, 782.

———, and Hahn, H. (1972), *Phys. Rev.* **A5**, 438.

Mason, E. A., and T. R. Marrero (1970), *Adv. Atom. Mol. Phys.* **6**, 155.

Mason, E. A., R. J. Munn, and F. J. Smith (1966), *Adv. Atom. Mol. Phys.* **2**, 33.

Mason, E. A., H. O'Hara, and F. J. Smith (1972), *J. Phys.* **B5**, 169.

Mason, E. A., and H. W. Schamp (1958), *Ann. Phys. (N.Y.)* **4**, 233.

Mason, E. A., and J. T. Vanderslice (1959), *Phys. Rev.* **114**, 497.

———, and J. M. Yos (1959), *Phys. Fluids* **2**, 688.

Massey, H. S. W., and C. B. O. Mohr (1934), *Proc. Roy. Soc. (London)* **A144**, 188.

Massey, H. S. W., and R. A. Smith (1933), *Proc. Roy. Soc. (London)* **A142**, 142.

Maxwell, J. C. (1860), *Phil. Mag.* **20**, 21. Reprinted in *The Scientific Papers of James Clerk Maxwell*, Vol. 1, Dover, New York, 1962, pp. 392–409.

——— (1867), *Phil. Trans. Roy. Soc. (London)* **157**, 49. Reprinted in *The Scientific Papers of James Clerk Maxwell*, Vol. 2, Dover, New York, 1962, pp. 26–78.

Miller, T. M., J. T. Moseley, D. W. Martin, and E. W. McDaniel (1968), *Phys. Rev.* **173**, 115.

Monchick, L. (1959), *Phys. Fluids* **2**, 695.

——— (1962), *Phys. Fluids* **5**, 1393.

———, and E. A. Mason (1967), *Phys. Fluids* **10**, 1377.

Monchick, L., R. J. Munn, and E. A. Mason (1966), *J. Chem. Phys.* **45**, 3051.

Monchick, L., S. I. Sandler, and E. A. Mason (1968), *J. Chem. Phys.* **49**, 1178.

Monchick, L., K. S. Yun, and E. A. Mason (1963), *J. Chem. Phys.* **39**, 654.

Mott, N. F., and H. S. W. Massey (1965), *The Theory of Atomic Collisions*, 3rd ed., Oxford University Press, London.

Munn, R. J., E. A. Mason, and F. J. Smith (1964), *J. Chem. Phys.* **41**, 3978.

Orient, O. J. (1967), *Can. J. Phys.* **45**, 3915.

―――― (1971), *J. Phys.* **B4**, 1257.

Parker, J. H., Jr., and J. J. Lowke (1969), *Phys. Rev.* **181**, 290.

Patterson, P. L. (1970a), *J. Chem. Phys.* **53**, 696.

―――― (1970b), *Phys. Rev.* **A2**, 1154.

Persson, K.-B., and S. C. Brown (1955), *Phys. Rev.* **100**, 729.

Peyerimhoff, S. (1965), *J. Chem. Phys.* **43**, 998.

Present, R. D. (1958), *Kinetic Theory of Gases*, McGraw-Hill, New York.

―――――, and A. J. de Bethune (1949), *Phys. Rev.* **75**, 1050.

Rayleigh, Lord (1891), *Phil. Mag.* **32**, 424. Reprinted in "Scientific Papers," Vol. 3, Cambridge University Press, London, 1902, pp. 473–490.

―――― (1900) *Proc. Roy. Soc. (London)* **66**, 68. Reprinted in "Scientific Papers," Vol. 4, Cambridge University Press, London, 1902, pp. 452–458.

Sandler, S. I., and J. S. Dahler (1967), *J. Chem. Phys.* **47**, 2621.

Sandler, S. I., and E. A. Mason (1967), *J. Chem. Phys.* **47**, 4653.

―――― (1968), *J. Chem. Phys.* **48**, 2873.

She, R. S. C., and N. F. Sather (1967), *J. Chem. Phys.* **47**, 4978.

Skullerud, H. R. (1968), *J. Phys.* **D1**, 1567.

―――― (1969a), *J. Phys.* **B2**, 86.

―――― (1969b), *ibid.*, 696.

―――― (1972), *Tech. Rept. EIP 72-1*, Physics Dept., Norwegian Institute of Technology, Trondheim, Norway.

Smirnov, B. M. (1966), *Soviet Phys.-Doklady (English Transl.)* **11**, 429 [*Dokl. Akad. Nauk SSSR* **168**, 322 (1966)].

―――― (1967a), *Soviet Phys.-Tech. Phys. (English Transl.)* **11**, 1388 [*Zh. Tekhn. Fiz.* **36**, 1864 (1966)].

―――― (1967b), *Soviet Phys.-Usp. (English Transl.)* **10**, 313 [*Usp. Fiz. Nauk* **92**, 75 (1967)].

Smith, F. J. (1967), *Mol. Phys.* **13**, 121.

Snider, R. F. (1960), *J. Chem. Phys.* **32**, 1051.

Stefan, J. (1871), *Sitzber. Akad. Wiss. Wien* **63**, 63.

―――― (1872), *Sitzber. Akad. Wiss. Wien* **65**, 323.

Taxman, N. (1958), *Phys. Rev.* **110**, 1235.

Varney, R. N. (1960), *Phys. Rev. Letters* **5**, 559.

Waldmann, L. (1957), *Z. Naturforsch.* **12a**, 660.

―――― (1958), *Z. Naturforsch.* **13a**, 609.

―――― (1963), *Z. Naturforsch.* **18a**, 1033.

Wang Chang, C. S., G. E. Uhlenbeck, and J. de Boer (1964), *Studies Stat. Mech.* **2**, 241.

Wannier, G. H. (1953), *Bell System Tech. J.* **32**, 170. See also *Phys. Rev.* **83**, 281 (1951); **87**, 795 (1952).

—— (1971), *Am. J. Phys.* **39**, 281. See also F. J. McCormack (1971), *Am. J. Phys.* **39**, 1413.

Whealton, J. H., and E. A. Mason (1972), *Phys. Rev.* **A6**, 1939.

—— (1973), *Ann. Phys. (N.Y.)* (to be published).

Wolniewicz, L. (1965), *J. Chem. Phys.* **43**, 1087.

Wood, H. T. (1971), *J. Chem. Phys.* **54**, 977.

Zwanzig, R. (1965), *Ann. Rev. Phys. Chem.* **16**, 67.

6

INTERACTION POTENTIALS

AND MOBILITIES

The relations connecting mobility and the ion-neutral interaction potential were given in Chapter 5. In this chapter we consider these relations in more detail, emphasizing two closely connected problems: how the mobility is calculated if the interaction is known and how the interaction can be found in the first place. The chapter closes with a section that summarizes the recipes for calculating mobilities, and particular attention is paid to estimation procedures that can be used in the face of meager initial data.

6-1. MOBILITIES FROM INTERACTION POTENTIALS

In principle, to obtain the mobility all we need to do is calculate the differential cross section from the potential and all cross sections and collision integrals will follow by integration. In practice, this is a poor computational procedure, and it is better to use other methods, for which it is worthwhile to draw a distinction between classical and quantum-mechanical cases, and to consider resonant charge transfer separately. Except for simple forms of interactions, the calculations constitute a fairly formidable numerical task, so much so that anisotropic potentials and inelastic collisions are still awaiting a future generation of high-speed computers. This section is limited essentially to elastic collisions of spherically symmetric potentials.

A. CLASSICAL CALCULATIONS. Given the potential, three successive integrations are needed to obtain a mobility. The first integration determines the deflection angle in an ion-neutral collision as a function of impact parameter b and relative energy E:

$$\theta(b, E) = \pi - 2b \int_{r_a}^{\infty} \left[1 - \frac{b^2}{r^2} - \frac{V(r)}{E} \right]^{-1/2} \frac{dr}{r^2}, \qquad (6\text{-}1\text{-}1)$$

where the distance of closest approach r_a is the outermost root of

$$1 - \frac{b^2}{r_a^2} - \frac{V(r_a)}{E} = 0. \qquad (6\text{-}1\text{-}2)$$

The second integration averages over impact parameters to yield transport cross sections as a function of energy:

$$Q^{(l)}(E) = 2\pi \left[1 - \frac{1 + (-1)^l}{2(1 + l)} \right]^{-1} \int_{0}^{\infty} (1 - \cos^l \theta) b \, db. \qquad (6\text{-}1\text{-}3)$$

The third integration averages the cross sections over energy to produce collision integrals as a function of temperature:

$$\overline{\Omega}^{(l,\, s)}(T) = [(s + 1)! \, (kT)^{s+2}]^{-1} \int_{0}^{\infty} Q^{(l)}(E) e^{-E/kT} E^{s+1} \, dE. \qquad (6\text{-}1\text{-}4)$$

Only algebra is required to find the mobility from the collision integrals. For high fields the last integration may need to be slightly altered so that $\overline{\Omega}^{(l,\, s)}$ is a function of v_d as well as of T, as in (5-6-4). In any case the evaluation of the integral for $\overline{\Omega}^{(l,\, s)}$ is relatively easy because the integrand varies smoothly, and we need not discuss it further here. The other two integrations are much more difficult because of singularities and other irregularities that can occur in their integrands. Here we discuss only the general nature of the difficulties and their physical interpretation, and do not go into computational details. The most detailed general discussion presently available can be found in the treatise of Hirschfelder, Curtiss, and Bird (1964, Section 8.4), and the most efficient numerical techniques are discussed by O'Hara and Smith (1970, 1971). For the record, however, we mention that a number of successful computational methods have been described in the literature, notably by Maxwell (1867), Langevin (1905), Hassé (1926), Hassé and Cook (1927, 1929, 1931), Kotani (1942), Kihara and Kotani (1943), de Boer and van Kranendonk (1948), Hirschfelder, Bird, and Spotz (1948), Holleran and Hulburt (1951), Mason (1954), Mason and Schamp (1958), Monchick (1959), Kihara, Taylor, and Hirschfelder

(1960), Itean, Glueck, and Svehla (1961), Monchick and Mason (1961), Smith and Munn (1964), Barker, Fock, and Smith (1964), Samoilov and Tsitelauri (1964), Dymond, Rigby, and Smith (1966), Mason, Munn, and Smith (1967), and DeRocco, Storvick, and Spurling (1968).

The most troublesome integration is that for the deflection angle. The singularity in the integrand at r_a is easily removed by a variable change such as $r_a/r = \sin \alpha$ and causes no difficulty. Difficulty arises from a singularity that can occur at low energies for potentials with an attractive component and corresponds physically to orbiting collisions. The deflection angle approaches negative infinity and the orbit approaches an asymptotic circle. This

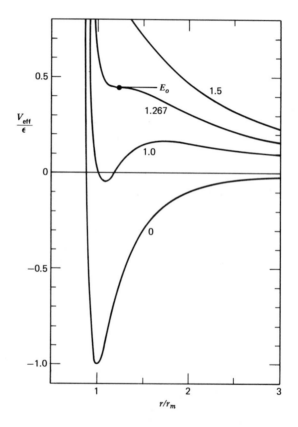

FIG. 6-1-1. Effective reduced potential for various values of reduced angular momentum $(E/\varepsilon)^{1/2}(b/r_m)$ for a (12-4) potential. The numbers on the curves are the reduced angular momenta. Orbiting occurs when the total energy equals the height of the centrifugal barrier. The barrier disappears at a "critical point" marked with the black dot and orbiting is impossible for energies above $E_o/\varepsilon = 5^{-1/2} = 0.447$. The effective potential is monotonic for reduced angular momentum equal to or greater than $(2/5)(3^{1/2})(5^{3/8}) = 1.267$.

behavior is conveniently described by means of the potential energy diagram for the radial motion, using the effective potential

$$V_{\text{eff}}(r) = V(r) + \frac{Eb^2}{r^2},$$ (6-1-5)

where Eb^2/r^2 is the centrifugal potential and $(2\mu E)^{1/2}b$ is the angular momentum. If the angular momentum is not too great, V_{eff} can have a maximum, as shown in Fig. 6-1-1, and an unstable circular orbit with $\theta = -\infty$ will result if the height of the maximum is equal to the energy E. As the angular momentum is increased, the height of the maximum in this centrifugal barrier increases, but eventually the maximum disappears entirely, degenerating to a horizontal inflection point of energy E_0, similar to the critical point in the pressure-volume isotherm of a gas. Thus orbiting is not possible for $E > E_0$, and instead of a singularity in θ we find a minimum. The behavior of θ as a function of b is shown in Fig. 6-1-2 for several energies. (Figures 6-1-1 and 6-1-2 are drawn for a (12-4) potential.) The minimum in θ for $E > E_0$ gives rise to intense scattering in the differential cross section; this is known as rainbow scattering because of the close analogy with optical rainbows (Ford and Wheeler, 1959a). Other optical analogies such as glories ($\theta = 0, \pi, 2\pi$) are

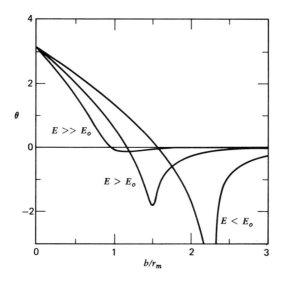

FIG. 6-1-2. Deflection angle as a function of impact parameter, showing the transition from orbiting ($E < E_0$) to large-angle rainbow scattering ($E > E_0$) to small-angle rainbow scattering ($E \gg E_0$). The three curves are for a (12-4) potential, with values of E/ε equal to 0.203 (= tan 0.2), 1.03 (= tan 0.8), and 14.1 (= tan 1.50).

also known, but none of these is of any great importance as far as mobility calculations are concerned. An elementary review of optical analogies in molecular scattering is available (Mason, Munn, and Smith, 1971).

The topology of the scattering is shown as a "phase diagram" in the plane of energy versus angular momentum in Fig. 6-1-3 (Ford and Wheeler, 1959b). The various regions in the figure are discussed in more detail in connection with quantum effects. The numerical integration for θ is rather troublesome in the neighborhood of orbiting or large-angle rainbow scattering because of the singularity or large maximum in the integrand; nothing very clever can be done about it and the numerical problem is usually solved by essentially brute-force methods.

The orbiting also creates difficulty in the integration for the cross sections because the factor $(1 - \cos^l \theta)$ causes the integrand to oscillate

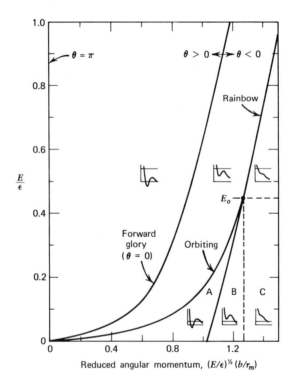

FIG. 6-1-3. Topology of scattering in the energy-angular momentum plane for a (12-4) potential. The black dot marks the "critical point" between orbiting and rainbow scattering, as in Fig. 6-1-1. In region A metastable bound states are possible, but in region B the energy is less than the minimum in V_{eff}. For reduced angular momentum greater than 1.267 (region C) the effective potential is monotonic.

violently near the orbiting value of the impact parameter. This can be dealt with in a number of ways. The simplest is to replace $(1 - \cos^l \theta)$ with its average value of $\frac{1}{2}$ in the region of violent oscillations. Another is to curve-fit θ versus b with a suitably simple function near orbiting and use analytical methods. (The orbiting singularity in θ is logarithmic.) Still another uses a change of variable that puts the oscillations at the end of the integration range, where they are less troublesome (O'Hara and Smith, 1970). However the matter is handled, the integration has a smoothing effect and the cross sections $Q^{(l)}(E)$ are smooth and well behaved in the orbiting region of $E < E_o$. In the large-angle rainbow region, in which E is not much larger than E_o, the factor $(1 - \cos^l \theta)$ has only a few oscillations, but these change rapidly with increasing E and cause the cross sections to show a few weak "rainbow undulations." Even these undulations are smoothed out by the final integration for the collision integrals.

Fortunately all the foregoing numerical work needs to be done only once for a given form of potential. With a suitable choice of dimensionless variables, the results can be enshrined as a set of numerical tables; the computational problem is thereby reduced to simple interpolation. Tables of reduced transport cross sections and collision integrals are given in Appendix I for (8-4), (∞-4), (12-6-4), and (12-4) core potentials, where the reduced quantities are defined as

$$Q^{(l)*}(E^*) \equiv \frac{Q^{(l)}(E)}{\pi r_m^2}, \tag{6-1-6}$$

$$E^* \equiv \frac{E}{\varepsilon}, \tag{6-1-7}$$

$$\Omega^{(l,\ s)*}(T^*) \equiv \frac{\overline{\Omega}^{(l,\ s)}(T)}{\pi r_m^2}, \tag{6-1-8}$$

$$T^* \equiv \frac{kT}{\varepsilon}, \tag{6-1-9}$$

$$A^* \equiv \frac{\Omega^{(2,\ 2)*}}{\Omega^{(1,\ 1)*}}, \tag{6-1-10}$$

$$C^* \equiv \frac{\Omega^{(1,\ 2)*}}{\Omega^{(1,\ 1)*}}. \tag{6-1-11}$$

The cross sections (rather than the collision integrals) are useful in calculations involving the Lorentz model, as mentioned in Sections 5-3-B and 5-5-B. Usually, however, only the collision integrals are needed. Appendix I also lists sources of cross sections and collision integrals for other potentials that are of less direct interest for the calculation of ion diffusion and mobility. To use these tables it is necessary to know values of ε, r_m and any other parameters in the potential model; how this knowledge is acquired is discussed in the later sections of this chapter. If the potential cannot be represented by one of the tabulated functions but a rough estimate of the mobility will suffice, then one of the approximations discussed in Section 5-2-B may be useful.

If the potential is orientation-dependent, a complete scattering calculation is usually prohibitively difficult, and some sort of approximation scheme is needed. In the simplest method we average the potential over all orientations to convert it to a spherically symmetric form. Although easy, this procedure may discard too much; for example, a dipole potential averages exactly to zero, and so all dipole effects are discarded by averaging over orientations. In a better method, which requires more work, we fix the relative orientation during the whole collision, so that the potential depends only on separa- tion, evaluate the cross sections for a number of orientations, and then average the cross sections or collision integrals over orientations (Monchick and Mason, 1961). The physical reasoning behind this averaging procedure is that the deflection angle in a collision is determined mostly by the interaction around the distance of closest approach, in which the orientation probably varies little during the collision. In mathematical language most of the contribution to the integral in (6-1-1) comes from the vicinity of the lower limit. This procedure has been applied so far only to the (12-6-3) and (12-6-5) potentials (see Table I-11), which are models for dipolar and quadrupolar neutral gases, and not to any potential of much direct interest for ion-neutral interactions.

B. SEMICLASSICAL AND QUANTUM-MECHANICAL CALCULATIONS.   As in the classical case, three successive integrations (or summations) are needed to obtain a mobility from a given potential, but the labor is much greater in the quantum-mechanical calculations. For one thing, an additional para- meter enters which corresponds to the de Broglie wavelength λ, a quantity implicitly taken as zero in a classical calculation. For another, the first integration is much more laborious in the quantum-mechanical case than in the classical.

The first integration determines the phase shifts δ_l as a function of wave- number $\kappa = 2\pi/\lambda = \mu v/\hbar = (2\mu E)^{1/2}/\hbar$; this is done by numerical integration of the radial wave equation,

$$\frac{d^2 G_l(r)}{dr^2} + \kappa^2 \left[1 - \frac{l(l+1)}{\kappa r^2} - \frac{V(r)}{E} \right] G_l(r) = 0, \qquad (6\text{-}1\text{-}12)$$

where $l = 0, 1, 2, \ldots$ is the angular-momentum quantum number. In principle, integration is to be carried out from the origin to a value of r large enough for $G_l(r)$ to acquire its asymptotic form

$$G_l(r) \sim \sin \left(\kappa r - \frac{l\pi}{2} + \delta_l \right), \qquad (6\text{-}1\text{-}13)$$

from which δ_l is identified. The most arduous part of the whole calculation is this numerical integration for $\delta_l(\kappa)$. A number of labor-saving tricks have been devised, some of which are discussed by Munn et al. (1964) and by Buckingham et al. (1965). In addition to the sheer numerical labor, difficulty develops from the same features that cause classical orbiting.

The transport cross sections are obtained by summing over the angular-momentum quantum numbers l, as discussed in Section 5-3-E; this corresponds to the integration over impact parameters in the classical case. The expression for the diffusion cross section is

$$Q^{(1)}(\kappa) = \frac{4\pi}{\kappa^2} \sum_{l=0}^{\infty} (l + 1) \sin^2 (\delta_l - \delta_{l+1}), \qquad (6\text{-}1\text{-}14)$$

and other transport cross sections are given by equations (5-3-34), (5-3-35), and (5-3-36).

The final integration for the collision integrals is exactly the same as in the classical case and is given by (6-1-4). The main difference is the occurrence of an extra parameter corresponding to the de Broglie wavelength; it is customary to take it as the dimensionless de Boer parameter Λ^*:

$$\Lambda^* = \frac{h}{r_m (2\mu\varepsilon)^{1/2}}; \qquad (6\text{-}1\text{-}15)$$

that is, the reduced collision integrals $\Omega^{(l,\,s)*}$ now depend on Λ^* as well as on T^*. Some extra difficulty of integration occurs because of orbiting resonances in the transport cross sections, which are absent in the classical case.

The foregoing description corresponds to a full quantum-mechanical calculation. If Λ^* is not too large, much numerical effort can be saved by judicious use of the semiclassical JWKB approximation for the phase shifts,

$$\delta_l(\kappa) \approx \delta(b, \kappa) = \kappa \int_{r_a}^{\infty} \left[1 - \frac{b^2}{r^2} - \frac{V(r)}{E} \right]^{1/2} dr - \kappa \int_{b}^{\infty} \left(1 - \frac{b^2}{r^2} \right)^{1/2} dr$$

$$(6\text{-}1\text{-}16)$$

$$= \frac{\kappa}{2} \int_{r_a}^{\infty} \left[1 - \frac{b^2}{r^2} - \frac{V(r)}{E} \right]^{1/2} \left[\frac{r}{E - V(r)} \right] \frac{dV(r)}{dr} dr; \qquad (6\text{-}1\text{-}17)$$

the second formula is due to F. T. Smith (1965). The Langer (1937) modification is incorporated into these formulas by the identification

$$l + \tfrac{1}{2} = \kappa b. \tag{6-1-18}$$

Efficient numerical methods for the evaluation of JWKB phase shifts are available (Kennedy and Smith, 1967); the labor required is about the same as that for the classical deflection angle, a result to be expected from the relation

$$\theta(b) = \frac{2}{\kappa} \frac{d\delta(b)}{db}. \tag{6-1-19}$$

The range of validity of the JWKB approximation for the phase shifts has been studied by Munn et al. (1964). Although a (12-6) potential was used for their numerical work, the main features they found are probably general. There is always a region in which the JWKB approximation is poor, and this region increases in extent as Λ^* increases. An illustration of the type of failure encountered is shown in Fig. 6-1-4, in which δ_l is plotted as a function of energy for values of $l = 5, 6, 7$ and $\Lambda^* = 1$. A fixed value of l corresponds to a fixed value of the angular momentum. The vertical arrows mark the energies for which classical orbiting occurs; the failure of the JWKB approximation is particularly obvious at energies close to the orbiting

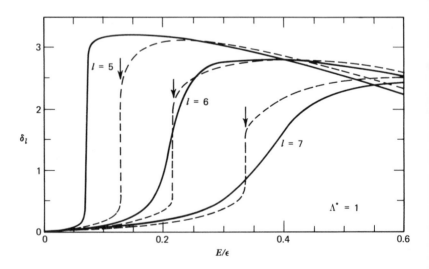

FIG. 6-1-4. Phase shifts as a function of energy for a (12-6) potential, showing the failure of the semiclassical JWKB results (broken curves), compared with the exact quantal values (solid curves). The vertical arrows mark the classical orbiting energies. An apparent case of resonant barrier tunneling is seen for $l = 5$.

energies, and it is easy to see why. For energies below the orbiting energy (region A in Fig. 6-1-3) penetration of the centrifugal barrier by tunneling occurs, and since the JWKB approximation contains no provision for tunneling the approximation is poor. For energies above the orbiting energy the incoming wave is partly reflected by the centrifugal barrier and the JWKB approximation is again rather poor.

Thus the failure of the JWKB approximation for the phase shift is largely associated with the centrifugal barrier in the effective potential; in the classical limit this barrier is responsible for orbiting. The connection between the "steps" in the phase shift in Fig. 6-1-4 and classical orbiting is shown by the relation (6-1-19) and is further illustrated in Fig. 6-1-5, which also illustrates a forward glory. In the classical orbiting region the

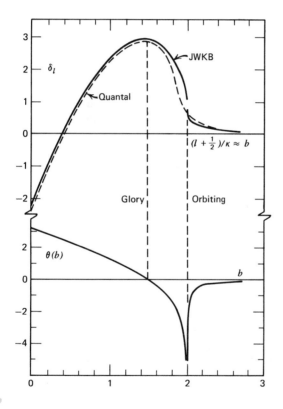

FIG. 6-1-5. The relation between phase shifts and the corresponding classical deflection angle, illustrated by results for a (12-6) potential ($E/\varepsilon = 0.4$, $\Lambda^* = 1$), showing the behavior around orbiting and a forward glory. The correct quantal phase shift is indicated by the dashed curve.

true phase shift changes rapidly over a small range of b at fixed E, and the JWKB phase shift exhibits an infinite slope and a discontinuity less than π (Ford and Wheeler, 1959a). In terms of Fig. 6-1-4 there is a rapid change over a small range of E at fixed l. Figure 6-1-5 also shows that JWKB phase-shift differences can be more accurate than the phase shifts themselves because the JWKB curve runs nearly parallel to the quantal curve over a large range. One consequence is that transport cross sections, which involve only differences of phase shifts, may be more accurate than expected on the basis of the errors in the JWKB phase shifts themselves.

All the difficulties in numerical integration encountered for the classical deflection angle in the vicinity of orbiting also occur for the phase shift, both in the JWKB approximation and in the full quantum-mechanical treatment.

Since the JWKB approximation requires so much less numerical effort than the complete quantal calculation, it is important to know the parameter values for which the JWKB results are unsatisfactory. These values can be represented by regions in the energy-angular momentum plane shown in Fig. 6-1-3. For small Λ^* the unsatisfactory region is a narrow loop around the orbiting curve. As Λ^* is increased, this loop expands to encompass region A, plus a large part of the area above the orbiting curve, as shown schematically in Fig. 6-1-6. For fixed reduced angular momentum the reduced energy at which the JWKB approximation becomes unsatisfactory is approximately proportional to $(\Lambda^*)^2$, except near the curve of classical orbiting (Munn et al, 1964). Knowledge of the qualitative behavior shown in Fig. 6-1-6 plus a few accurate computations serves to locate the boundaries of the loops and thus to minimize the effort needed to compute a set of phase shifts. In addition, tunneling corrections can be applied to some of the JWKB phase shifts (Livingston, 1966).

No very effective procedures have been described for evaluating the transport cross sections $Q^{(l)}$ once the phase shifts have been computed; direct summation over l seems to be most effective. If we try to ease the labor by replacing the summations with integrations, the completely classical result is obtained, as shown in Section 5-3-E.

The relation between classical orbiting and the low-energy resonances in the transport cross section, illustrated in Fig. 5-3-5, can be understood from the phase-shift curve shown in Fig. 6-1-5. A plot of δ_l versus l at constant E looks very much like the upper curve in Fig. 6-1-5, and as E is increased the steep portion of the curve (corresponding to classical orbiting) moves to higher l values. As the "step" crosses each integral values of l the phase shift for that l takes a sudden jump which is reflected as a jump in the cross section at that energy. These orbiting resonances persist until E is so high that orbiting is no longer possible, and only rainbow scattering occurs. Then the "step" in the δ_l versus l curve becomes less steep and there is only a small undulation in the cross section.

The best strategy for a quantum-mechanical calculation of the mobility therefore is probably as follows: the phase shifts can be calculated by the JWKB approximation when possible but some must always be calculated by numerical integration of the radial wave equation. The cross sections must be calculated by direct summation and the collision integrals can be found by straightforward numerical integration. As in the classical case, all this numerical work can be done once and preserved as a set of numerical tables, which now contain the additional parameter Λ^*. Unfortunately such tables are available only for the (12-6) potential (Imam-Rahajoe et al., 1965; Munn et al., 1965) but not for any ion-neutral potentials.

C. CHARGE-TRANSFER CALCULATIONS. The effect of resonant charge transfer on mobility was discussed in some detail in Section 5-7. The type

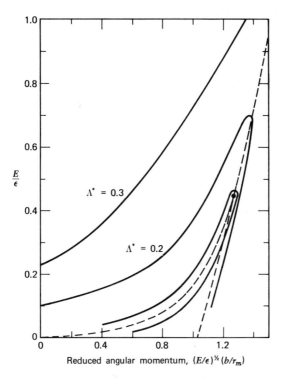

FIG. 6-1-6. Diagram of regions in which the JWKB approximation is unsatisfactory, exhibited as loops in the same energy-angular momentum plane shown in Fig. 6-1-3.

of numerical work needed in a calculation of the mobility from the interaction potentials is the same as that described in the preceding section, except that phase shifts must be calculated for two or more potentials instead of for only one. There is, of course, no complete counterpart to the classical calculations described in Section 6-1-A, since resonant charge transfer is not describable classically. The various approximations that are useful for making reasonable estimates with little labor were discussed in Section 5-7-E and need not be elaborated here.

6-2. THEORY OF ION-ATOM AND ION-MOLECULE INTERACTIONS

We almost never know the true interaction potential and instead must be satisfied with mathematical models that hopefully mimic the true potential in a reasonable way. Such models are ordinarily concocted to reproduce various known asymptotic forms of the true potential and to behave in a qualitatively correct way in intermediate regions. In this section we summarize briefly the available information on the theory of the interaction potential. We do not go into much detail because several excellent reviews are readily available (Hirschfelder, Curtiss, and Bird, 1964; Hirschfelder, 1967; Margenau and Kestner, 1969). It is convenient, although arbitrary, to divide intermolecular forces into long-range, short-range, and intermediate-range forces.

A. LONG-RANGE FORCES.   By far the most important long-range component of the interaction between an ion and a neutral molecule is the polarization or induction potential. At ordinary temperatures this component may even dominate the mobility, as discussed in Section 5-3-D. There are terms in the polarization energy, however, other than the spherically symmetric r^{-4}. In the first place, if the molecule is not spherically symmetric there will be an angle-dependent term in the potential due to the anisotropy of the molecular polarizability. Second, the ion induces not only a dipole moment in the molecule but also quadrupole and higher moments, which interact with the ionic charge. The complete expression for the polarization energy is thus an infinite series containing angle-dependent terms. For a molecule with cylindrical symmetry the first few terms are

$$V_p(r) = -\frac{e^2\bar{\alpha}}{2r^4}[1 + \kappa(3\cos^2\theta - 1)] - \frac{e^2\alpha_q}{2r^6} + \cdots, \qquad (6\text{-}2\text{-}1)$$

in which the r^{-4} term results from the induced dipole and the r^{-6} term from the induced quadrupole. The average dipole polarizability $\bar{\alpha}$ and the anisotropy of the polarizability κ (not to be confused with wavenumber, for which we have used the same symbol) are defined in terms of the dipole polarizabilities parallel and perpendicular to the molecular axis as

$$\bar{\alpha} = \tfrac{1}{3}(\alpha_\| + 2\alpha_\perp),\qquad (6\text{-}2\text{-}2)$$

$$\kappa = \frac{(\alpha_\| - \alpha_\perp)}{(\alpha_\| + 2\alpha_\perp)}. \qquad (6\text{-}2\text{-}3)$$

The molecular axis makes an angle θ with the line drawn from the charge to the center of the molecule. The quadrupole polarizability is denoted as α_q. The angle-dependent term is usually ignored in mobility calculations because of the difficulty of evaluating the appropriate collision integrals (Section 6-1-A), but the induced quadrupole term can be included and is often significant (Margenau, 1941; Mason and Schamp, 1958).

The expression (6-2-1) for the polarization energy is valid in both classical and quantum mechanics. The only way in which quantum phenomena would enter (6-2-1) is through *ab initio* calculations of $\bar{\alpha}$, κ, or α_q; such calculations are often the best source for α_q, but the best values of $\bar{\alpha}$ and κ usually come from experiment or from a judicious combination of theory and experiment. Summaries of numerical values of $\bar{\alpha}$, κ, and α_q for some common gases are given in Appendix II. Even if an accurate experimental or theoretical value is not available, estimates based on simple models can be made that are often adequate for mobility calculations. The dipole polarizability for an atom or ion can be estimated from the formula (Kirkwood, 1932)

$$\alpha = \frac{4}{9N_e a_0}\left(\sum_{i=1}^{N_e}\langle r_i^2\rangle\right)^2, \qquad (6\text{-}2\text{-}4)$$

where N_e is the number of electrons in the atom, a_0, the Bohr atomic radius (0.529 Å), and $\langle r_i^2\rangle$, the mean square distance of the ith electron from the nucleus. Values of $\langle r_i^2\rangle$ can be calculated very simply with the aid of Slater screening constants (Hirschfelder, Curtiss, and Bird, 1964, Section 13.2d),

$$\langle r_i^2\rangle = (2n_i^* + 1)(2n_i^* + 2)\left[\frac{n_i^*}{2(Z - S_i)}\right]^2 a_0^2, \qquad (6\text{-}2\text{-}5)$$

where n_i^* is the effective principal quantum number of the orbital containing the ith electron, Z is the charge on the nucleus, and S_i, the amount the ith electron is screened from the nuclear charge by the other electrons. The values of n_i^* and S_i are given by simple rules. Calculated values of α for some representative atoms and ions are compared with accurate values in Table 6-2-1. The accuracy for atoms or ions with full or nearly full outer

TABLE 6-2-1. Dipole polarizabilities calculated from Slater screening constants according to (6-2-4) and (6-2-5), compared with accurate values

Atom or Ion	$\alpha(\text{Å}^3)$	
	Calculated	Accurate
He	0.142	0.205
Ar	2.00	1.64
Xe	4.31	4.04
Li	7.6	24.3
K	19	43.4
Cs	23	64.0
Li^+	0.0223	0.0286
K^+	1.17	1.2
Cs^+	2.9	3.1

electronic shells is quite good (on the order of 25%) but rather poor for the one-electron alkali atoms (too low by a factor of 2 or 3).

Values of $\bar{\alpha}$ for complicated molecules can be estimated by adding up the contributions from each of the bonds, using a table of empirical bond polarizabilities (Hirschfelder, Curtiss, and Bird, 1964, Section 13.2c).

Values of α_q can be estimated by use of a model to relate α_q to α and other experimental quantities. Using a harmonic-oscillator model, Margenau (1941) obtained the relation

$$\alpha_q = \frac{3}{2}\alpha^2\frac{h\nu}{e^2f},\qquad(6\text{-}2\text{-}6)$$

where $h\nu$ is the oscillator frequency or some mean excitation energy of the atom and f is the oscillator strength. The usual simplest guess takes $h\nu$ equal to the ionization potential and $f = 1$. Some values of α_q calculated in this way from experimental dipole polarizabilities and ionization potentials are compared in Table 6-2-2 with more accurate values for some representative atoms. Sometimes the agreement is surprisingly good, but errors amounting to a factor of 4 or 5 are obviously possible. Equation 6-2-6 could also be used to estimate α_q for molecules.

Further small contributions to the polarization energy occur if either the ion or the neutral has permanent dipole or higher multipole moments. These contributions all involve the interaction of a permanent multipole

TABLE 6-2-2. Quadrupole polarizabilities calculated from the oscillator
model (6-2-6), compared with accurate
values

Atom	$\alpha_q(\text{Å}^5)$ Calculated	Accurate
H	0.630	0.622
He	0.108	0.101
Ne	0.351	0.370
Ar	4.43	2.19
Li	332	60.0
K	852	212
Cs	1660	441

on one partner with an induced multipole on the other (Hirschfelder,
Curtiss, and Bird, 1964, Section 13.5). Most of them vanish on averaging
over all orientations. The only nonvanishing interaction of interest here is
that between a permanent dipole on one partner and the dipole it induces
in the other; averaged over all orientations, this is

$$\overline{V}(\mu, ind\ \mu) = -\frac{1}{r^6}(\mu_n^2\overline{\alpha}_i + \mu_i^2\overline{\alpha}_n), \qquad (6\text{-}2\text{-}7)$$

where μ is the dipole moment (not to be confused with reduced mass),
the subscript i denotes ion, and n denotes neutral molecule. Note that this
dipole-induced dipole energy has the same r^{-6} behavior as the charge-
induced quadrupole energy. Few dipole moments are known for ions, but
values for some common neutral gases are given in Appendix II, as are
polarizabilities for some ions.

Two further important contributions to the long-range interaction between
an ion and a molecule are the dispersion energy and the electrostatic
energy. The dispersion energy operates even between spherically symmetric
ions and molecules, whereas the electrostatic energy is angle-dependent
and vanishes when averaged over all orientations.

The dispersion energy is entirely quantum mechanical in origin but can
be given a simple physical interpretation in terms of a semiclassical description. Although the average motion of the electrons about a single nucleus
is spherically symmetric, at any instant of time there may be a temporary
accumulation of negative charge in one region, hence an instantaneous
dipole moment of the atom. This instantaneous dipole induces corresponding

instantaneous dipole, quadrupole, etc., moments in another nearby atom. Although the instantaneous dipole moment averages to zero over a period of time, the interaction energy between it and its induced multipoles does not average to zero because the two moments are in phase. The dispersion energy can thus be written as a series,

$$V_{\text{dis}}(r) = -\frac{C^{(6)}}{r^6} - \frac{C^{(8)}}{r^8} - \cdots, \qquad (6\text{-}2\text{-}8)$$

in which the first term represents the dipole-induced dipole interaction, the second term the dipole-induced quadrupole interaction, and so on. Values of the coefficients $C^{(6)}$, $C^{(8)}$, etc., can, in principle, be calculated *ab initio* from quantum mechanics, but the most reliable values come from the semiempirical combination of theoretical formulas with experimental oscillator strengths (Dalgarno, 1967). We are concerned here only with $C^{(6)}$, since we have already neglected terms of the order of r^{-8} in the polarization energy. An excellent way of consolidating and summarizing the available information on dispersion energies employs the approximate Slater-Kirkwood (1931) formula for the interaction of atoms 1 and 2:

$$C_{12}^{(6)} = \frac{3}{2} \frac{\alpha_1 \alpha_2}{(\alpha_1/N_1)^{1/2} + (\alpha_2/N_2)^{1/2}} e^2 a_0^5, \qquad (6\text{-}2\text{-}9)$$

where α_1 and α_2 are in units of a_0^3 and N_1 and N_2 are the number of equivalent electron oscillators in the atoms. By choosing N_1 and N_2 to reproduce the accurately known values of $C_{11}^{(6)}$ and $C_{22}^{(6)}$, respectively, we can also calculate the value of $C_{12}^{(6)}$ with remarkable accuracy (Wilson, 1965; Kramer and Herschbach, 1970); that is, treatment of N_1 and N_2 as empirical parameters is equivalent to use of the combination rule

$$C_{12}^{(6)} = \frac{2C_{11}^{(6)} C_{22}^{(6)}}{(\alpha_2/\alpha_1)C_{11}^{(6)} + (\alpha_1/\alpha_2)C_{22}^{(6)}}. \qquad (6\text{-}2\text{-}10)$$

Values of N for a number of atoms, molecules, and ions are given in Appendix II. Except for atoms in excited electronic states, the empirical values of N are rather close to the number of electrons in the outermost subshell of the atom or ion. For molecules the value of N is fairly close to the number of electrons involved in the chemical bonds. These two rules enable reasonably good estimates of $C^{(6)}$ to be made by (6-2-9) from knowledge of polarizabilities alone.

The electrostatic energy is entirely orientation-dependent and has so far played no role in mobility calculations, presumably because of the difficulty of the collision integrals. For cylindrically symmetric ions and neutrals the first few terms in the electrostatic energy are (Hirschfelder, Curtiss, and Bird, 1964; Sections 1.3 and 12.1)

$$V_{el}(r) = -\frac{e\mu_n}{r^2}\cos\theta_n + \frac{e\theta_n}{2r^3}(3\cos^2\theta_n - 1)$$

$$-\frac{\mu_i\mu_n}{r^3}[2\cos\theta_i\cos\theta_n - \sin\theta_i\sin\theta_n\cos(\phi_i - \phi_n)]$$

$$+\frac{3\mu_i\theta_n}{2r^4}[\cos\theta_i(3\cos^2\theta_n - 1)$$

$$-2\sin\theta_i\sin\theta_n\cos\theta_n\cos(\phi_i - \phi_n)] + \cdots, \qquad (6\text{-}2\text{-}11)$$

where θ_i and θ_n are the angles made by the ionic and molecular axes, respectively, with the line drawn between centers, ϕ_i and ϕ_n are the azimuthal rotation angles about the line of centers, and θ is a permanent quadrupole moment. Some values of θ for neutral molecules are given in Appendix II. It should be noted that the definition of θ we are using here is equal to one-half that used by Hirschfelder, Curtiss, and Bird (1964).

The expressions given in this section, plus the numerical values tabulated in Appendix II, allow estimates to be made of long-range forces that are adequate for most mobility calculations. To give an idea of relative magnitudes we show in Table 6-2-3 the values of the two r^{-6} coefficients (ion-induced quadrupole polarization and dipole-dipole dispersion) for some representative ion-neutral pairs. It can be seen that the polarization energy dominates for the light ions but that the dispersion energy may dominate for heavy ions. In Table 6-2-4 we show the ratio of the total r^{-6} energy to the r^{-4} polarization energy, at the position of the potential minimum, for a few ion-neutral pairs. We can conclude from these results that it would seldom be safe to ignore the r^{-6} energy completely.

TABLE 6-2-3. Comparison of charge-induced quadrupole polarization energy with dipole-dipole dispersion energy for some ion-neutral pairs

		$C^{(6)}$		
	$\frac{1}{2}e^2\alpha_q$	Li^+	K^+	Cs^+
He	0.727	0.200	4.26	8.97
Ar	15.7	1.20	28.3	61.2
Xe	148.	2.15	55.6	124.

Units are eV-$Å^6$.

TABLE 6-2-4. Ratio of total r^{-6} energy to r^{-4} polarization energy for a few ion-neutral pairs, taken at the position of the potential energy minimum

	$[\frac{1}{2}e^2\alpha_q + C^{(6)}]/(\frac{1}{2}e^2\alpha r_m^2)$		
	Li^+	K^+	Cs^+
He	0.15		0.97
Ar		0.3	
Xe			0.3

Data from Mason and Schamp (1958).

B. SHORT-RANGE FORCES. Accurate information for the short-range forces is much scarcer than for the long-range forces. Their quantum-mechanical origin is clear enough but accurate calculations are difficult. When an ion and atom or molecule are brought close enough together that their electronic charge clouds can overlap, large distortions are produced because of the requirements of the Pauli exclusion principle. If the partners originally had closed electronic shells, their electrons tend to avoid each other according to the Pauli principle, and there is a decrease of charge density in the region between the atoms. This decrease reduces the screening of the nuclear charges from each other, and the net effect is one of repulsion between the pair. If the original partners did not have closed shells, there may be an increase in charge density between them because of electron spin pairing, leading to the formation of a chemical bond. Short-range repulsive forces thus have the same origin as chemical bonds and are often called overlap or valence forces.

From a theoretical point of view short-range forces already appear in a first-order perturbation calculation. Unfortunately most of the electrons have to be treated individually in setting up appropriate wavefunctions that will satisfy the Pauli principle and the calculations thus become complicated (Margenau and Kestner, 1969, Chapter 3). Even for a simple repulsive interaction about all that can be seen in a general way in the final energy expression is exponential terms in r multiplied by polynomials in r. The simplest function that might reasonably be expected to mimic this behavior is a single exponential, and the repulsive overlap energy is therefore often written as

$$V_{\text{rep}}(r) = Ae^{-ar}, \qquad (6\text{-}2\text{-}12)$$

where A and a are constants. Empirically it is found that quantum-mechanical calculations of $V_{rep}(r)$ can usually be represented accurately over a large range of r by such an expression (see, for example, Matcha and Nesbet, 1967). For mathematical convenience $V_{rep}(r)$ is sometimes represented by an inverse power

$$V_{rep}(r) \approx \frac{B}{r^n},$$ (6-2-13)

where B and n are constants. This is often found empirically to be satisfactory for a limited range of r but it has no particular theoretical basis.

Because of the lack of direct information, it is often desirable to estimate an unknown short-range interaction from data on known interactions. Examination of the quantum-mechanical formulas leading to (6-2-12) suggests the following combination rule for the parameter a (Zener, 1931):

$$a_{12} = \tfrac{1}{2}(a_{11} + a_{22}).$$ (6-2-14)

The formulation of a combination rule for the parameter A is less clear, although simple analogy with a would suggest a geometric-mean rule. The best theoretical justification at present is based on calculations that start from the united atom at $r = 0$ and work outward; these suggest an equation for $V_{rep}(r)$ of the form

$$V_{rep}(r) = \frac{Z_1 Z_2}{r} P(r) e^{-ar},$$ (6-2-15)

where Z_1 and Z_2 are the nuclear charges and $P(r)$ is a slowly varying function of r (Byers-Brown and Power, 1970, and previous papers referred to therein). This form suggests a geometric rule for A:

$$A_{12} = (A_{11} A_{22})^{1/2}.$$ (6-2-16)

Taken together, (6-2-14) and (6-2-16) imply that $V_{rep}(r)$ itself obeys a geometric-mean combination rule, as has been approximately verified experimentally in a number of cases by the scattering of fast atomic beams in gases (e.g., Colgate et al., 1969). Presumably the rule would still work if one of the partners were an ion instead of a neutral. The geometric mean rule for $V_{rep}(r)$ gives the following combination formulas for the parameters of the inverse-power potential:

$$n_{12} = \tfrac{1}{2}(n_{11} + n_{22}), \qquad B_{12} = (B_{11} B_{22})^{1/2}.$$ (6-2-17)

Another combination rule based on the distortion of the individual charge clouds of the partners has been suggested by F. T. Smith (1972) and leads to the expressions

$$a_{12}^{-1} = \tfrac{1}{2}(a_{11}^{-1} + a_{22}^{-1}),$$ (6-2-18)

$$(a_{12} A_{12})^{2/a_{12}} = (a_{11} A_{11})^{1/a_{11}}(a_{22} A_{22})^{1/a_{22}}.$$ (6-2-19)

These formulas usually give results similar to those from the geometric mean rule of (6-2-14) and (6-2-16), and it has not yet been possible to choose between the two on the basis of available experimental results (Smith, 1972).

In electron spin pairing leading to strongly attractive short-range valence forces it is found that the interaction can often be empirically represented by exponential functions, as in simple repulsion. An example is the Morse potential, a sum of two exponentials,

$$V_{val}(r) = D_e[e^{-2\beta(r-r_e)} - 2e^{-\beta(r-r_e)}],$$ (6-2-20)

where D_e is the dissociation energy, r_e, the equilibrium bond length, and β, a parameter related to the fundamental vibrational frequency. Many other similar empirical potential functions have also been used, primarily for spectroscopic work (Varshni, 1957; Steele, Lippincott, and Vanderslice, 1962). Usually many molecular states are possible from the interaction of two ground-state particles (Mason and Monchick, 1967).

C. INTERMEDIATE-RANGE FORCES.   Although the short-range forces appear in a first-order perturbation calculation, the long-range polarization and dispersion forces appear only in a second-order calculation. Moreover, properly symmetrized wavefunctions must be used for the short-range calculations, whereas the long-range results discussed in Section 6-2-A are based on unsymmetrized wavefunctions. Clearly it would be theoretically inconsistent simply to add the two results together. A consistent calculation would use a properly symmetrized wavefunction and proceed as far as second order. This will give rise to a new set of energy terms, usually called second-order exchange terms, which are not important at small separations because the first-order terms are much larger. They die away fast enough at large separations to become unimportant compared with the polarization and dispersion terms, but they are not necessarily negligible at intermediate separations. Second-order exchange forces are discussed in Chapter 4 of the review by Margenau and Kestner (1969).

Despite this admonition, second-order exchange is usually ignored and the short-range first-order energies are simply added to the long-range second-order energies. The result is usually surprisingly successful, especially if a parameter is left adjustable. This procedure is illustrated in Section 6-4 with the calculation of the mobility of Li^+ in He.

6-3. DETERMINATION OF INTERACTION POTENTIALS

From the theoretical results sketched in the preceding section, we can usually put together a mathematical model of the interaction potential. With rare exceptions such a model will contain one or more parameters whose numerical values must be found in some way, usually by comparison with experimental data. Suitable data are the results on ion mobilities and the scattering of ion beams in gases, which are discussed in the first three subsections to follow. The last two subsections discuss methods by which the potential can be found more directly: namely by quantum-mechanical calculations and by spectroscopic observations.

A. MOBILITY AND DIFFUSION DATA.   We have already pointed out in Section 5-3-D that the temperature dependence of the low-field mobility is sensitive enough to the ion-neutral interaction to furnish information on it. Combination of one or more attraction terms with a short-range repulsion term yields a simple model containing one or more adjustable parameters whose values can be determined by comparison with mobility data. This adjustability hopefully compensates for completely ignoring the inter-mediate-range forces in setting up the model. Some early attempts along this line used the (∞-4) and (8-4) models (Tyndall, 1938, Chapter 6). Dalgarno, McDowell, and Williams (1958) used an exponential repulsion plus polarization attraction model and evaluated the parameters for Li^+, Na^+, and Cs^+ ions in He. Their numerical calculation of the collision integrals was, however, only approximate. A more elaborate model was put together from polarization and dispersion attraction terms plus an inverse twelfth power repulsion (Mason and Schamp, 1958) and the para-meters for this (12-6-4) model found for a number of ion-neutral pairs. In practice, three adjustable parameters are too many to be determined from mobility data, but the three can be reduced effectively to one para-meter freely adjustable and one parameter adjustable only within rather narrow limits by using an accurate theoretical value for the coefficient of the r^{-4} term and an approximate value for the coefficient of the r^{-6} term. Parameter values so determined are given in Appendix II (Table II-6). They can be used to calculate mobility values over a fairly wide temperature range and the dependence of the mobility on field strength.

If more accurate data were available on the variation of mobility with field strength, they could also be used to determine ion-neutral potential parameters (Section 5-4-E).

B. ELASTIC SCATTERING OF ION BEAMS.   The scattering of a beam of ions by neutral gas obviously depends on the ion-neutral interaction, and a study of such scattering can lead to information of the potential. Many such experiments furnish a useful source of material for ion-neutral potentials. The energy range of the experiments runs from a few electron volts into the megaelectron-volt regime and thus obviously probes the short-range portion of the potential. The experiments do not extend to lower energies because of the difficulty of producing and manipulating ion beams at energies below about 1 eV; at such low energies the techniques change to swarm methods and drift tubes and in effect measure mobilities.

The magnitude of the short-range potential in the range of interest for mobilities is only of the order of 1 eV and lower. It may seem at first thought surprising that such information can be obtained from the scattering of ion beams of kinetic energies greater than 1 eV—up to about 10^3 eV, in fact. The trick is to study only the small-angle scattering of an energetic beam; that is, to deflect a beam particle through a small angle θ requires that the ratio of the potential energy at the distance of closest approach to the kinetic energy be of order θ. Study of deflections of 10^{-3} rad in a beam of 10^3-eV kinetic energy thus gives information on the potential energy in the range of about $(10^{-3})(10^3 \text{ eV}) = 1$ eV. The upper limit of ion beam energies of interest here is thus about 10^3 eV because of the difficulty of observing angular deflections of less than about 10^{-3} rad.

A detailed description of the analysis of ion-beam elastic scattering would require at least a chapter of its own. The subject is treated briefly in a number of monographs (Massey and Burhop, 1952, Chapter 8; Hasted, 1964, Sections 10.1 and 10.2; McDaniel, 1964, Sections 4-8 and 4-9). A more detailed review has been given by Mason and Vanderslice (1962), who also review work on neutral beams. Other reviews of neutral beam work have been published by Amdur and Jordan (1966), by Amdur, Jordan, and Mason (1969), and by Jordan, Mason, and Amdur (1972). Neutral beam studies may be helpful because a short-range ion-neutral interaction can sometimes be approximated by an analogous neutral-neutral interaction, suitably scaled (e.g., $K^+ - Ar$ by $Ar - Ar$).

A resumé of available work on short-range ion-neutral potentials appears in Appendix II (Table II-7). The use of these results is illustrated in Section 6-4 with the calculation of the mobility of Li^+ in He.

C. RESONANT CHARGE-TRANSFER DATA.   Although charge transfer data come from measurements with ion beams, the topic deserves individual mention because of the special relation of the diffusion cross section to the charge transfer cross section

$$Q^{(1)} \approx 2Q_T. \tag{6-3-1}$$

This relation was discussed in some detail in Section 5-7, where it was also pointed out that $Q_T^{1/2}$ was very nearly linear in $\ln E$,

$$Q_T^{1/2} \approx a_1 - a_2 \ln E, \tag{6-3-2}$$

where a_1 and a_2 are constants. Thus, if Q_T is measured experimentally as a function of E by beam techniques, it is possible to pass directly to the mobility without having to calculate the ion-neutral interaction at all. It is necessary only to fit the data by an equation of the form of (6-3-2), and the mobility is then given in terms of a_1 and a_2 by (5-7-54).

Resonant charge transfer is treated in a number of standard references (Massey and Burhop, 1952, Chapter 8; Hasted, 1964, Chapter 12; McDaniel, 1964, Section 6-2; Hasted, 1962, 1968). A review has also been given by Fite (1964). When no experimental data exist, recourse may be made to various semiempirical schemes for predicting the values of a_1 and a_2 (Firsov, 1951; Gurnee and Magee, 1957; Rapp and Francis, 1962; Hasted, 1968). Some references for specific systems are given in Table II-7 of Appendix II.

D. QUANTUM-MECHANICAL CALCULATIONS. Interaction potentials are calculable in principle from the Schrödinger equation and the fundamental constants, but such calculations soon bog down in computational difficulties. Approximations are made and three levels may be distinguished: *ab initio*, semiempirical, and model calculations. A review is available (Mason and Monchick, 1967).

In *ab initio* calculations the only experimental numbers used are the values of the physical constants. Accurate calculations are available for systems with a small number of electrons, such as H^+-He and Li^+-He, and less accurate calculations for a few systems with more electrons, such as H^+-Ne and Na^+-He. References for specific systems are listed in Table II-7 of Appendix II.

Semiempirical calculations judiciously mix in extra experimental information, such as polarizabilities or oscillator strengths. Most of the best calculations of long-range interactions discussed in Section 6-2-A and summarized in the tables of Appendix II fall in this category. Semiempirical calculations of short-range interactions have been most useful when incomplete electron shells are involved, and many potential energy curves arise from a single ground-state ion-neutral pair. We use a simple form of first-order perturbation theory to derive relationships among the various potential curves. If several of the curves are already known, say from spectroscopic data, the relations can be used to predict the remaining curves. This procedure has been used rather extensively for neutral-neutral interactions (Mason and Monchick, 1967), but only a few ion-neutral systems, included in the summary in Table II-7 of Appendix II, have been treated in this way.

In model calculations the original physical problem is replaced by one that is more tractable mathematically. A good model accurately mimics all the important features of the original problem and discards only the useless complications; needless to say, such models are few and far between. An example of a model calculation has been given in Section 6-2-A, in which a harmonic-oscillator model was used to estimate quadrupole polarizabilities. Two models have proved to be of some use for short-range interactions: delta-function models and statistical models. In a delta-function model the Coulomb wells at the nuclei are replaced by square wells, which are eventually collapsed into delta functions. Electron-electron interactions are ignored and a one-dimensional treatment is used. Surprisingly good results have been obtained for short-range interactions between neutral atoms (Mason and Vanderslice, 1958), but the extension to ions has so far included only two-electron ions and atoms (Weber, 1964). In a statistical model the electrons are treated as a Fermi gas. The method has been given only limited application to ion-neutral interactions (Sida, 1957) but has been extensively applied to atom-atom interactions (Abrahamson, 1969). Recent improvements in the method by Gordon and Kim (1972) have yielded promising results in both the short-range and intermediate-range regions.

E. SPECTROSCOPIC OBSERVATIONS. When an ion-atom interaction involves a valence attraction, it may be possible to form a stable molecular ion whose vibration-rotation bands can be observed spectroscopically. The location of the vibrational-rotational energy levels is determined by the shape of the potential energy curve, which can be determined by inversion of the energy level data by the Rydberg-Klein-Rees (RKR) method (Mason and Monchick, 1967). The RKR method has been applied to a large number of neutral species but to only a few ionic species (Gilmore, 1965), which are listed in Table II-7 of Appendix II.

6-4. ESTIMATION OF MOBILITIES FROM MEAGER DATA

In the preceding sections we discussed the calculation of mobilities when the ion-neutral interaction was rather accurately known. It frequently happens that the interaction is known only poorly or must even be guessed. In this section we show by three examples how mobilities can be estimated in the face of meager information. The methods are also useful when the ion-neutral interaction is accurately known, but only an approximate mobility value is needed, for which the full computational techniques discussed in Section 6-1 are too laborious.

The examples in this section also serve to indicate the sort of accuracy that can be obtained in mobility estimates. Since complete theoretical calculations for the same systems are available in the literature, we obtain further insight concerning the errors introduced by the simplification of the mobility calculations, compared with those introduced by uncertainties in the interactions.

The most important step is to find the nature of the short-range ion-neutral interaction. This controls the mobility at high temperatures and determines the nature of the temperature dependence at intermediate temperatures (Section 5-3-D). The three major short-range interactions are repulsion, resonant charge transfer, and valence attraction, illustrated, respectively, by Li^+ in He, He^+ in He, and H^+ in He. Although all these interactions are in reality known accurately, we pretend to have only limited information for purposes of illustration.

A. SHORT-RANGE REPULSION: Li^+ IN He. The nature of the interaction here is indicated by the fact that both Li^+ and He have closed electronic shells and by the fact that the room-temperature mobility is higher than the polarization limit. The polarization limit is readily calculated from (5-3-29), taking the polarizability of He from Table II-1 of Appendix II; it is found to be $K_p = 19.2$ cm^2/V-sec. First we pretend that no measured mobility values are available and that the only information on the short-range forces comes from ion-beam scattering measurements or from *ab initio* quantum-mechanical calculations. It happens that there are a number of these, all in good agreement (summarized in Table II-7 of Appendix II); we choose the calculations of Fischer (1968), who fitted his results by the formula

$$V_{rep}(r) = 580e^{-5.4r} \text{ eV}, \qquad r \text{ in Å}, \qquad (6\text{-}4\text{-}1)$$

to which we add the dominant term of the polarization energy to produce the potential

$$V(r) = 580e^{-5.4r} - \frac{1.48}{r^4} \text{ eV}, \qquad r \text{ in Å}. \qquad (6\text{-}4\text{-}2)$$

It is probably not worth the trouble to include the r^{-6} attraction term, since of necessity we are ignoring all the second-order exchange energy anyway. Even this (exp-4) potential presents a large numerical challenge, since the collision integrals for it have never been accurately evaluated. We therefore replace it with an $(n\text{-}4)$ potential matched near the potential minimum. The depth and position of the minimum of (6-4-2) are $\varepsilon = 0.123$ eV and $r_m = 1.59$ Å; we keep the value of ε and the coefficient of the r^{-4} term correct in the $(n\text{-}4)$ potential, thereby letting the value of r_m shift to 1.68 Å for a (12-4) potential and to 1.57 Å for an (8-4) potential. Clearly the (8-4) potential is a better representation; the values of K_0 as a function of temperature are now easily calculated from (5-3-18) with the aid of the collision integrals in Table I-4 of Appendix I. The results are shown in Fig. 6-4-1; comparison with measured mobilities shows agreement within 10%.

If there had been no direct results on the short-range repulsion of Li^+

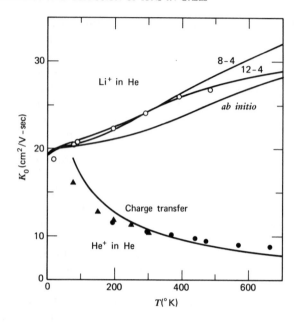

FIG. 6-4-1. Examples of estimation methods for ion mobilities. Short-range repulsion is illustrated by Li^+ in He; the *ab initio* curve is based on quantum-mechanical calculations and the other two curves use the measured point near 300°K. Resonant charge transfer is illustrated by He^+ in He; the calculated curve is based on ion beam measurements of the charge transfer cross section. Experimental points are from Hoselitz ($\bigcirc$, 1941), Orient ($\bullet$, 1967), and Patterson ($\blacktriangle$, 1970).

in He (as there probably would not have been for more complicated particles), we could still have estimated it from one mobility measurement not too close to the polarization limit. Let us pretend that one measurement at 291°K had yielded a value of $K_0 = 24.2$ cm²/V-sec (Hoselitz, 1941). Its ratio to the polarization value is 1.258. Interpolating in the collision integral tables of Appendix I, we find that this ratio corresponds to kT/ε of 0.298 for the (8-4) potential and 0.364 for the (12-4) potential. From these values plus the coefficient of the r^{-4} polarization potential we find $\varepsilon = 0.084$ eV and $r_m = 1.72$ Å for the (8-4) potential and $\varepsilon = 0.069$ eV and $r_m = 1.95$ Å for the (12-4) potential. The mobilities are now easily calculated as before and are shown in Fig. 6-4-1; the agreement with experiment is now within 5%.

The accuracy obtained by these simple estimates is comparable to that of the complete theoretical calculations (Catlow et al., 1970).

B. RESONANT CHARGE TRANSFER: He^+ IN He. The nature of the interaction here is indicated by the electronic structures of He^+ and He and by the fact that the room-temperature mobility is lower than the polarization limit. The polarization limit is readily calculated to be $K_p = 21.7$ cm²/V-sec.

The charge-transfer cross section has been measured for ion beam energies between 4 and 400 eV by Cramer and Simons (1957), and we find that their results can be fitted by the expression

$$Q_T^{1/2} = 5.00 - 0.71 \log_{10} E, \qquad (6\text{-}4\text{-}3)$$

where Q_T is in square angstroms and E is the relative energy in electron volts. From this expression the mobilities are easily calculated according to (5-7-54), with the results shown in Fig. 6-4-1; the worst disagreement with experiment is about 16% at 76°K. Comparison should be made with the full quantum-mechanical calculations in Fig. 5-7-5; the accuracies are comparable.

C. VALENCE ATTRACTION: H^+ IN He.   The nature of the interaction is indicated by simple considerations of electronic structure (e.g., HeH$^+$ is isoelectronic with H_2) and by the fact that HeH$^+$ appears as a stable ion in the mass spectrograph. The polarization limit is calculated to be $K_p = 34.1$ cm^2/V-sec. Although the references in Appendix II show that the potential energy curve is accurately known, for the sake of illustration let us pretend that we know only that the well depth is about $\varepsilon = 2.0$ eV. If we fit a (12-4) potential to this well depth and keep the correct coefficient of the r^{-4} polarization energy, we calculate the curve marked "classical" in Fig. 5-3-6. Actually we expect the curve at 300°K to be a few percent lower than the polarization limit, according to the discussion in Section 5-3-D, but the crude calculation we are allowing ourselves here does not admit this. Even so, the agreement with experiment is within 10% and within 5% of the full quantum-mechanical calculations in Fig. 5-3-6 except at very low temperatures.

Further examples of mobility estimates, involving more complicated systems, are given by Mason (1970) and Dukowicz (1970).

REFERENCES

Abrahamson, A. A. (1969), *Phys. Rev.* **178**, 76.

Amdur, I., and J. E. Jordan (1966), *Adv. Chem. Phys.* **10**, 29.

———, and E. A. Mason (1969), *Entropie* **30**, 135.

Barker, J. A., W. Fock, and F. Smith (1964), *Phys. Fluids* **7**, 897.

Buckingham, R. A., J. W. Fox, and E. Gal (1965), *Proc. Roy. Soc. (London)* **A284**, 237.

Byers-Brown, W., and J. D. Power (1970), *Proc. Roy. Soc. (London)* **A317**, 545.

Catlow, G. W., M. R. C. McDowell, J. J. Kaufman, L. M. Sachs, and E. S. Chang (1970), *J. Phys.* **B3**, 833.

Colgate, S. O., J. E. Jordan, I. Amdur, and E. A. Mason (1969), *J. Chem. Phys.* **51**, 968.

Cramer, W. H., and J. H. Simons (1957), *J. Chem. Phys.* **26**, 1272.

Dalgarno, A. (1967), *Adv. Chem. Phys.* **12**, 143.

264   THE MOBILITY AND DIFFUSION OF IONS IN GASES

Dalgarno, A., M. R. C. McDowell, and A. Williams, (1958), *Phil. Trans. Roy. Soc. (London)* **A250**, 411.

de Boer, J., and J. van Kranendonk (1948), *Physica* **14**, 442.

De Rocco, A. G., T. S. Storvick, and T. H. Spurling (1968), *J. Chem. Phys.* **48**, 997.

Dukowicz, J. K. (1970), *AIAA J.* **8**, 827.

Dymond, J. H., M. Rigby, and E. B. Smith (1966), *Phys. Fluids* **9**, 1222.

Firsov, O. B. (1951), *Zh. Eksper. Teor. Fiz.* **21**, 1001.

Fischer, C. R. (1968), *J. Chem. Phys.* **48**, 215.

Fite, W. L. (1964), *Ann. Géophysique* **20**, 47.

Ford, K. W., and J. A. Wheeler (1959a), *Ann. Phys. (N. Y.)* **7**, 259.

——— (1959b), *ibid.*, 287.

Gilmore, F. R. (1965), *J. Quant. Spectry. Radiat. Transfer* **5**, 369.

Gordon, R. G., and Y. S. Kim (1972), *J. Chem. Phys.* **56**, 3122.

Gurnee, E. F., and J. L. Magee (1957), *J. Chem. Phys.* **26**, 1237.

Hassé, H. R. (1926), *Phil. Mag.* **1**, 139.

———, and W. R. Cook (1927), *Phil. Mag.* **3**, 977.

——— (1929), *Proc. Roy. Soc. (London)* **A125**, 196.

——— (1931), *Phil. Mag.* **12**, 554.

Hasted, J. B. (1962), "Charge Transfer and Collisional Detachment," in *Atomic and Molecular Processes*, D. R. Bates, Ed., Chapter 18, Academic, New York.

——— (1964), *Physics of Atomic Collisions*, Butterworths, Washington, D.C.

——— (1968), *Adv. Atom. Mol. Phys.* **4**, 237.

Hirschfelder, J. O., Ed. (1967), *Intermolecular Forces*, Wiley, New York.

———, R. B. Bird, and E. L. Spotz (1948), *J. Chem. Phys.* **16**, 968.

Hirschfelder, J. O., C. F. Curtiss, and R. B. Bird (1964), *Molecular Theory of Gases and Liquids*, Wiley, New York.

Holleran, E. M., and H. M. Hulburt (1951), *J. Chem. Phys.* **19**, 232.

Hoselitz, K. (1941), *Proc. Roy. Soc. (London)* **A177**, 200.

Iman-Rahajoe, S., C. F. Curtiss, and R. B. Bernstein (1965), *J. Chem. Phys.* **42**, 530.

Itean, E. C., A. R. Glueck, and R. A. Svehla (1961), *NASA Tech. Note* **D-481**.

Jordan, J. E., E. A. Mason, and I. Amdur (1972), "Molecular Beams in Chemistry," in *Physical Methods of Chemistry, Part III*, A. Weissberger and B. W. Rossiter, Eds., Chapter 6, Wiley, New York.

Kennedy, M., and F. J. Smith (1967), *Mol. Phys.* **13**, 443.

Kihara, T., and M. Kotani (1943), *Proc. Phys.-Math. Soc. Japan* **25**, 602.

Kihara, T., M. H. Taylor, and J. O. Hirschfelder (1960), *Phys. Fluids* **3**, 715.

Kirkwood, J. G. (1932), *Physik. Z.* **33**, 57.

Kotani, M. (1942), *Proc. Phys.-Math. Soc. Japan* **24**, 76.

Kramer, H. L., and D. R. Herschbach (1970), *J. Chem. Phys.* **53**, 2792.

Langer, R. E. (1937), *Phys. Rev.* **51**, 669.

Langevin, P. (1905), *Ann. Chim. Phys.* **5**, 245. A translation is given in E. W. McDaniel, *Collision Phenomena in Ionized Gases*, Wiley, New York 1964, Appendix II.

Livingston, P. M. (1966), *J. Chem. Phys.* **45**, 601.

Margenau, H. (1941), *Phil. Sci.* **8**, 603.

—— and N. R. Kestner (1969), *Theory of Intermolecular Forces*, Pergamon, New York.

Mason, E. A. (1954), *J. Chem. Phys.* **22**, 169.

—— (1970), *Planet. Space Sci.* **18**, 137.

—— and L. Monchick (1967), *Adv. Chem. Phys.* **12**, 329.

Mason, E. A., R. J. Munn, and F. J. Smith (1967), *Phys. Fluids* **10**, 1827.

—— (1971), *Endeavour* **30**, 91.

Mason, E. A., and H. W. Schamp (1958), *Ann. Phys. (N. Y.)* **4**, 233.

Mason, E. A., and J. T. Vanderslice (1958), *J. Chem. Phys.* **28**, 432.

—— (1962), "High-Energy Elastic Scattering of Atoms, Molecules, and Ions," in *Atomic and Molecular Processes*, D. R. Bates, Ed., Chapter 17, Academic, New York.

Matcha, R. L., and R. K. Nesbet (1967), *Phys. Rev.* **160**, 72.

Massey, H. S. W., and E. H. S. Burhop (1952), *Electronic and Ionic Impact Phenomena*, Oxford University Press, London.

Maxwell, J. C. (1867), *Phil. Trans. Roy. Soc. (London)* **157**, 49. Reprinted in *The Scientific Papers of James Clerk Maxwell*, Vol. 2, Dover, New York, 1962, pp. 26–78, and in S. G. Brush, *Kinetic Theory*, Vol. 2, Pergamon, New York, 1966, pp. 23–87.

McDaniel, E. W. (1964), *Collision Phenomena in Ionized Gases*, Wiley, New York.

Monchick, L. (1959), *Phys. Fluids* **2**, 695.

—— and E. A. Mason (1961), *J. Chem. Phys.* **35**, 1676.

Munn, R. J., E. A. Mason, and F. J. Smith (1964), *J. Chem. Phys.* **41**, 3978.

Munn, R. J., F. J. Smith, E. A. Mason, and L. Monchick (1965), *J. Chem. Phys.* **42**, 537.

O'Hara, H., and F. J. Smith (1970), *J. Comput. Phys.* **5**, 328.

—— (1971), *Comput. Phys. Comm.* **2**, 47.

Orient, O. J. (1967), *Can. J. Phys.* **45**, 3915.

—— and E. A. Mason (1961), *J. Chem. Phys.* **35**, 1676.

Patterson, P. L. (1970), *Phys. Rev.* **A2**, 1154.

Rapp, D., and W. E. Francis (1962), *J. Chem. Phys.* **37**, 2631.

Samoilov, E. V., and N. N. Tsitelauri (1964), *High Temp. (English Transl.)* **2**, 509 [*Tepl. Vys. Temp.* **2**, 565 (1964)].

Sida, D. W. (1957), *Phil. Mag.* **2**, 761.

Slater, J. C., and J. G. Kirkwood (1931), *Phys. Rev.* **37**, 682.

Smith, F. J., and R. J. Munn (1964), *J. Chem. Phys.* **41**, 3560.

Smith, F. T. (1965), *J. Chem. Phys.* **42**, 2419.

—— (1972), *Phys. Rev.* **A5**, 1708.

Steele, D., E. R. Lippincott, and J. T. Vanderslice (1962), *Rev. Mod. Phys.* **34**, 239.

Tyndall, A. M. (1938), *The Mobility of Positive Ions in Gases*, Cambridge University Press, London.

Varshni, Y. P. (1957), *Rev. Mod. Phys.* **29**, 664.

Weber, G. G. (1964), *J. Chem. Phys.* **40**, 1762.

Wilson, J. N. (1965), *J. Chem. Phys.* **43**, 2564.

Zener, C. (1931), *Phys. Rev.* **37**, 556.

7

EXPERIMENTAL DATA ON

MOBILITIES AND

DIFFUSION COEFFICIENTS

This chapter contains a collection of experimental data on the mobility and diffusion of ions in gases. The data compilation is by no means exhaustive—in fact, an effort has been made in most cases to include only those data that are considered reliable in the light of recent developments. Omitted are the results of many experiments that were useful in the development of the subject but have been superseded by later results or proved to be difficult to interpret. Many data were excluded because the identity of the ions was not at least reasonably well established or because the effects of reactions could not be properly assessed. The following abbreviations denote the experimental apparatus used to obtain the data: DT, drift tube; DTMS, drift-tube mass spectrometer; SA, stationary afterglow; SAMS, stationary-afterglow mass spectrometer. The reader is cautioned not to accept the indicated identification of the ions as definitely established unless a mass spectrometer had been employed in the measurements.

It should be pointed out that several different notations for molecular ions appear in the literature; for example, N_4^+, $K^+ \cdot N_2$, and $SF_6^-(SF_6)_2$. In general here in each instance we use the notation employed by the authors of the original papers from which the data were taken.

266

7-1. THE MOBILITY OF IONS IN PURE GASES AT OR NEAR ROOM TEMPERATURE; THE MOBILITY OF IONS IN VAPORS

In this section we present a large amount of data on mobilities and drift velocities of ions in pure gases at or near room temperature and a smaller collection of mobilities of ions in vapors. The temperatures at which the latter data were taken were determined by vapor pressure considerations. The data are arranged according to the gas or vapor in which the measurements were made. Mobilities of ions in gas mixtures are presented in Section 7-2, and the variation of mobilities with gas temperature is treated in Section 7-3.

Most of the data presented here are in the form of the *reduced mobility* K_0, which is defined by (1-3-1):

$$K_0 = \frac{p}{760} \frac{273.16}{T} K.$$

Here p is the gas pressure in Torr and T is the gas temperature in degrees Kelvin at which the measured mobility K was obtained. Most of these data were obtained from drift-tube measurements. Some of the zero-field reduced mobilities are derived from measurements of the ambipolar diffusion coefficient D_a under conditions in which the electron and ion temperatures were equal to the gas temperature. In these cases K_0 was calculated from (1-12-10):

$$K_0 = \frac{D_a p}{T^2} 2.086 \times 10^3.$$

The data displayed here are of uneven quality. In many cases accurate measurements have been made with apparatus incorporating mass spectrometers to ensure that the identity of the ions is certain and that no difficulties of interpretation will develop because of reactions of the ions with the gas molecules. In others no mass spectrometer was employed but the reaction pattern was simple enough to permit little doubt concerning the identity of the ions and the fact that the data refer to a single ionic species. In a few in which experiments are difficult, such as in metal vapors, the accuracy and precision of the data leave something to be desired and the identity of the ions may be in doubt. These data, however, are so important that they warrant inclusion. The experimenter's estimate of probable error is provided whenever available.

In most instances mobilities obtained directly with drift tubes are to be preferred to those derived from ambipolar diffusion measurements. However, data from the latter kind of measurements are presented here when drift-tube data are unavailable or to illustrate the quality of agreement that can be obtained with data from the two kinds of experiment.

The results of theoretical calculations of mobilities are interspersed throughout Chapters 5 and 6.

A. HELIUM

TABLE 7-1-A-1. Experimental zero-field reduced mobility K_0 in square centimeters per volt per second of ions in helium (the data were taken at a temperature of $300°K$ unless otherwise specified)

Ion	K_0	$T(°K)$	Method	Reference
He^+	10.40 ± 0.10		DT	Beaty and Patterson (1965); Patterson (1970); see Figs. 7-1-A-1, 7-1-A-2
He_2^+	16.70 ± 0.17		DT	Beaty and Patterson (1965); Patterson (1970); see Fig. 7-1-A-1
$He_2^{+m}(^4\Sigma_u^+)$ (metastable)	19.6 ± 0.3		DT	Beaty and Patterson (1965); see Fig. 7-1-A-1
Ne^+	24 ± 3	380	SAMS	Sauter, Gerber, and Oskam (1966)
$HeNe^+$	20 ± 2	380	SAMS	Sauter, Gerber, and Oskam (1966)
Ne_2^+	17.5		DT	Biondi and Chanin (1961)
Ar^+	19.5		SAMS	Veatch and Oskam (1970)
Kr^+	20.2		SA	Chen (1963)
Xe^+	18		SA	Chen (1963)
H^+	31.8		DTMS	Orient (1971)
D^+	24.9		DTMS	Orient (1972)
Hg^+	19.6		DT	Chanin and Biondi (1957)
	19.4 ± 0.5	292	DTMS	Johnsen and Biondi (1972)
	19.8	291	SA	Biondi (1953)
N^+	20 ± 2		DTMS	Johnsen, Brown, and Biondi (1970)

TABLE 7-1-A-1 (*continued*)

Ion	K_0	$T(°K)$	Method	Reference
N_2^+	19 ± 2		DTMS	Johnsen, Brown, and Biondi (1970)
O^+	22 ± 2		DTMS	Johnsen, Brown, and Biondi (1970)
O_2^+	21 ± 2		DTMS	Johnsen, Brown, and Biondi (1970)
Li^+	24.2	291	DT	Tyndall (1938)
Na^+	22.7	291	DT	Tyndall (1938)
	20.7 ± 1	300	DTMS	Johnsen, Brown, and Biondi (1971)
K^+	21.5	291	DT	Tyndall (1938)
	19.3 ± 1		DTMS	Johnsen, Brown, and Biondi (1972)
Rb^+	20.1	291	DT	Tyndall (1938)
Cs^+	18.4	291	DT	Tyndall (1938)
	18.5		SA	Chen and Raether (1962)
U^+	16.0 ± 0.5	305	DTMS	Johnsen and Biondi (1972)
SF_6^-	9.2		DTMS	Patterson (1972)
SF_5^-	11		DTMS	Patterson (1972)

REFERENCES

Beaty, E. C., and P. L. Patterson (1965), *Phys. Rev.* **137**, A346.
Biondi, M. A. (1953), *Phys. Rev.* **90**, 730.
———, and L. M. Chanin (1961), *Phys. Rev.* **122**, 843.
Chanin, L. M., and M. A. Biondi (1957), *Phys. Rev.* **107**, 1219.
Chen, C. L. (1963), *Phys. Rev.* **131**, 2550.
———, and M. Raether (1962), *Phys. Rev.* **128**, 2679.
Johnsen, R., H. L. Brown, and M. A. Biondi (1970), *J. Chem. Phys.* **52**, 5080.
——— (1971), *J. Chem. Phys.* **55**, 186.
——— (1972), private communication.
Johnsen, R., and M. A. Biondi (1972), *J. Chem. Phys.* **57**, 5292.
Orient, O. J. (1971), *J. Phys. B* **4**, 1257.
——— (1972), *J. Phys. B* **5**, 1056.
Patterson, P. L. (1970), *Phys. Rev. A* **2**, 1154.
——— (1972), *J. Chem. Phys.* **56**, 3943.
Sauter, G. F., R. A. Gerber, and H. J. Oskam (1966), *Physica* **32**, 1921.
Tyndall, A. M. (1938), *The Mobility of Positive Ions in Gases*, Cambridge University Press, Cambridge, p. 92.
Veatch, G. E., and H. J. Oskam (1970), *Phys. Rev. A* **1**, 1498.

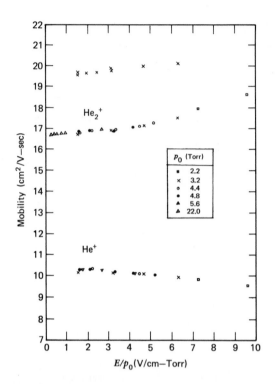

FIG. 7-1-A-1. The reduced mobilities of three helium ions in helium at 300°K plotted as functions of E/p_0. [E. C. Beaty and P. L. Patterson (1965), *Phys. Rev.* **137**, A346. DT]. The mass of the slowest ion was shown to be 4 amu and the mass of both the faster ions was shown to be 8 amu by J. M. Madson, H. J. Oskam, and L. M. Chanin (1965), *Phys. Rev. Letters* **15**, 1018-DTMS. It has been suggested that the fastest ion is the metastable $^4\Sigma_u^+$ state of He_2^+ by E. C. Beaty, J. C. Browne, and A. Dalgarno (1966), *Phys. Rev. Letters* **16**, 723.

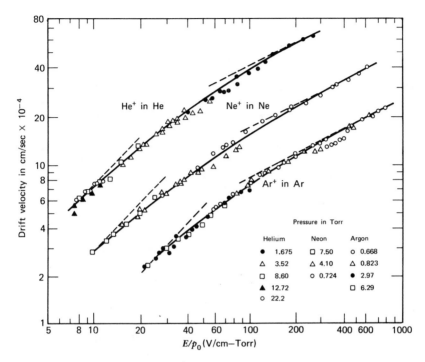

FIG. 7-1-A-2. The drift velocity of He$^+$ ions in helium at 300°K as a function of E/p_0. Data on Ne$^+$ in neon and Ar$^+$ in argon are included for comparison [J. A. Hornbeck (1951), *Phys. Rev.* **84**, 615. DT]. The broken lines at the left of each experimental curve have slope = 1, whereas the broken lines at the right have slope = $\frac{1}{2}$. The high-field behavior is consistent with Wannier's predictions for a constant mean free path situation [see the discussion following (5-2-14)]. These data are not so accurate as the later data shown in Figs. 7-1-A-1, 7-1-B-1, and 7-1-C-1, but they are of special interest because they extend to high values of E/p_0.

B. NEON

TABLE 7-1-B-1. Experimental zero-field reduced mobility K_0 in square centimeters per volt per second of ions in neon (the data were taken at a temperature of 300°K unless otherwise specified)

Ion	K_0	$T(°K)$	Method	Reference
Ne^+	4.07		DT	Beaty and Patterson (1968); see Fig. 7-1-B-1
	4.2 ± 0.3		SAMS	Smith et al. (1972)
	4.18 ± 0.23		SA	Biondi (1954)
	4.1 ± 0.3		SAMS	Märk and Oskam (1971)
Ne_2^+	6.14		DT	Beaty and Patterson (1968); see Fig. 7-1-B-1
He^+	17.2		DT	Courville and Biondi (1962)
He_2^+	9.3		DT	Biondi and Chanin (1961)
Hg^+	5.95		DT	Chanin and Biondi (1957)
Li^+	11.1	291	DT	Tyndall (1938)
Na^+	8.16	291	DT	Tyndall (1938)
K^+	7.42	294	DT	Crompton and Elford (1959)
	7.50	291	DT	Tyndall (1938)
Rb^+	6.73	291	DT	Tyndall (1938)
Cs^+	6.1	291	DT	Tyndall (1938)
N_2^+	8.9 ± 0.6		SAMS	Märk and Oskam (1971)

REFERENCES

Beaty, E. C., and P. L. Patterson (1968), *Phys. Rev.* **170**, 116.
Biondi, M. A. (1954), *Phys. Rev.* **93**, 1136.
———, and L. M. Chanin (1961), *Phys. Rev.* **122**, 843.
Chanin, L. M., and M. A. Biondi (1957), *Phys. Rev.* **107**, 1219.
Courville, G. E., and M. A. Biondi (1962), *J. Chem. Phys.* **37**, 616.
Crompton, R. W., and M. T. Elford (1959), *Proc. Phys. Soc. (London)* **74**, 497.
Märk, T. D., and H. J. Oskam (1971), *Z. Physik* **247**, 84.
Smith, D., A. G. Dean, and N. G. Adams (1972), *Z. Physik* **253**, 191.
Tyndall, A. M. (1938), *The Mobility of Positive Ions in Gases*, Cambridge University Press, Cambridge, p. 92.

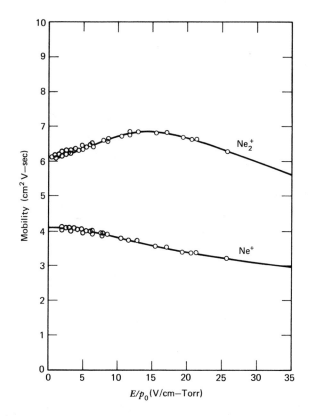

FIG. 7-1-B-1. The reduced mobilities of Ne$^+$ and Ne$_2^+$ ions in neon at 300°K plotted as functions of E/p_0 [E. C. Beaty and P. L. Patterson (1968), *Phys. Rev.* **170**, 116. DT]. Less accurate data on Ne$^+$ ions in neon which extend to high values of E/p_0 are shown in Fig. 7-1-A-2.

C. ARGON

TABLE 7-1-C-1. Experimental zero-field reduced mobility K_0 in square centimeters per volt per second of ions in argon (the data were taken at a temperature of $300°K$ unless otherwise specified)

Ion	K_0	$T(°K)$	Method	Reference
Ar^+	1.535 ± 0.007	296	DT	Beaty (1962); see Fig. 7-1-C-1
	1.63 ± 0.11		SA	Biondi (1954)
	1.41 ± 0.12		SAMS	Smith et al. (1972)
Ar_2^+	1.833 ± 0.008	296	DT	Beaty (1962); see Fig. 7-1-C-1
Ar^{2+}	2.60 ± 0.02	296	DT	Beaty (1962); see Fig. 7-1-C-1
Hg^+	1.84		DT	Chanin and Biondi (1957)
H^+	5.75		DTMS	McAfee, Sipler, and Edelson (1967)
ArH^+	1.70		DTMS	McAfee, Sipler, and Edelson (1967)
Kr^+	2.30		DTMS	McAfee, Sipler, and Edelson (1967)
Li^+	4.68	291	DT	Tyndall (1938)
Na^+	3.02	291	DT	Tyndall (1938)
K^+	2.66 ± 0.05		DTMS	James et al. (1973); see Fig. 7-1-C-2
	2.63	291	DT	Tyndall (1938)
	2.73 ± 0.17	310	DTMS	Keller et al. (1973)
Rb^+	2.25	291	DT	Tyndall (1938)
Cs^+	2.10	291	DT	Tyndall (1938)
	2.1		DTMS	McKnight and Sawina (1972)
H_3O^+	3.0	337	DTMS	Young, Edelson, and Falconer (1970)
$H_3O^+ \cdot H_2O$	2.5	337	DTMS	Young, Edelson, and Falconer (1970)
$H_3O^+ \cdot 2H_2O$	2.2	337	DTMS	Young, Edelson, and Falconer (1970)
$H_3O^+ \cdot 3H_2O$	2.0	337	DTMS	Young, Edelson, and Falconer (1970)
ReO_3^-	1.94	295	DTMS	Center (1972)
ReO_4^-	1.67	295	DTMS	Center (1972)
WO_3^-	1.96	295	DTMS	Center (1972)

REFERENCES

Beaty, E. C. (1962), *Proc. 5th Intern. Conf. Ionization Phenomena Gases* (Munich, 1961), Vol. 1, p. 183, North-Holland, Amsterdam.

Biondi, M. A. (1954), *Phys. Rev.* **93**, 1136.

Center, R. E. (1972), *J. Chem. Phys.* **56**, 371.

Chanin, L. M., and M. A. Biondi (1957), *Phys. Rev.* **107**, 1219.

James, D. R., E. Graham, G. M. Thomson, I. R. Gatland, and E. W. McDaniel (1973), *J. Chem. Phys.*, **58**.

Keller, G. E., R. A. Beyer, and L. M. Colonna-Romano (1973) *Phys. Rev. A.* to be published.

McAfee, K. B., D. Sipler, and D. Edelson (1967), *Phys. Rev.* **160**, 130.

McKnight, L. G., and J. M. Sawina (1972), *J. Chem. Phys.* **57**, 5176.

Smith, D., A. G. Dean, and N. G. Adams (1972), *Z. Physik* **253**, 191.

Tyndall, A. M. (1938), *The Mobility of Positive Ions in Gases*, Cambridge University Press, Cambridge, p. 92.

Young, C. E., D. Edelson, and W. E. Falconer (1970), *J. Chem. Phys.* **53**, 4295.

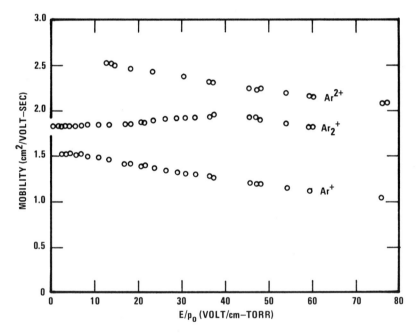

FIG. 7-1-C-1. The reduced mobilities of Ar^+, $Ar_2{}^+$, and Ar^{2+} ions in argon at 296°K plotted as functions of E/p_0 [E. C. Beaty (1962), *Proc. 5th Int. Conf. Ionization Phenomena Gases* (Munich, 1961), Vol. 1, p. 183, North-Holland, Amsterdam. DT]. The identities of the ions were established by K. B. McAfee, D. Sipler, and D. Edelson (1967), *Phys. Rev.* **160**, 130—DTMS and by J. M. Madson and H. J. Oskam (1967), *Physics Letters* **25A**, 407—DTMS. Less accurate data on Ar^+ ions in argon which extend to high values of E/p_0 are shown in Fig. 7-1-A-2.

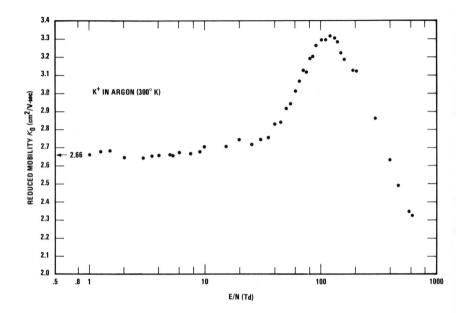

FIG. 7-1-C-2. The reduced mobility of K⁺ ions in argon at 300°K plotted as a function of *E/N* [D. R. James, E. Graham, G. M. Thomson, I. R. Gatland and E. W. McDaniel (1973) *J. Chem. Phys.*, **58**—DTMS].

D. KRYPTON

TABLE 7-1-D-1. Experimental zero-field reduced mobility K_0 in square centimeters per volt per second of ions in krypton (the data were taken at a temperature of 300°K unless otherwise specified)

Ion	K_0	$T(°K)$	Method	Reference
Kr^+	0.90		DT	Biondi and Chanin (1954); see Fig. 7-1-D-1
	0.96 ± 0.09		SAMS	Smith, D. et al. (1972)
	0.90–0.95		DT	Varney, R. N. (1952); see Fig. 7-1-D-2
Kr_2^+	1.2		DT	Biondi and Chanin (1954); see Fig. 7-1-D-1
	1.1–1.2		DT	Varney (1952); K_0 obtained by long extrapolation from high E/p_0.
	1.0		DT	Beaty (1956); K_0 obtained by long extrapolation from high E/p_0
Li^+	3.72	291	DT	Tyndall (1938)
Na^+	2.19	291	DT	Tyndall (1938)
K^+	1.86	291	DT	Tyndall (1938)
Rb^+	1.47	291	DT	Tyndall (1938)
Cs^+	1.33	291	DT	Tyndall (1938)

REFERENCES

Beaty, E. C. (1956), *Phys. Rev.* **104**, 17.
Biondi, M. A., and L. M. Chanin (1954), *Phys. Rev.* **94**, 910.
Smith, D., A. G. Dean, and N. G. Adams (1972), *Z. Physik* **253**, 191.
Tyndall, A. M. (1938), *The Mobility of Positive Ions in Gases*, Cambridge University Press, Cambridge, p. 92.
Varney, R. N. (1952), *Phys. Rev.* **88**, 362.

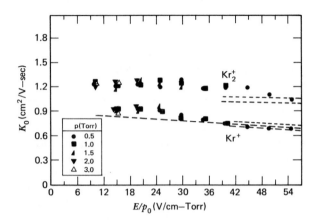

FIG. 7-1-D-1. The reduced mobilities of Kr^+ and Kr_2^+ ions in krypton at $300°K$ plotted as functions of E/p_0 [M. A. Biondi and L. M. Chanin (1954), *Phys. Rev.* **94**, 910—DT]. The short-dashed lines refer to the measurements of R. N. Varney (1952), *Phys. Rev.* **88**, 362—DT. The long-dashed line refers to the measurements of A. M. Tyndall and R. J. Munson (1940), *Proc. Roy. Soc.* (*London*) **A177**, 187—DT.

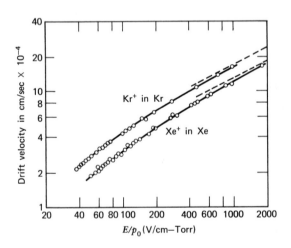

FIG. 7-1-D-2. The drift velocity of Kr^+ ions in krypton at $300°K$ as a function of E/p_0. Data on Xe^+ in xenon are included for comparison. [R. N. Varney (1952), *Phys. Rev.* **88**, 362—DT]. The broken lines at the right of each curve have a slope of $\frac{1}{2}$. The high-field behavior is consistent with Wannier's predictions for a constant mean free path situation [see the discussion following (5-2-14)].

E. XENON

TABLE 7-1-E-1. Experimental zero-field reduced mobility K_0 in square centimeters per volt per second of ions in xenon (the data were taken at a temperature of 300°K unless otherwise specified)

Ion	K_0	$T(°K)$	Method	Reference
Xe$^+$	0.58		DT	Biondi and Chanin (1954); see Fig. 7-1-E-1
	0.57 $\pm$ 0.05		SA	Chantry (1964)
	0.6–0.65		DT	Varney (1952); see Fig. 7-1-D-2
Xe$_2^+$	0.79		DT	Biondi and Chanin (1954); see Fig. 7-1-E-1
	0.67–0.77		DT	Varney (1952); K_0 obtained by long extrapolation from high E/p_0
Li$^+$	2.85	291	DT	Tyndall (1938)
Na$^+$	1.69	291	DT	Tyndall (1938)
K$^+$	1.35	291	DT	Tyndall (1938)
Rb$^+$	1.03	291	DT	Tyndall (1938)
Cs$^+$	0.91	291	DT	Tyndall (1938)

REFERENCES

Biondi, M. A., and L. M. Chanin (1954), *Phys. Rev.* **94**, 910.
Chantry, P. J. (1964), *Atomic Collision Processes*, M. R. C. McDowell, Ed., North-Holland, Amsterdam, p. 565.
Tyndall, A. M. (1938), *The Mobility of Positive Ions in Gases*, Cambridge University Press, Cambridge, p. 92.
Varney, R. N. (1952), *Phys. Rev.* **88**, 362.

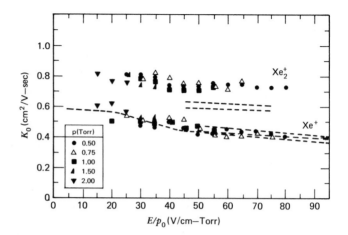

FIG. 7-1-E-1. The reduced mobilities of Xe$^+$ and Xe$_2$$^+$ ions in xenon at 300°K plotted as functions of E/p_0 [M. A. Biondi and L. M. Chanin (1954), *Phys. Rev.* **94**, 910—DT]. The short-dashed lines refer to the measurements in R. N. Varney (1952), *Phys. Rev.* **88**, 362—DT. The long-dashed line refers to the measurements in A. M. Tyndall and R. J. Munson (1940), *Proc. Roy. Soc. (London)* **A177**, 187—DT. Data on Xe$^+$ which extend to high values of E/p_0 are presented in Fig. 7-1-D-2.

F. METAL VAPORS

TABLE 7-1-F-1. Experimental zero-field reduced mobility K_0 in square centimeters per volt per second of ions in metal vapors

Ion/Vapor	K_0	$T(°K)$	Method	Reference
Rb^+ in Rb	0.18	621	DT	Lee and Mahan (1965)
Rb_2^+ in Rb	0.29	621	DT	Lee and Mahan (1965)
Cs^+ in Cs	0.075	579–679	DT	Chanin and Steen (1963)
	0.12	628	DT	Lee and Mahan (1965)
	0.15	527	DT	Popescu and Niculescu (1969)
	0.14–0.15	433–493	DT	Musa et al. (1969)
Cs_2^+ in Cs	0.21	579–679	DT	Chanin and Steen (1963)
	0.20	625	DT	Lee and Mahan (1965)
	0.28–0.30	527	DT	Popescu and Niculescu (1969)
Hg^+ in Hg	0.24	500	DT	Kovar (1964); K_0 obtained by long extrapolation from high E/p_0
	0.22	350	SA	Biondi (1953)
Hg_2^+ in Hg	0.45	500	DT	Kovar (1964); K_0 obtained by long extrapolation from high E/p_0

REFERENCES

Biondi, M. A. (1953), *Phys. Rev.* **90**, 730.
Chanin, L. M., and R. D. Steen (1963), *Phys. Rev.* **132**, 2554.
Kovar, F. R. (1964), *Phys. Rev.* **133**, A681.
Lee, Y., and B. H. Mahan (1965), *J. Chem. Phys.* **43**, 2016.
Musa, G., A. Baltog, L. Năstase, and I. Mustată (1969), *Proc. 9th Intern. Conf. Phenomena Ionized Gases* (Bucharest, 1969), p. 10, Academiei Republicii Socialiste Romania, Bucharest.
Popescu, A., and N. D. Niculescu (1969), *Proc. 9th Intern. Conf. Phenomena Ionized Gases* (Bucharest, 1969), p. 8, Academiei Republicii Socialiste Romania, Bucharest.

G. HYDROGEN AND DEUTERIUM

TABLE 7-1-G-1. Experimental zero-field reduced mobility K_0 in square centimeters per volt per second of ions in hydrogen (H_2) (the data were taken at a temperature of 300°K unless otherwise specified)

Ion	K_0	$T(°K)$	Method	Reference
H^+	15.7 ± 0.6	302	DTMS	Miller, Moseley, Martin, and McDaniel (1968); see Fig. 7-1-G-1
	16.0 ± 0.3		DTMS	Graham (1974)
H_3^+	11.1 ± 0.5	302	DTMS	Miller et al. (1968); see Fig. 7-1-G-1
H_3O^+	12.6		DT	Fleming, Tunnicliffe, and Rees (1969)
Li^+	12.3 ± 0.6	304	DTMS	Miller et al. (1968); see Fig. 7-1-G-2
	12.5	291	DT	Tyndall (1938)
Na^+	12.2 ± 0.6	304	DTMS	Miller et al. (1968); see Fig. 7-1-G-2
	12.8	291	DT	Tyndall (1938)
K^+	12.8 ± 0.6	304	DTMS	Miller et al. (1968); see Fig. 7-1-G-2
	12.7	291	DT	Tyndall (1938)
	12.75	293	DT	Elford (1967)
	12.70		DT	Fleming, Tunnicliffe, and Rees (1969)
Rb^+	12.6	291	DT	Tyndall (1938)
Cs^+	12.6	291	DT	Tyndall (1938)

REFERENCES

Elford, M. T. (1967), *Australian J. Phys.* **20**, 471.
Fleming, I. A., R. J. Tunnicliffe, and J. A. Rees (1969), *J. Phys. B* **2**, 780.
Graham, E. (1974), Ph.D. thesis, Georgia Institute of Technology, Atlanta.
Miller, T. M., J. T. Moseley, D. W. Martin, and E. W. McDaniel (1968), *Phys. Rev.* **173**, 115.
Tyndall, A. M. (1938), *The Mobility of Positive Ions in Gases*, Cambridge University Press, Cambridge, p. 92.

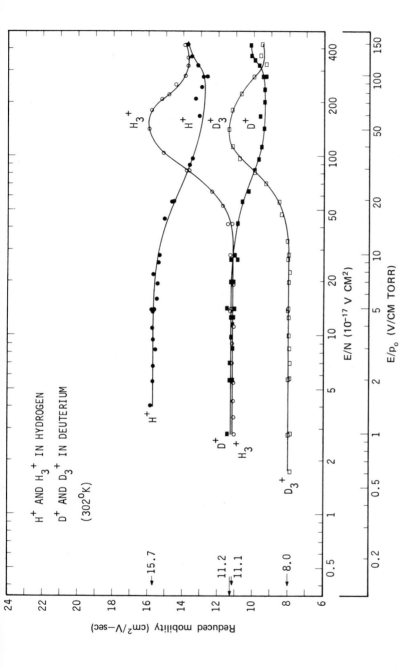

FIG. 7-1-G-1. The reduced mobilities of H^+ and H_3^+ ions in H_2 and D^+ and D_3^+ ions in D_2 at 302°K plotted as functions of E/N and E/p_0 [T. M. Miller, J. T. Moseley, D. W. Martin, and E. W. McDaniel (1968), *Phys. Rev.* **173**, 115. DTMS].

TABLE 7-1-G-2. Experimental zero-field reduced mobility K_0 in square centimeters per volt per second of ions in deuterium (D_2)

Ion	K_0	$T(^\circ K)$	Method	Reference
D^+	11.2 ± 0.5	302	DTMS	Miller, Moseley, Martin, and McDaniel (1968); see Fig. 7-1-G-1
D_3^+	8.0 ± 0.3	302	DTMS	Miller et al. (1968); see Fig. 7-1-G-1
Li^+	9.6 ± 0.5	304	DTMS	Miller et al. (1968); see Fig. 7-1-G-2
Na^+	8.9 ± 0.4	304	DTMS	Miller et al. (1968); see Fig. 7-1-G-2
K^+	9.4 ± 0.5	304	DTMS	Miller et al. (1968); see Fig. 7-1-G-2

REFERENCE

Miller, T. M., J. T. Moseley, D. W. Martin, and E. W. McDaniel (1968), *Phys. Rev.* **173**, 115.

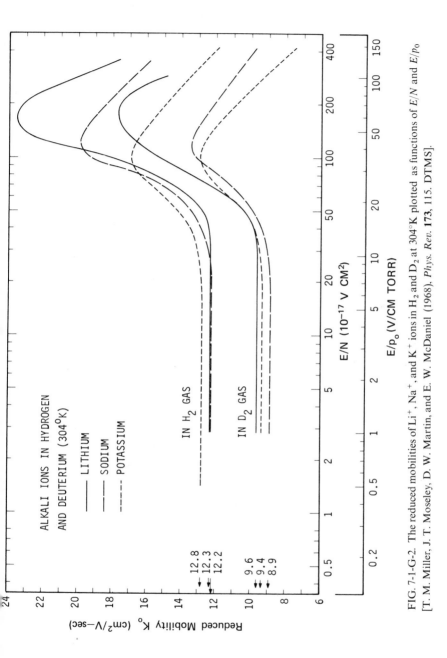

FIG. 7-1-G-2. The reduced mobilities of Li^+, Na^+, and K^+ ions in H_2 and D_2 at 304°K plotted as functions of E/N and E/p_0 [T. M. Miller, J. T. Moseley, D. W. Martin, and E. W. McDaniel (1968). *Phys. Rev.* **173**, 115. DTMS].

H. NITROGEN

TABLE 7-1-H-1. Experimental zero-field reduced mobility K_0 in square centimeters per volt per second of ions in nitrogen (N_2) (the data were taken at a temperature of 300°K unless otherwise specified)

Ion	K_0	$T(°K)$	Method	Reference
N^+	2.97 ± 0.09		DTMS	Moseley, Snuggs, Martin, and McDaniel (1969); see Fig. 7-1-H-1
N_2^+	1.87 ± 0.06		DTMS	Moseley, Snuggs, Martin, and McDaniel (1969); see Fig. 7-1-H-2
N_3^+	2.26 ± 0.06		DTMS	Moseley, Snuggs, Martin, and McDaniel (1969); see Fig. 7-1-H-1
N_4^+	2.33 ± 0.06		DTMS	Moseley, Snuggs, Martin, and McDaniel (1969); see Fig. 7-1-H-2
Li^+	3.95	291	DT	Tyndall (1938)
Na^+	2.85	291	DT	Tyndall (1938)
K^+	2.55 ± 0.05		DTMS	Moseley, Gatland, Martin, and McDaniel (1969); Thomson et al. (1973); see Fig. 7-1-H-3
	2.54	294	DT	Crompton and Elford (1959)
	2.53		DT	Fleming, Tunnicliffe, and Rees (1969)
	2.55 ± 0.07		DT	Davies, Dutton, and Llewellyn-Jones (1966)
	2.53	291	DT	Tyndall (1938)
	2.55 ± 0.09	310	DTMS	Keller and Beyer (1971)
Rb^+	2.24	291	DT	Tyndall (1938)
Cs^+	2.21	291	DT	Tyndall (1938)
	2.2		DTMS	McKnight and Sawina (1972)
Ba^+	2.03		DTMS	Johnsen and Biondi (1973)

REFERENCES

Crompton, R. W., and M. T. Elford (1959), *Proc. Phys. Soc. (London)* **74**, 497.
Davies, P. G., J. Dutton, and F. Llewellyn-Jones (1966), *Phil. Trans. Roy. Soc. London* **259**, 321.
Fleming, I. A., R. J. Tunnicliffe, and J. A. Rees (1969), *Brit. J. Appl. Phys. (J. Phys. D)* **2**, 551.

Johnsen, R., and M. A. Biondi (1973), to be published.

Keller, G. E., and R. A. Beyer (1971), private communication.

McKnight, L. G., and J. M. Sawina (1972), *J. Chem. Phys.* **57**, 5156.

Moseley, J. T., I. R. Gatland, D. W. Martin, and E. W. McDaniel (1969), *Phys. Rev.* **178**, 234.

——, R. M. Snuggs, D. W. Martin, and E. W. McDaniel (1969), *Phys. Rev.* **178**, 240.

Thomson, G. M., J. H. Schummers, D. R. James, E. Graham, I. R. Gatland, M. R. Flannery, and E. W. McDaniel (1973), *J. Chem. Phys.* **58**.

Tyndall, A. M. (1938), *The Mobility of Positive Ions in Gases*, Cambridge University Press, Cambridge, p. 92.

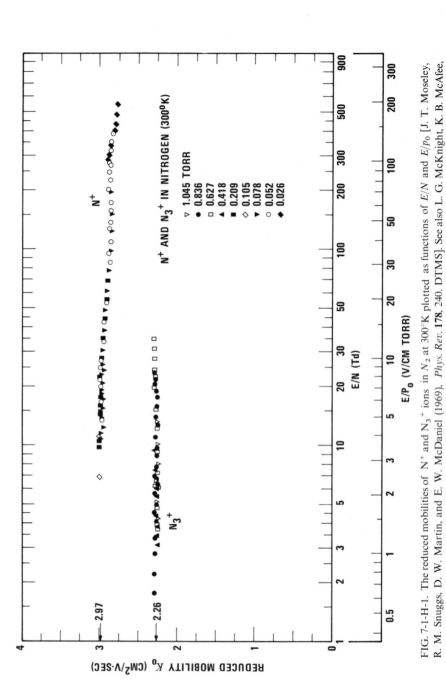

FIG. 7-1-H-1. The reduced mobilities of N^+ and N_3^+ ions in N_2 at 300°K plotted as functions of E/N and E/p_0 [J. T. Moseley, R. M. Snuggs, D. W. Martin, and E. W. McDaniel (1969), *Phys. Rev.* **178**, 240. DTMS]. See also L. G. McKnight, K. B. McAfee, and D. P. Sipler (1967), *Phys. Rev.* **164**, 62—DTMS.

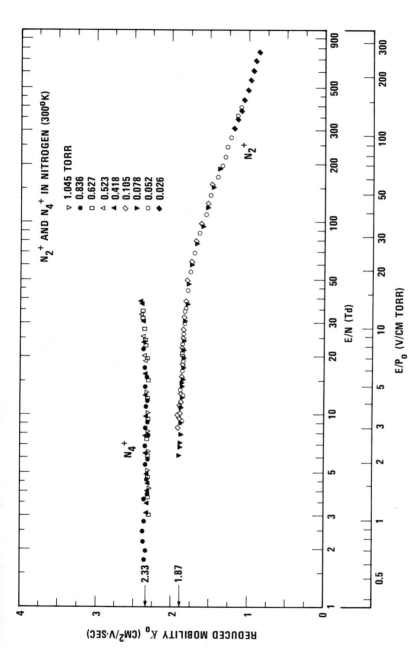

FIG. 7-1-H-2. The reduced mobilities of N_2^+ and N_4^+ ions in N_2 at 300°K plotted as functions of E/N and E/p_0 [J. T. Moseley, R. M. Snuggs, D. W. Martin, and E. W. McDaniel (1969), *Phys. Rev.* **178**, 240. DTMS]. See also L. G. McKnight, K. B. McAfee, and D. P. Sipler (1967), *Phys. Rev.* **164**, 62—DTMS.

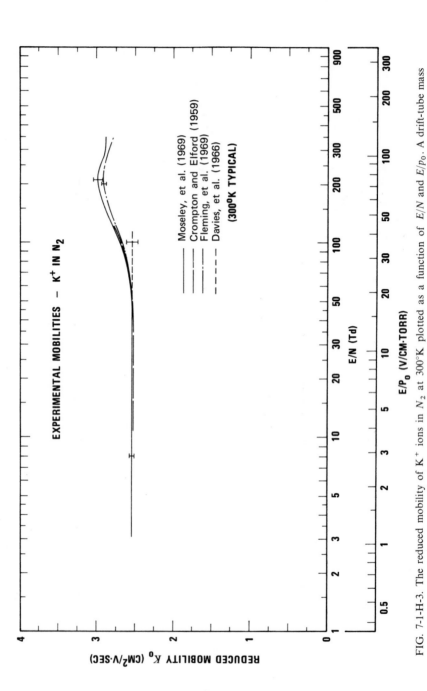

FIG. 7-1-H-3. The reduced mobility of K$^+$ ions in N_2 at 300°K plotted as a function of E/N and E/p_0. A drift-tube mass spectrometer was used to obtain the data in J. T. Moseley, I. R. Gatland, D. W. Martin, and E. W. McDaniel (1969), *Phys. Rev.* 178, 234. The other papers are cited in the references for Table 7-1-H-1.

I. OXYGEN

TABLE 7-1-I-1. Experimental zero-field reduced mobility K_0 in square centimeters per volt per second of ions in oxygen (O_2) (the data were taken at a temperature of 300°K unless otherwise specified)

Ion	K_0	$T(°K)$	Method	Reference
O_2^+	2.24 ± 0.07		DTMS	Snuggs, Volz, Schummers, Martin, and McDaniel (1971); see Fig. 7-1-I-1
O_4^+	2.16 ± 0.08		DTMS	Snuggs et al. (1971)
O^-	3.20 ± 0.09		DTMS	Snuggs et al. (1971); see Fig. 7-1-I-2
O_2^-	2.16 ± 0.07		DTMS	Snugg et al. (1971); see Fig. 7-1-I-2
O_3^-	2.55 ± 0.08		DTMS	Snuggs et al. (1971); see Fig. 7-1-I-3
O_4^-	2.14 ± 0.08		DTMS	Snuggs et al. (1971)
	2.19		DTMS	McKnight and Sawina (1971)
CO_3^-	2.50 ± 0.07		DTMS	Snuggs et al. (1971)
CO_4^-	2.45 ± 0.07		DTMS	Snuggs et al. (1971)
K^+	2.68 ± 0.07		DTMS	Snuggs et al. (1971); see Fig. 7-1-I-4
Cs^+	2.2	315	DTMS	McKnight and Sawina (1972)
SF_6^-	2.0		DTMS	Patterson (1972)
SF_5^-	2.2		DTMS	Patterson (1972)
Ba^+	1.88	298	DTMS	Johnsen and Biondi (1973)

REFERENCES

Johnsen, R., and M. A. Biondi (1973), to be published.
McKnight, L. G., and J. M. Sawina (1971), *Phys. Rev. A* **4**, 1043.
——— (1972), private communication.
Patterson, P. L. (1972), *J. Chem. Phys.* **56**, 3943.
Snuggs, R. M., D. J. Volz, J. H. Schummers, D. W. Martin, and E. W. McDaniel (1971), *Phys. Rev. A* **3**, 477.

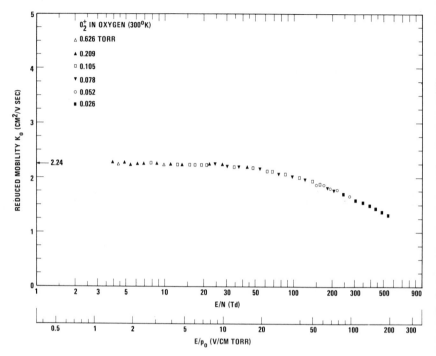

FIG. 7-1-I-1. The reduced mobility of $O_2{}^+$ ions in O_2 at 300°K plotted as a function of E/N and E/p_0 [R. M. Snuggs, D. J. Volz, J. H. Schummers, D. W. Martin, and E. W. McDaniel (1971), *Phys. Rev. A* **3**, 477—DTMS].

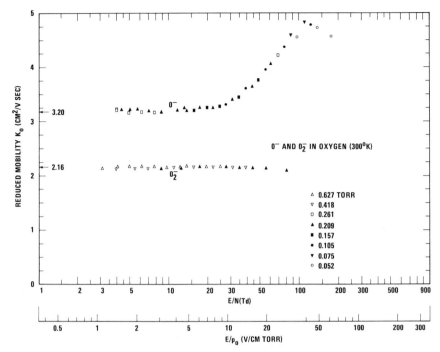

FIG. 7-1-I-2. The reduced mobilities of O^- and O_2^- ions in O_2 at 300°K plotted as functions of E/N and E/p_0 [R. M. Snuggs, D. J. Volz, J. H. Schummers, D. W. Martin, and E. W. McDaniel (1971), *Phys. Rev. A* **3**, 477—DTMS]. For high-field data on O_2^- see L. G. McKnight (1970), *Phys. Rev. A* **2**, 762—DTMS.

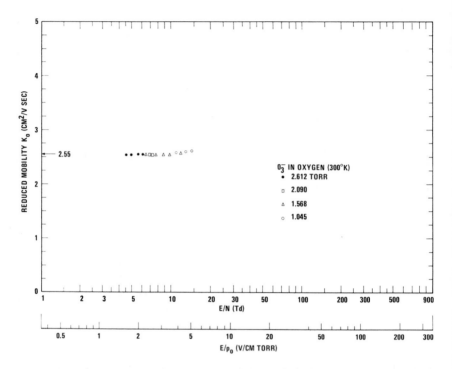

FIG. 7-1-I-3. The reduced mobility of O_3^- ions in O_2 at 300°K plotted as a function of E/N and E/p_0 [R. M. Snuggs, D. J. Volz, J. H. Schummers, D. W. Martin, and E. W. McDaniel (1971), *Phys. Rev. A* **3**, 477—DTMS].

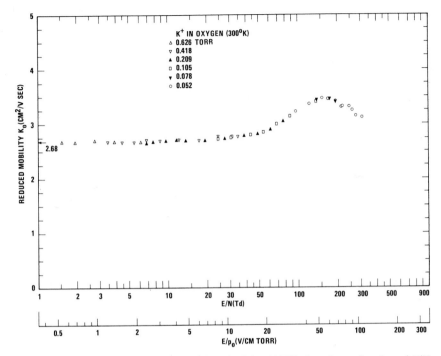

FIG. 7-1-I-4. The reduced mobility of K$^+$ ions in O$_2$ at 300°K plotted as a function of E/N and E/p_0 [R. M. Snuggs, D. J. Volz, J. H. Schummers, D. W. Martin, and E. W. McDaniel (1971), *Phys. Rev. A* **3**, 477—DTMS].

J. NITRIC OXIDE

TABLE 7-1-J-1. Experimental zero-field reduced mobility K_0 in square centimeters per volt per second of ions in nitric oxide (NO)

Ion	K_0	$T(^\circ K)$	Method	Reference
NO^+	1.91 ± 0.06	300	DTMS	Volz, Schummers, Laser, Martin, and McDaniel (1971); see Fig. 7-1-J-1
$NO^+ \cdot NO$	1.78 ± 0.05	300	DTMS	Volz et al. (1971); see Fig. 7-1-J-2
K^+	2.245 ± 0.067	300	DTMS	Volz et al. (1971); see Fig. 7-1-J-2

REFERENCE

Volz, D. J., J. H. Schummers, R. D. Laser, D. W. Martin, and E. W. McDaniel (1971), *Phys. Rev. A* **4**, 1106.

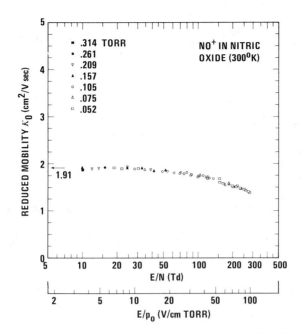

FIG. 7-1-J-1. The reduced mobility of NO⁺ ions in NO at 300°K plotted as a function of E/N and E/p_0 [D. J. Volz, J. H. Schummers, R. D. Laser, D. W. Martin, and E. W. McDaniel (1971), *Phys. Rev. A.* **4**, 1106—DTMS].

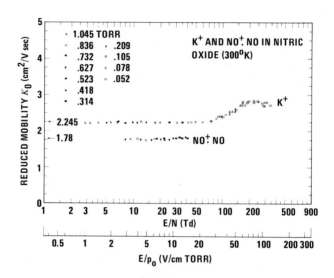

FIG. 7-1-J-2. The reduced mobilities of $NO^+ \cdot NO$ and K^+ ions in NO at 300°K plotted as functions of E/N and E/p_0 [D. J. Volz, J. H. Schummers, R. D. Laser, D. W. Martin, and E. W. McDaniel (1971), *Phys. Rev. A* **4**, 1106—DTMS].

K. CARBON MONOXIDE

TABLE 7-1-K-1. Experimental zero-field reduced mobility K_0 in square centimeters per volt per second of ions in carbon monoxide (CO)

Ion	K_0	$T(^\circ K)$	Method	Reference
$CO^+ \cdot CO$	1.90 ± 0.03	300	DTMS	Schummers, Thomson, James, Gatland, and McDaniel (1973); see Fig. 7-1-K-1
C^+	2.7 ± 0.1	300	DTMS	Schummers, Thomson, James, Gatland, and McDaniel (1973); see Fig. 7-1-K-2
K^+	2.30 ± 0.04	300	DTMS	Thomson, Schummers, James, Graham, Gatland, Flannery, and McDaniel (1973); see Fig. 7-1-K-3

REFERENCES

Schummers, J. H., G. M. Thomson, D. R. James, I. R. Gatland, and E. W. McDaniel (1973), *Phys. Rev. A* **7**, 683.

Thomson, G. M., J. H. Schummers, D. R. James, E. Graham, I. R. Gatland, M. R. Flannery, and E. W. McDaniel (1973), *J. Chem. Phys.* **58**.

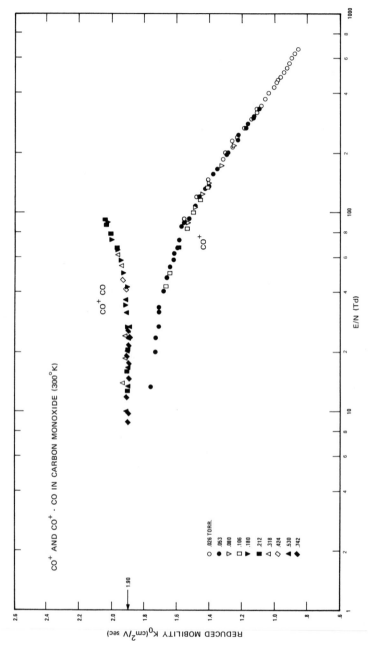

FIG. 7-1-K-1. The reduced mobility of CO^+ and $CO^+ \cdot CO$ ions in CO at 300°K plotted as a function of E/N [J. H. Schummers, G. M. Thomson, D. R. James, I. R. Gatland, and E. W. McDaniel (1973), *Phys. Rev. A* **7**, 683—DTMS].

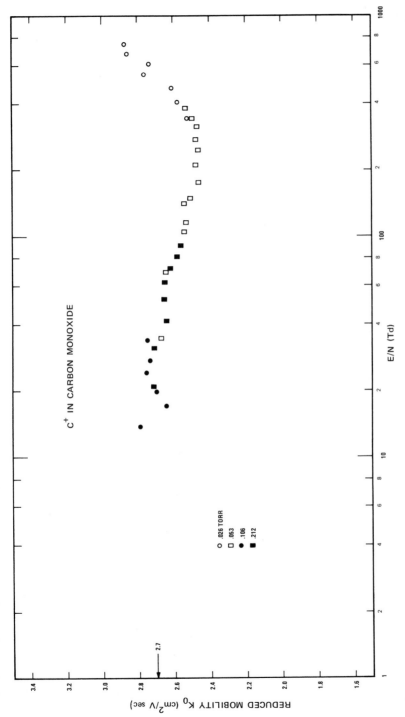

FIG. 7-1-K-2. The reduced mobility of C^+ ions in CO at 300°K plotted as a function of E/N. The data points are badly scattered at low E/N because of the extremely low intensity of the C^+ ions. [J. H. Schummers, G. M. Thomson, D. R. James, I. R. Gatland, and E. W. McDaniel (1973), *Phys. Rev. A* **7**, 683—DTMS].

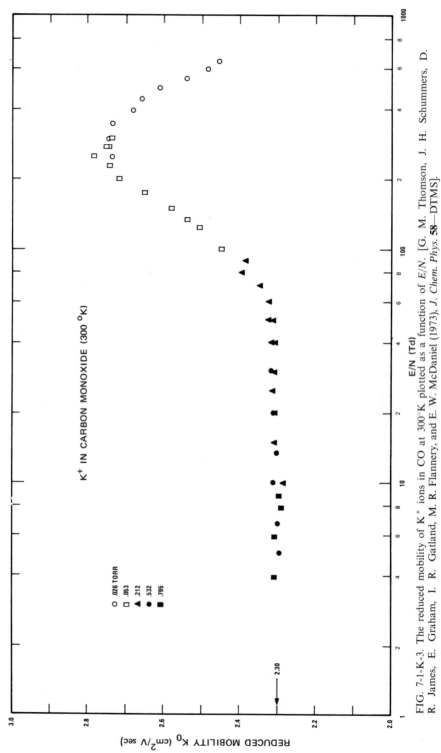

FIG. 7-1-K-3. The reduced mobility of K^+ ions in CO at 300°K plotted as a function of E/N. [G. M. Thomson, J. H. Schummers, D. R. James, E. Graham, I. R. Gatland, M. R. Flannery, and E. W. McDaniel (1973), *J. Chem. Phys.* **58**—DTMS].

302

L. SULFUR HEXAFLUORIDE

TABLE 7-1-L-1. Experimental zero-field reduced mobility K_0 in square centimeters per volt per second of ions in sulfur hexafluoride (SF_6)

Ion	K_0	$T(^\circ K)$	Method	Reference
SF_5^-	0.595 ± 0.007	300	DTMS	Patterson (1970); see Fig. 7-1-L-1
SF_6^-	0.542 ± 0.007	300	DTMS	Patterson (1970); see Fig. 7-1-L-1
$SF_6^- (SF_6)$	0.470 ± 0.010	300	DTMS	Patterson (1970); see Fig. 7-1-L-1
$SF_6^- (SF_6)_2$	0.420 ± 0.010	300	DTMS	Patterson (1970); see Fig. 7-1-L-1

REFERENCES

Patterson, P. L. (1970), *J. Chem. Phys.* **53**, 696.
See also P. L. Patterson (1972), *J. Chem. Phys.* **56**, 3943.

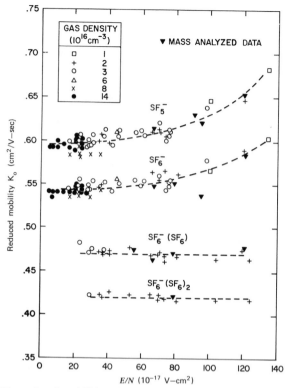

FIG. 7-1-L-1. The reduced mobilities of ions in SF_6 at $300°K$ plotted as functions of E/N [P. L. Patterson (1970), *J. Chem. Phys.* **53**, 696—DTMS].

7-2. THE MOBILITY OF IONS IN MIXTURES OF GASES—BLANC'S LAW

A. BINARY MIXTURES OF OXYGEN WITH OTHER GASES.   Drift-tube data obtained by McDaniel and Crane on a negative ion in binary mixtures of oxygen with other gases are displayed in Figs. 7-2-A-1 and 7-2-A-2. The ion is thought to be O_3^-, but a mass spectrometer was not employed and the identity of the ion is by no means certain. In each figure the reciprocal of the zero-field reduced mobility of the ion is plotted as a function of the percentage of oxygen in the mixture to facilitate comparison with the pre- dictions of Blanc's law (Section 5-3-G). To within experimental error, each plot is linear, and Blanc's law appears to have been obeyed.

B. HELIUM-NEON MIXTURES.   Blanc's law also seems to hold for the data on He^+ and Ne_2^+ ions in helium-neon mixtures, presented in Figs. 7-2-B-1

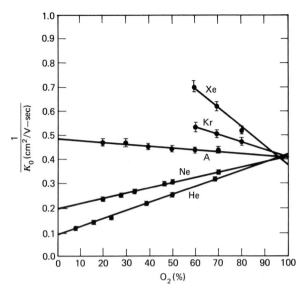

FIG. 7-2-A-1. Reciprocal zero-field reduced mobilities for a negative ion believed to be O_3^- in mixtures of oxygen and each of the noble gases. [E. W. McDaniel and H. R. Crane (1957), *Rev. Sci. Instr.* **28**, 684. DT.]

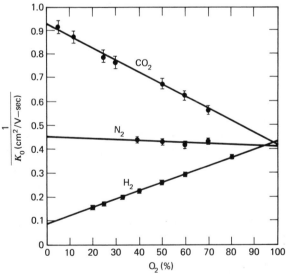

FIG. 7-2-A-2. Reciprocal zero-field reduced mobilities for a negative ion believed to be O_3^- in mixtures of oxygen and each of the gases CO_2, N_2, and H_2. [E. W. McDaniel and H. R. Crane (1957), *Rev. Sci. Instr.* **28**, 684. DT.]

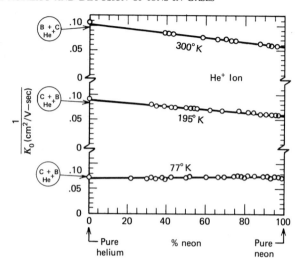

FIG. 7-2-B-1. Reciprocal zero-field reduced mobilities for He^+ in helium-neon mixtures at gas temperatures of 77, 195, and 300°K. [G. E. Courville and M. A. Biondi (1962), *J. Chem. Phys.* **37**, 616. DT.] The circled symbols indicate values previously determined in single-gas studies by M. A. Biondi and L. M. Chanin (1954). *Phys. Rev.* **94**, 910 (DT) and L. M. Chanin and M. A. Biondi (1957), *Phys. Rev.* **106**, 473 (DT).

and 7-2-B-2. These data were obtained by Courville and Biondi with a drift tube that lacked a mass spectrometer, but there is little doubt about the identities of the ions.

7-3. THE VARIATION OF IONIC MOBILITIES WITH GAS TEMPERATURE

A fairly substantial amount of data is available on the dependence of ionic mobilities on gas temperature. Many however, are vitiated by the uncertainty in the identification of the ions, the likelihood that in some cases the ions may have been clustered to different extents over different parts of the temperature range covered or by inaccurate measurements of the mobilities. In this section an effort has been made to restrict the discussion to those ion-gas combinations for which these problems did not appear.

A. IONS IN HELIUM.   Experimental data on the temperature variation of the zero-field reduced mobility of He^+, He_2^+, and Ne_2^+ ions in helium are presented in Figs. 7-3-A-1 through 7-3-A-3. Some of the experimental data on He^+ and He_2^+ in He are also displayed in Figs. 5-7-5, 5-8-2, and 6-4-1. The experimental temperature variation of the mobility of Li^+ ions in helium is shown in Figs. 5-3-3, 5-7-5, and 6-4-1. The reader may wish to

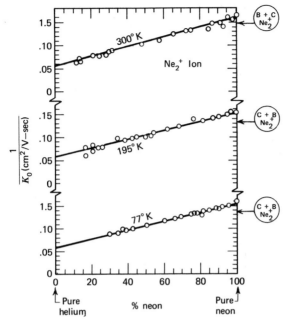

FIG. 7-2-B-2. Reciprocal zero-field reduced mobilities for Ne_2^+ in helium-neon mixtures at gas temperatures of 77, 195, and 300°K. [G. E. Courville and M. A. Biondi (1962), *J. Chem. Phys.* **37**, 616. DT.] The circled symbols are explained in the legend for Fig. 7-2-B-1.

TABLE 7-3-A-1. Zero-field reduced mobility of Na^+ and Cs^+ ions in helium as a function of the gas temperature[a]

$T(°K)$	K_0 (cm²/V-sec)	
	Na^+	Cs^+
79		16.4
92	17.3	16.9
195	19.6	18.0
290	21.4	17.7
392		17.0
405	22.5	
477	23.0	
492		16.3

[a] A. F. Pearce (1936), *Proc. Roy. Soc.* **A155**, 490-DT.

307

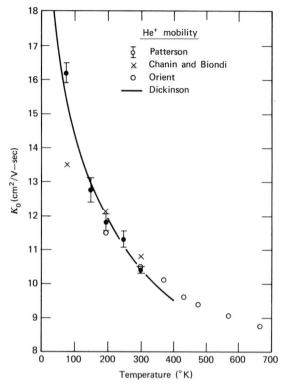

FIG. 7-3-A-1. Zero-field reduced mobility of He$^+$ ions in helium as a function of the gas temperature. The symbols refer to experimental data in P. L. Patterson (1970), *Phys. Rev. A* **2**, 1154—DT, L. M. Chanin and M. A. Biondi (1957), *Phys. Rev.* **106**, 473—DT, and O. J. Orient (1967), *Can. J. Phys.* **45**, 3915—DT. The curve refers to calculations by A. S. Dickinson (1968), *J. Phys. B* **1**, 387.

refer to the theoretical discussions accompanying these figures. Experimental data on Na$^+$ and Cs$^+$ ions in helium appear in Table 7-3-A-1. All of these data were obtained with drift tubes that lacked mass spectrometers, but the identification of the ions over the entire temperature range covered is probably secure. The small polarizability of helium militates against cluster formation in this gas.

B. IONS IN NEON.   Figures 7-3-B-1 and 7-3-B-2 display experimental data on Ne$^+$, Ne$_2$$^+$, and He$^+$ ions in neon. Again, mass spectrometric analysis of the ions was not performed, but the ions can probably be considered to be of the types indicated.

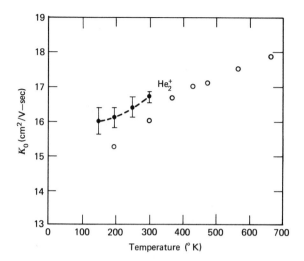

FIG. 7-3-A-2. Zero-field reduced mobility of He_2^+ ions in helium as a function of the gas temperature. The dots refer to experimental data in P. L. Patterson (1970), *Phys. Rev. A* **2**, 1154—DT, the circles, to experimental data in O. J. Orient (1967), *Can. J. Phys.* **45**, 3915—DT. At temperatures below about 200°K and at relatively high pressures Patterson observed the formation of an ion believed to be He_3^+ from the He_2^+ ion. This ion has a well-defined mobility of 18.0 cm²/V-sec at 76°K.

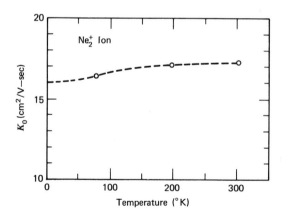

FIG. 7-3-A-3. Zero-field reduced mobility of Ne_2^+ ions in helium as a function of the gas temperature. The circles refer to experimental data in G. E. Courville and M. A. Biondi (1962), *J. Chem. Phys.* **37**, 616—DT. The heavy bar at 0°K gives the theoretical value of K_0 in the polarization limit.

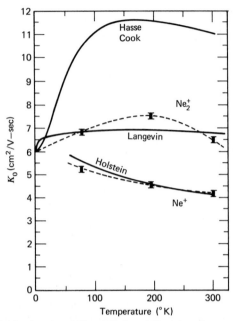

FIG. 7-3-B-1. Zero-field reduced mobility of Ne^+ and Ne_2^+ ions in neon as a function of the gas temperature. The symbols refer to experimental data in L. M. Chanin and M. A. Biondi (1957), *Phys. Rev.* **106**, 473—DT. The curves refer to various theoretical predictions. Slightly lower mobilities for the Ne_2^+ ion were observed by G. E. Courville and M. A. Biondi (1962), *J. Chem. Phys.* **37**, 616—DT, but the temperature dependence was similar to that indicated here.

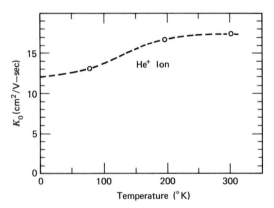

FIG. 7-3-B-2. Zero-field reduced mobility of He^+ ions in neon as a function of the gas temperature. The circles refer to experimental data in G. E. Courville and M. A. Biondi (1962), *J. Chem. Phys.* **37**, 616—DT. The heavy bar at 0°K gives the theoretical value of K_0 in the polarization limit.

C. IONS IN ARGON.   Experimental data on Ar^+ and Ar^{2+} ions in argon are presented in Fig. 7-3-C-1. In the light of the data in Fig. 7-1-C-1, the indicated identification of the ions at 300°K appears certain, and it is unlikely that the identities of the ions change over the temperature range covered. Data on K^+ in argon are displayed in Table 7-3-C-1.

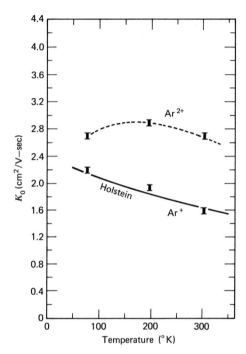

FIG. 7-3-C-1. Zero-field reduced mobility of Ar^+ and Ar^{2+} ions in argon as a function of the gas temperature. The symbols refer to experimental data in L. M. Chanin and M. A. Biondi (1957), *Phys. Rev.* **106**, 473—DT. The solid curve shows the theoretical prediction by Holstein for Ar^+. The ion of higher mobility was originally believed to be Ar_2^+ by Chanin and Biondi but subsequent work indicates that it was probably Ar^{2+} (see Fig. 7-1-C-1).

D. IONS IN KRYPTON.   Experimental data on the temperature variation of the mobility of Rb^+ ions in krypton appear in Table 7-3-D-1.

E. IONS IN XENON.   Data on the variation with gas temperature of the mobility of Cs^+ ions in xenon are displayed in Fig. 5-7-6.

F. IONS IN HYDROGEN.   Table 7-3-F-1 displays Chanin's experimental data on the temperature variation of the mobility of H_3^+ ions in hydrogen.

TABLE 7-3-C-1. Zero-field reduced mobility of K^+ ions in argon as a function of the gas temperature[a]

$T(°K)$	$K_0(cm^2/V\text{-sec})$
291	2.63
400	2.88
460	2.77

[a] K. Hoselitz (1941), *Proc. Roy. Soc.* **A177**, 200-DT.

TABLE 7-3-D-1. Zero-field reduced mobility of Rb^+ ions in krypton as a function of the gas temperature[a]

$T(°K)$	$K_0(cm^2/V\text{-sec})$
195	1.47
273	1.47
291	1.48
370	1.49
455	1.54

[a] K. Hoselitz (1941), *Proc. Roy. Soc.* **A177**, 200-DT.

TABLE 7-3-F-1. Zero-field reduced mobility of H_3^+ ions in hydrogen as a function of the gas temperature[a]

$T(°K)$	$K_0(cm^2/V\text{-sec})$
77	13.0
195	13.3
300	12.3

[a] L. M. Chanin (1961), *Phys. Rev.* **123**, 526-DT.

His mobility at 300°K is slightly higher than the drift-tube mass spectrometer value listed in Table 7-1-G-1, but there is little doubt that the ion is indeed H_3^+.

7-4. THE DIFFUSION OF IONS IN GASES

This section deals mainly with the results of direct drift-tube measurements of longitudinal and transverse diffusion coefficients D_L and D_T as a function of E/N. Accurate measurements of this type are intrinsically difficult to make because they require close correspondence between the experimental arrangement and a soluble mathematical model of the drift tube with realistic initial and boundary conditions. In addition, they require the detailed and accurate mapping of the arrival-time spectrum of the drifting ions in the case of D_L (see Section 2-3) or of the spatial distribution of the drifting swarm in the case of D_T (see Chapter 3). By contrast the determination of drift velocities is comparatively simple, since it is based merely on the evaluation of a properly defined *average* time of transit through the drift tube. Reliable, directly measured diffusion coefficients for ions of known composition have only recently become available. These data are few in number and are much less accurate, in general, than the corresponding drift velocities. Many of the data presented here are believed to be known only within 10 or 20%. It is possible, however, to obtain much more accurate values for diffusion coefficients *at low E/N* (thermal energies) by computing them from mobilities accurately measured with drift tubes. [The reader will recall that at low E/N, $D_L = D_T$ and the single diffusion coefficient D required to characterize the diffusion in all directions is related in a simple manner to the mobility by the Einstein equation (1-1-5).]

The bulk of the data in this section was obtained with drift-tube mass spectrometers; therefore it is certain that only a single species of ion was involved in each measurement and the identity of the species is known. The same claims cannot be made with absolute confidence for the data derived from measurements not involving mass spectrometers, but those data selected for presentation here almost certainly refer to single known ionic species.

The only values of D_L available at present appear to be those obtained with the Georgia Tech drift-tube mass spectrometer. The precision and accuracy of the data published before 1972 are not high. During 1972, however, improvements in the apparatus and methods of analyzing the experimental arrival-time spectra resulted in data of much higher quality. The values of D_L for K^+ ions in N_2, CO, and Ar displayed in Fig. 7-4-F-1 illustrate the degree of precision and accuracy now attainable.

The extrapolated low-field experimental data for K^+ ions in N_2 agree within 0.6% with the value of D_L calculated from the zero-field mobility by use of the Einstein equation (1-1-5). The agreement is within 2% in the case of K^+ ions in CO and Ar.

The measurements of D_T by the Georgia Tech group discussed here all involved the use of a drift-tube mass spectrometer and the attenuation method (Section 3-1). This method suffers from the disadvantage of requiring a constant ion source output over long periods of time as the ion source is moved along the axis of the drift tube. The other D_T data presented here were obtained by the Townsend method (Section 3-2), which does not entail this requirement, and usually a smaller scatter in the data points results.

Diffusion coefficients vary inversely with the gas number density N. For this reason it is convenient to display the data in the form of the product of D_L or D_T with N. Of course, mobilities also vary inversely with N, but a different convention prevails with respect to the presentation of mobility data, which are usually displayed in the modern literature in the form of the mobility reduced to 760 Torr pressure and 0°C [see (1-3-1)].

Comparisons between the experimental diffusion data and certain predictions of Wannier's theory (Section 5-2-A) are made for those ion-gas combinations in which resonant charge transfer does not occur. An interesting and useful feature of Wannier's analysis of the motion of ions in gases is the derivation of equations for D_L and D_T based on the solution of the Boltzmann equation for two different scattering models. In one model the scattering was assumed to be governed by the polarization force between an ion and a neighboring molecule. This force gives rise to a constant mean free time between collisions. The other model assumed constant mean free time and isotropic scattering. The results for the two models do not differ appreciably, which is as expected, since polarization scattering is very nearly isotropic.

Wannier's original equations for D_L and D_T based on the *polarization model*, numbered (149) and (150) in his paper, are

$$D_L = \frac{M + m}{Mm} 0.905\tau_s kT + \frac{1}{3} \frac{(M + m)^3(M + 3.72m)}{M^2 m(M + 1.908m)} \left(\frac{eE}{m}\right)^2 (0.905\tau_s)^3 \quad (7\text{-}4\text{-}1)$$

and

$$D_T = \frac{M + m}{Mm} 0.905\tau_s kT + \frac{1}{3} \frac{(M + m)^4}{M^2 m(M + 1.908m)} \left(\frac{eE}{m}\right)^2 (0.905\tau_s)^3. \quad (7\text{-}4\text{-}2)$$

Here the mass of the ion is denoted by m, that of the molecule by M; τ_s is the mean free time for spiraling collisions. These equations may be put into the form

$$ND_L(E) = ND(0) + \frac{(M + 3.72m)M}{3(M + 1.908m)e} \frac{v_d^3}{E/N},$$ (7-4-3)

$$ND_T(E) = ND(0) + \frac{(M + m)M}{3(M + 1.908m)e} \frac{v_d^3}{E/N},$$ (7-4-4)

where we have transformed from Wannier's original quantity τ_s to the observable variable v_d. Here $D(0)$ is the zero-field value of D, which can be calculated from the experimental value of $K(0)$ by the Einstein equation (1-1-5). Equations 7-4-3 and 7-4-4 predict that $D_L > D_T$ and that both ND_L and ND_T should vary as $v_d^3(E/N)^{-1}$ at high E/N.

In the development of these equations τ_s is treated as a variable and related to the drift velocity v_d at any E/N by the equation

$$\xi\tau_s \frac{M + m}{Mm} = \frac{v_d}{eE}.$$ (7-4-5)

[Cf. (5-2-2).] The constant ξ may be determined by invoking the Einstein relation at low E/N:

$$\xi \tau_s(0) \frac{M + m}{Mm} = \frac{v_d}{eE}\bigg|_{E\to 0} = \frac{D(0)}{kT},$$ (7-4-6)

where the last form, $D(0)/kt$, is given by the Wannier equations (7-4-1) and (7-4-2) in the limit $E \to 0$ as

$$\frac{D(0)}{kT} = 0.905\, \tau_s(0) \frac{M + m}{Mm}.$$ (7-4-7)

The same technique used above to obtain the modified version of Wannier's equations for the polarization model, (7-4-3) and (7-4-4), may be employed to obtain the modified version of his equations for the isotropic scattering model. Wannier's original equations for this model are numbered (151) and (152) in his paper. When transformed to depend explicitly on the drift velocity, they assume the form of (5-2-30) and (5-2-31) in Chapter 5.

Equations 7-4-3 and 7-4-4, or (5-2-30) and (5-2-31), are used here for direct comparison with experiment because v_d has been measured as a function of E/N for all of the ions for which diffusion coefficient data are available.

Comparisons are not made with Wannier's predictions for those cases in which the ion and gas molecule differ in composition only by an electron. In such ion-gas combinations (e.g., N_2^+ in N_2 and O_2^- in O_2) the ionic motion is dominated by resonant charge transfer and would not be expected to be described realistically by Wannier's predictions, which did not take this scattering mechanism into account.

We shall see that there is a significant measure of agreement between the experimental data and Wannier's theoretical predictions, although the agreement in absolute magnitude is not close in all cases. However, when we consider the approximate nature of the theory, the wide range of ion-molecule systems considered, and the rather large probable errors (up to about 25%) in the measured values, the agreement is surprisingly good.

A. IONS IN HYDROGEN.   Data on the transverse diffusion of K^+ and H_3^+ ions in H_2 at room temperature are displayed in Fig. 7-4-A-1. The number at the left side of each curve refers to the value of ND at zero field, which is calculated from the measured low-field mobility by the Einstein equation. Similarly placed numbers on the other figures presented here have the same significance.

The value of ND at temperature T may be calculated from the measured zero-field reduced mobility K_0 by the equation

$$ND = 2.315 \times 10^{15} T K_0 \qquad (7\text{-}4\text{-}8)$$

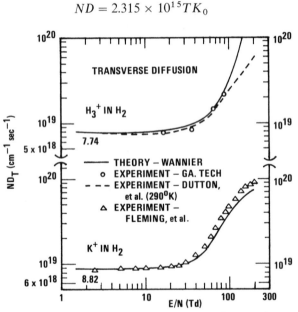

FIG. 7-4-A-1. Experimental results on transverse diffusion of K^+ and H_3^+ ions in hydrogen at room temperature, compared with the predictions of Wannier's modified theory. N is the gas number density and D_T is the transverse diffusion coefficient. The circles refer to the data in T. M. Miller, J. T. Moseley, D. W. Martin, and E. W. McDaniel (1968), *Phys. Rev.* **173**, 115—DTMS. The dashed line indicates the data in J. Dutton, F. Llewellyn-Jones, W. D. Rees, and E. M. Williams (1966), *Phil. Trans. Roy. Soc. (London)* A-259, 339—DT. The triangles refer to the data in I. A. Fleming, R. J. Tunnicliffe, and J. A. Rees (1969), *J. Phys. B* **2**, 780—DT. Data on D_L for ions in H_2 appear in the Ph.D. theses of E. Graham and D. R. James (1974) Ga. Inst. of Technology, Atlanta.

where T, the temperature at which the mobility was measured, is expressed in degrees Kelvin and N is in molecules per cubic centimeter.

B. IONS IN NITROGEN.   Figures 7-4-B-1 through 7-4-B-3 display data on the longitudinal and transverse diffusion of various ions in N_2. The case of N_2^+ in N_2, for which resonant charge transfer occurs, has some interesting features. Figure 7-4-B-3 shows that D_L for N_2^+ in N_2 rises much less sharply with increasing E/N than D_L does for K^+ and N^+ in N_2. Resonant charge transfer does not occur in the last two cases. Furthermore, D_T for N_2^+ in N_2 remains essentially constant out to very high values of E/N.

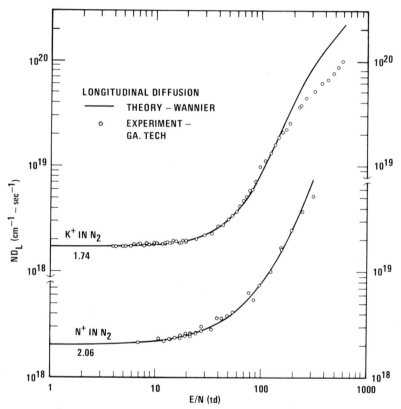

FIG. 7-4-B-1. Experimental results on longitudinal diffusion of N^+ and K^+ ions in nitrogen at room temperature, compared with the predictions of Wannier's modified theory. N is the gas number density and D_L is the longitudinal diffusion coefficient. The N^+ experimental data are those in J. T. Moseley, I. R. Gatland, D. W. Martin, and E. W. McDaniel (1969), *Phys. Rev.* **178**, 234—DTMS. The K^+ data are those in G. M. Thomson, J. H. Schummers, D. R. James, E. Graham, I. R. Gatland, M. R. Flannery, and E. W. McDaniel (1973), *J. Chem. Phys.* **58**—DTMS.

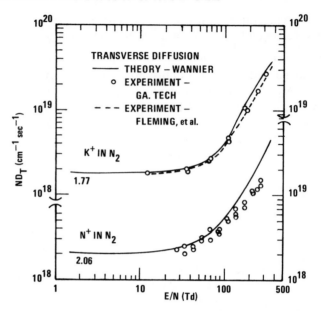

FIG. 7-4-B-2. Experimental results on transverse diffusion of N^+ and K^+ ions in nitrogen at room temperature, compared with the predictions of Wannier's modified theory. N is the gas number density and D_T is the transverse diffusion coefficient. The circles refer to the experimental data in J. T. Moseley, R. M. Snuggs, D. W. Martin, and E. W. McDaniel (1968), *Phys. Rev. Letters* **21**, 873—DTMS, and J. T. Moseley, I. R. Gatland, D. W. Martin, and E. W. McDaniel (1969), *Phys. Rev.* **178**, 234—DTMS. The dashed line indicates the experimental data in I. A. Fleming, R. J. Tunnicliffe, and J. A. Rees (1969), *J. Phys. D* **2**, 551—DT.

C. IONS IN OXYGEN. Longitudinal diffusion data for K^+ and O^- in O_2 are presented in Fig. 7-4-C-1. The K^+ data are displayed again in Fig. 7-4-C-2 for comparison with data on O_2^+ in O_2 to dramatize the strong effect of resonant charge transfer in the latter case. Similarly, in Fig. 7-4-C-3 a comparison is made between the behavior of O^- in O_2 and that of O_2^- in O_2. The inhibiting effect of resonant charge transfer on longitudinal diffusion in the latter case is apparent. Transverse diffusion data for O_2^+ in O_2 appear in Fig. 7-4-C-4.

D. IONS IN NITRIC OXIDE. Figure 7-4-D-1 provides data on D_L for NO^+ and K^+ ions in NO. As expected, the effect of resonant charge transfer on D_L for NO^+ in the parent gas is dramatic.

E. IONS IN CARBON MONOXIDE. Data on D_L for CO^+ and K^+ ions in CO are shown in Fig. 7-4-E-1.

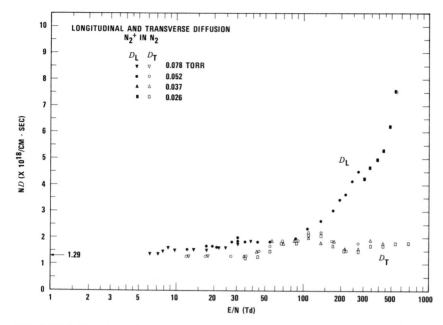

FIG. 7-4-B-3. Experimental results on longitudinal and transverse diffusion of N_2^+ ions in nitrogen at room temperature. N is the gas number density, D_L is the longitudinal diffusion coefficient, and D_T is the transverse diffusion coefficient. The data are those in J. T. Moseley, R. M. Snuggs, D. W. Martin, and E. W. McDaniel (1968), *Phys. Rev. Letters* **21**, 873—DTMS, and J. T. Moseley, R. M. Snuggs, D. W. Martin, and E. W. McDaniel (1969), *Phys. Rev.* **178**, 240—DTMS.

F. IONS IN ARGON.   Fig. 7-4-F-1 shows experimental data for K^+ ions in argon along with data on these ions in N_2 and CO for purposes of comparison. The solid curves give the values of ND_L calculated from the modified Wannier equation (7-4-3), expressed in terms of the variable v_d. The dashed curves are plots of the original Wannier equation (7-4-1), which may be recovered from (7-4-3) by replacing $(v_d/E) = K$ with $K(0)$; that is to say, the use of the zero-field mobility $K(0)$ for (v_d/E) at all E/N is tantamount to reverting to (7-4-1), which contains the constant τ_s rather than the variable v_d.

Better agreement between theory and experiment, especially at high E/N, is provided by the first-order correction to the modified Wannier equations which has been developed by Whealton and Mason. The Whealton-Mason theory is discussed in the Note Added in Proof in Section 5-6, pg. 208.

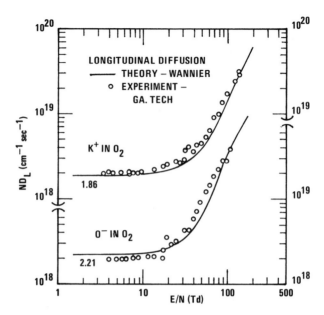

FIG. 7-4-C-1. Experimental results on longitudinal diffusion of O^- and K^+ ions in oxygen at room temperature, compared with the predictions of Wannier's modified theory. N is the gas number density and D_L is the longitudinal diffusion coefficient. The experimental data are those of R. M. Snuggs, D. J. Volz, J. H. Schummers, D. W. Martin, and E. W. McDaniel (1971), *Phys. Rev. A* **3**, 477—DTMS.

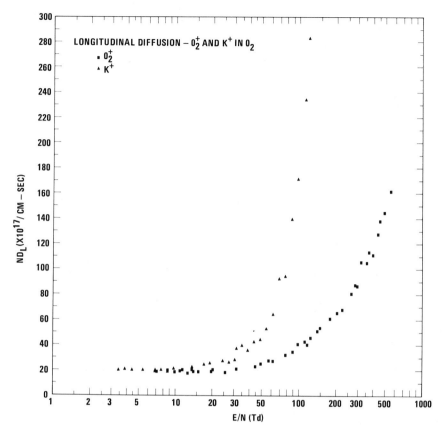

FIG. 7-4-C-2. Experimental results on longitudinal diffusion of O_2^+ and K^+ ions in oxygen at room temperature. N is the gas number density and D_L is the longitudinal diffusion coefficient. The data are those of R. M. Snuggs, D. J. Volz, J. H. Schummers, D. W. Martin, and E. W. McDaniel (1971), *Phys. Rev. A* **3**, 477—DTMS.

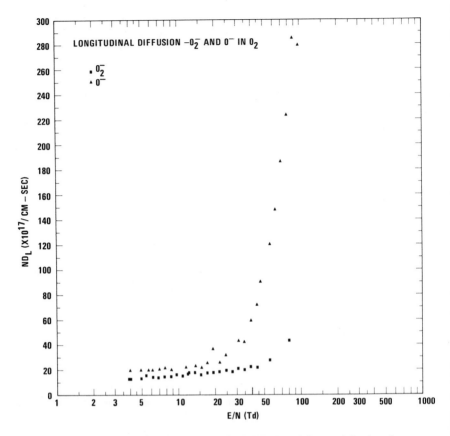

FIG. 7-4-C-3. Experimental results on longitudinal diffusion of O_2^- and O^- ions in oxygen at room temperature. N is the gas number density and D_L is the longitudinal diffusion coefficient. The data are those of R. M. Snuggs, D. J. Volz, J. H. Schummers, D. W. Martin, and E. W. McDaniel (1971), *Phys. Rev. A* **3**, 477—DTMS.

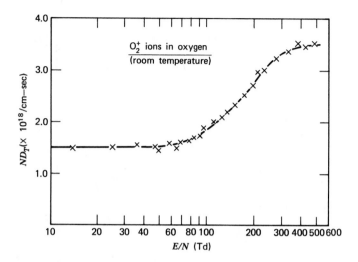

FIG. 7-4-C-4. Experimental data on transverse diffusion of O_2^+ ions in oxygen obtained with the apparatus described in Section 3-2-C [D. R. Gray and J. A. Rees (1972), *J. Phys. B* **5**, 1048—DTMS].

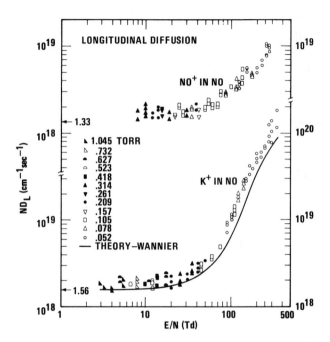

FIG. 7-4-D-1. Experimental results on longitudinal diffusion of NO$^+$ and K$^+$ ions in nitric oxide at room temperature. The predictions of Wannier's modified theory are shown for K$^+$ in NO. N is the gas number density and D_L is the longitudinal diffusion coefficient. The experimental data are those of D. J. Volz, J. H. Schummers, R. D. Laser, D. W. Martin, and E. W. McDaniel (1971), *Phys. Rev. A* **4**, 1106—DTMS.

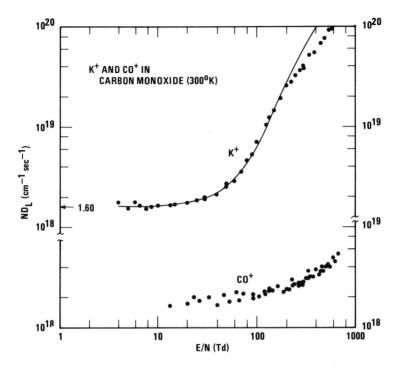

FIG. 7-4-E-1. Experimental results on longitudinal diffusion of CO^+ and K^+ ions in CO at room temperature. The curve shows the predictions of Wannier's modified theory for K^+. N is the gas number density and D_L is the longitudinal diffusion coefficient. The CO^+ data are those of J. H. Schummers, G. M. Thomson, D. R. James, I. R. Gatland, and E. W. McDaniel (1973), *Phys. Rev. A* **7**, 683—DTMS. The K^+ data are those of G. M. Thomson, J. H. Schummers, D. R. James, E. Graham, I. R. Gatland, M. R. Flannery, and E. W. McDaniel (1973), *J. Chem. Phys.* **58**—DTMS.

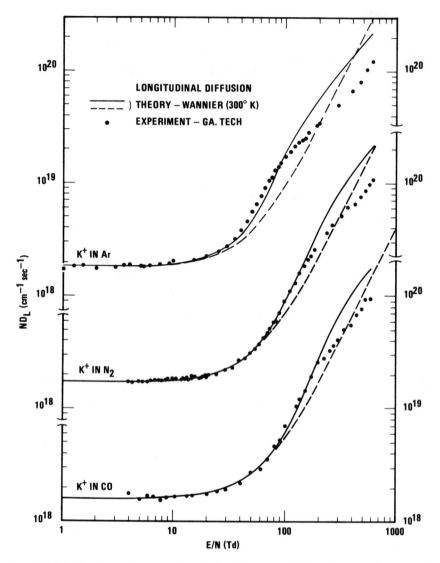

FIG. 7-4-F-1. Experimental results on longitudinal diffusion of K^+ ions in argon, nitrogen, and carbon monoxide at room temperature. The data on N_2 and CO are the same as those presented in Figs. 7-4-B-1 and 7-4-E-1. The data on Ar are those of D. R. James, E. Graham, G. M. Thomson, I. R. Gatland, and E. W. McDaniel (1973), *J. Chem. Phys.*, **58**. The solid curves are plots of the modified Wannier equation (7-4-3). The dashed curves are plots of the original Wannier equation (7-4-1).

TABLES OF

TRANSPORT CROSS SECTIONS

AND COLLISION INTEGRALS

I-1   Cross Sections $Q^{(1)*}$ and $Q^{(2)*}$ for the (8-4) Potential

I-2   Cross Sections $Q^{(1)*}$ and $Q^{(2)*}$ for the (∞-4) Potential

I-3   Cross Sections $Q^{(1)*}$ and $Q^{(2)*}$ for (12-6-4) Potentials

I-4   Collision Integrals $\Omega^{(1,1)*}$, A^*, and $6C^*$-5 for the (8-4) and (∞-4) Potentials

I-5   Collision Integrals $\Omega^{(1,1)*}$ for (12-6-4) Potentials

I-6   Collision Integral Ratios A^* for (12-6-4) Potentials

I-7   Collision Integral Ratios $6C^*$-5 for (12-6-4) Potentials

I-8   Collision Integrals $\Omega^{(1,1)*}$ for (12-4) Core Potentials

I-9   Collision Integral Ratios A^* for (12-4) Core Potentials

I-10 Collision Integral Ratios $6C^*$-5 for (12-4) Core Potentials

I-11 Summary of Classical Collision Integrals Available for other Potentials

TABLE I-1. Cross sections for the (8-4) potential[a]

$(E^*)^{-1/2}$	E^*	$Q^{(1)*}$	$Q^{(2)*}$
∞	0	$3.1260/(E^*)^{1/2}$	$3.2680/(E^*)^{1/2}$
5	0.040000	15.075	16.387
4	0.062500	11.986	13.112
3	0.111111	8.8748	9.8356
2	0.250000	5.7812	6.5591
tan 60°	0.333333	4.9491	5.6757
tan 55°	0.490291	4.1811	4.5406
tan 50°	0.704088	3.0330	4.1616
1.000	1.00000	2.0732	3.4581
0.875	1.30612	1.5769	2.7200
0.750	1.77778	1.2003	1.9967
0.625	2.56000	0.92985	1.4198
0.500	4.00000	0.74132	1.0235
0.375	7.11111	0.60573	0.77573
0.250	16.00000	0.49167	0.61243
0.125	64.00000	0.35947	0.45324
0	∞	$1.0771/(E^*)^{1/4}$	$1.4018/(E^*)^{1/4}$

[a] Hassé and Cook (1929). An accuracy of five figures is claimed. The tabulated functions $I(k)$ and $Y(k)$ are related to the cross sections as follows:

$$Q^{(1)*} = 2k^{1/2}\frac{I_1(k)}{2\pi} = 2^{3/2}kY_1(k),$$

$$Q^{(2)*} = 6k^{1/2}\frac{I_2(k)}{2\pi} = 3 \cdot 2^{3/2}kY_2(k),$$

$$k = 2\tan\alpha = 2(E^*)^{-1/2}.$$

TABLE I-2. Cross sections for the (∞-4) potential[a]

$(E^*)^{1/2}$	E^*	$Q^{(1)*}$	$Q^{(2)*}$
0	0	$2.2104/(E^*)^{1/2}$	$2.3108/(E^*)^{1/2}$
1/16	0.003906	33.768	37.150
1/8	0.015625	16.540	18.627
1/4	0.062500	8.0376	9.3638
1/3	0.111111	5.9447	7.0492
1/2	0.250000	3.8834	4.7366
3/4	0.562500	2.5454	3.2002
1	1.00000	1.8965	2.4353
1.15470...	1.33333	1.6326	2.1353
4/3	1.77778	1.4255	1.8437
2	4.	1.1438	1.2796
3	9.	1.0530	1.0930
4	16.	1.0278	1.0456
6	36.	1.0117	1.0181
8	64.	1.0064	1.0097
∞	∞	1.0000	1.0000

[a] Hassé (1926) and Hassé and Cook (1927). An accuracy of five figures is claimed. The tabulated functions x and y are related to the cross sections as follows: $Q^{(1)*} = 4y/(z\lambda)$; $Q^{(2)*} = 12x/(z\lambda)$; $z\lambda = (E^*)^{1/2}$.

TABLE I-3. Cross sections for (12-6-4) potentials[a]

$\tan^{-1} E^*$	E^*	$Q^{(1)*}$			$Q^{(2)*}$		
		$\gamma = 0$	$\gamma = 0.25$	$\gamma = 0.50$	$\gamma = 0$	$\gamma = 0.25$	$\gamma = 0.50$
0	0	$2.7072[(1 - \gamma)/E^*)]^{1/2}$			$2.8301[(1 - \gamma)/E^*]^{1/2}$		
0.2	0.20271	5.501	4.918	4.306	6.491	5.148	5.579
0.4	0.42279	3.719	3.432	3.137	4.354	3.970	3.440
0.6	0.68414	2.892	2.730	2.506	3.454	3.178	3.019
0.7	0.84229	2.394	2.333	2.258	3.263	2.935	2.630
0.8	1.0296	1.974	1.950	1.926	3.002	2.799	2.583
0.9	1.2602	1.636	1.627	1.620	2.610	2.504	2.392
1.0	1.5574	1.368	1.368	1.368	2.182	2.131	2.080
1.1	1.9648	1.155	1.160	1.164	1.778	1.757	1.737
1.2	2.5722	0.9842	0.9919	0.9984	1.425	1.420	1.415
1.3	3.6021	0.8438	0.8540	0.8625	1.134	1.137	1.141
1.4	5.7979	0.7224	0.7346	0.7447	0.9019	0.9111	0.9188
1.5	14.101	0.5951	0.6088	0.6203	0.7089	0.7218	0.7326
1.51	16.428	0.5789	0.5927	0.6042	0.6883	0.7016	0.7126
1.52	19.670	0.5609	0.5748	0.5864	0.6663	0.6798	0.6911
1.53	24.498	0.5402	0.5542	0.5658	0.6417	0.6556	0.6671
1.54	32.461	0.5154	0.5294	0.5410	0.6129	0.6270	0.6388
1.55	48.078	0.4833	0.4970	0.5086	0.5760	0.5903	0.6024
1.56	92.621	0.4344	0.4476	0.4588	0.5199	0.5343	0.5464
1.57	1255.8	0.2831	0.2932	0.3016	0.3421	0.3538	0.3637
$\pi/2$	∞	$0.9322[(1 + \gamma)/E^*]^{1/6}$			$1.1270[(1 + \gamma)/E^*]^{1/6}$		

[a] Mason and Schamp (1958). An accuracy of 0.1% is claimed; $Q^{(3)*}$, $Q^{(4)*}$ and $Q^{(5)*}$ are also given. Results for $\gamma = 1$ correspond to a (12-6) potential and are available elsewhere (Kihara and Kotani, 1943; Hirschfelder, Bird, and Spotz, 1948).

TABLE I-4. Collision integrals for two (n-4) potentials[a]

		$\Omega^{(1,1)*}$		A^*		$6C^*$-5	
$(T^*)^{1/2}$	T^*	8-4	∞-4	8-4	∞-4	8-4	∞-4
0	0	$1.4691\left[\left(1-\dfrac{4}{n}\right)T^*\right]^{-1/2}$		0.8713	0.8713	0	0
0.1	0.01	20.20	13.67	0.898	0.946	-0.02	-0.05
0.2	0.04	9.878	6.640	0.928	0.981	-0.04	-0.05
0.3	0.09	6.453	4.343	0.937	1.017	-0.08	-0.05
0.4	0.16	4.665	3.213	0.984	1.033	-0.22	-0.04
0.5	0.25	3.454	2.548	1.047	1.052	-0.44	-0.03
0.6	0.36	2.603	2.117	1.093	1.062	-0.60	0.00
0.7	0.49	2.019	1.823	1.115	1.064	-0.66	0.06
0.8	0.64	1.619	1.617	1.122	1.055	-0.63	0.15
0.9	0.81	1.344	1.472	1.122	1.050	-0.54	0.25
1.0	1.00	1.150	1.368	1.120	1.040	-0.43	0.35
1.1	1.21	1.010	1.292	1.117	1.032	-0.30	0.44
1.2	1.44	0.9076	1.236	1.116	1.025	-0.18	0.53
1.3	1.69	0.8293	1.194	1.115	1.019	-0.08	0.60
1.4	1.96	0.7688	1.161	1.116	1.014	0.02	0.66
1.5	2.25	0.7208	1.136	1.117	1.011	0.10	0.70
1.6	2.56	0.6818	1.116	1.120	1.008	0.17	0.74
1.7	2.89	0.6497	1.100	1.122	1.006	0.23	0.78
1.8	3.24	0.6225	1.087	1.124	1.005	0.28	0.81
1.9	3.61	0.5996	1.076	1.127	1.004	0.32	0.83
2.0	4.00	0.5796	1.067	1.130	1.003	0.35	0.85
2.2	4.84	0.5465	1.054	1.135	1.002	0.41	0.88
2.4	5.76	0.5200	1.044	1.139	1.001	0.44	0.90
2.6	6.76	0.4980	1.036	1.143	1.001	0.47	0.92
2.8	7.84	0.4794	1.031	1.147	1.001	0.49	0.94
3.0	9.00	0.4631	1.026	1.150	1.000	0.50	0.94
3.2	10.24	0.4487	1.023	1.153	1.000	0.52	0.95
3.4	11.56	0.4359	1.020	1.155	1.000	0.52	0.96
3.6	12.96	0.4243	1.017	1.158	1.000	0.53	0.97
3.8	14.44	0.4138	1.015	1.159	1.000	0.54	0.97
4.0	16.00	0.4040	1.014	1.161	1.000	0.54	0.98
4.4	19.36	0.3866		1.165		0.54	
4.8	23.04	0.3715		1.167		0.54	
5.2	27.04	0.3582		1.169		0.54	
5.6	31.36	0.3463		1.171		0.54	
6.0	36.00	0.3354		1.173		0.54	
6.4	40.96	0.3257		1.174		0.54	
6.8	46.24	0.3167		1.175		0.54	
7.2	51.84	0.3085		1.176		0.54	
7.6	57.76	0.3009		1.177		0.54	
8.0	64.00	0.2938		1.178		0.53	
∞	∞	$0.8661/(T^*)^{1/4}$	1.0000	1.1929	1.0000	0.5000	1.0000

[a] Hassé (1926) and Hassé and Cook (1927, 1929, 1931). An accuracy of about 0.2% is claimed for $\Omega^{(1,1)*}$ and A^*. Values of $6C^*$-5 have been calculated by numerical differentiation according to (5-3-26) and thus have uncertainties of the order of 0.02. The tabulated functions $I(s)$ for the (8-4) potential and $A(\lambda)$ and $X(\lambda)$ for the (∞-4) potential are related to the collision integrals as follows: $\Omega^{(l,l)*} = 2I(s)/s^{1/2}$ for $l = 1, 2$; $T^* = 4s^2$; $\Omega^{(1,1)*} = 3/4\lambda A(\lambda)$, $\Omega^{(2,2)*} = 4X(\lambda)/\lambda$, $T^* = \lambda^2$.

331

TABLE I-5. $\Omega^{(1, 1)*}$ for (12-6-4) potentials[a]

			$\Omega^{(1, 1)*}$		
T^*	$\gamma = 0$	$\gamma = 0.25$	$\gamma = 0.50$	$\gamma = 0.75$	$\gamma = 1$
0			$1.7994[(1 - \gamma)/T^*]^{1/2}$		∞
0.1	5.234	4.704	4.133	3.63	3.181
0.2	3.527	3.263	2.972	2.72	2.484
0.3	2.713	2.562	2.392	2.25	2.103
0.4	2.216	2.123	2.018	1.93	1.837
0.5	1.885	1.826	1.757	1.70	1.640
0.6	1.653	1.614	1.568	1.53	1.490
0.7	1.483	1.457	1.425	1.40	1.373
0.8	1.354	1.337	1.315	1.30	1.280
0.9	1.254	1.243	1.228	1.22	1.204
1.0	1.174	1.168	1.158	1.15	1.143
1.2	1.056	1.055	1.052	1.05	1.048
1.4	0.9727	0.9758	0.9766	0.977	0.9791
1.6	0.9114	0.9170	0.9203	0.922	0.9270
1.8	0.8645	0.8718	0.8768	0.881	0.8862
2.0	0.8275	0.8359	0.8422	0.848	0.8535
2.5	0.7615	0.7718	0.7800	0.787	0.7942
3.0	0.7176	0.7290	0.7383	0.747	0.7540
3.5	0.6861	0.6981	0.7080	0.716	0.7247
4	0.6621	0.6745	0.6848	0.694	0.7021
5	0.6273	0.6403	0.6511	0.661	0.6689
6	0.6026	0.6159	0.6269	0.637	0.6452
7	0.5837	0.5971	0.6083	0.618	0.6268
8	0.5685	0.5820	0.5933	0.604	0.6120
9	0.5558	0.5694	0.5808	0.591	0.5997
10	0.5450	0.5587	0.5701	0.581	0.5891
12	0.5275	0.5412	0.5526	0.563	0.5716
14	0.5137	0.5273	0.5388	0.549	0.5576
∞			$0.8039[(1 + \gamma)/T^*]^{1/6}$		

[a] Mason and Schamp (1958). An accuracy of about 0.2% is estimated for $\gamma = 0$, 0.25, 0.50. Results for $\gamma = 1$ correspond to a (12-6) potential and are claimed to be accurate to better than 0.1% (Monchick and Mason, 1961). Results for $\gamma = 0.75$ are interpolated, and are accordingly less accurate.

TABLE I-6. A^* for (12-6-4) potentials[a]

T^*	A^*				
	$\gamma = 0$	$\gamma = 0.25$	$\gamma = 0.50$	$\gamma = 0.75$	$\gamma = 1$
0	0.8713	0.8713	0.8713	0.8713	1.0065
0.1	0.964	0.919	1.010	1.020	1.023
0.2	1.014	0.998	1.010	1.022	1.042
0.3	1.061	1.049	1.049	1.056	1.072
0.4	1.088	1.078	1.076	1.081	1.094
0.5	1.100	1.093	1.091	1.095	1.105
0.6	1.106	1.100	1.098	1.102	1.110
0.7	1.107	1.102	1.101	1.104	1.111
0.8	1.107	1.103	1.102	1.104	1.110
0.9	1.105	1.102	1.101	1.103	1.109
1.0	1.103	1.100	1.100	1.102	1.106
1.2	1.100	1.098	1.097	1.099	1.102
1.4	1.098	1.096	1.095	1.097	1.098
1.6	1.097	1.095	1.094	1.096	1.096
1.8	1.096	1.094	1.094	1.094	1.094
2.0	1.096	1.094	1.093	1.093	1.093
2.5	1.097	1.095	1.094	1.094	1.093
3.0	1.099	1.097	1.095	1.094	1.093
3.5	1.101	1.099	1.098	1.097	1.095
4	1.104	1.102	1.100	1.098	1.096
5	1.108	1.105	1.104	1.102	1.100
6	1.111	1.108	1.107	1.105	1.102
7	1.113	1.111	1.110	1.108	1.105
8	1.115	1.113	1.112	1.110	1.107
9	1.117	1.115	1.114	1.112	1.109
10	1.119	1.117	1.116	1.114	1.111
12	1.122	1.120	1.119	1.117	1.113
14					1.115
∞	1.1419	1.1419	1.1419	1.1419	1.1419

[a] Mason and Schamp (1958) for $\gamma = 0$, 0.25, 0.50, with an estimated accuracy of 0.4%. Monchick and Mason (1961) for $\gamma = 1$, with an estimated accuracy better than 0.2%. Results for $\gamma = 0.75$ by interpolation.

TABLE I-7. $(6C^*-5)$ for (12-6-4) potentials[a]

T^*	$6C^*-5$				
	$\gamma = 0$	$\gamma = 0.25$	$\gamma = 0.75$	$\gamma = 0.75$	$\gamma = 1$
0	0	0	0	0	0.3333
0.1	−0.099	−0.031	0.063	0.18	0.315
0.2	−0.222	−0.124	−0.007	0.11	0.236
0.3	−0.367	−0.264	−0.139	−0.02	0.111
0.4	−0.439	−0.340	−0.221	−0.11	0.013
0.5	−0.451	−0.360	−0.252	−0.15	−0.042
0.6	−0.428	−0.344	−0.247	−0.16	−0.062
0.7	−0.385	−0.307	−0.221	−0.14	−0.059
0.8	−0.331	−0.261	−0.183	−0.11	−0.041
0.9	−0.275	−0.210	−0.140	−0.08	−0.013
1.0	−0.219	−0.159	−0.095	−0.04	0.018
1.2	−0.112	−0.060	−0.007	0.04	0.086
1.4	−0.016	0.029	0.075	0.11	0.151
1.6	0.067	0.108	0.147	0.18	0.210
1.8	0.138	0.174	0.209	0.24	0.263
2.0	0.197	0.231	0.262	0.29	0.309
2.5	0.310	0.337	0.362	0.382	0.398
3.0	0.388	0.411	0.431	0.448	0.462
3.5	0.444	0.465	0.482	0.496	0.508
4	0.486	0.506	0.521	0.533	0.542
5	0.542	0.557	0.570	0.580	0.588
6	0.575	0.588	0.600	0.609	0.616
7	0.596	0.609	0.619	0.628	0.636
8	0.612	0.624	0.633	0.642	0.649
9	0.624	0.635	0.642	0.650	0.658
10	0.633	0.642	0.651	0.658	0.665
12					0.674
14					0.680
∞	0.6667	0.6667	0.6667	0.6667	0.6667

[a] Mason and Schamp (1958) for $\gamma = 0$, 0.25, 0.50, with the third decimal uncertain. Monchick and Mason (1961) for $\gamma = 1$. Results for $\gamma = 0.75$ by interpolation.

TABLE I-8. $\Omega^{(1,1)*}$ for (12-4) core potentials[a]

$\Omega^{(1,1)*}$

T^*	$a^* = 0$	$a^* = 0.1$	$a^* = 0.2$	$a^* = 0.3$	$a^* = 0.4$	$a^* = 0.5$	$a^* = 0.6$	$a^* = 0.7$	$a^* = 0.8$
0.01	17.19	14.52	12.07	9.868	7.893	6.158	4.661	3.401	2.375
0.02	12.04	10.27	8.640	7.159	5.823	4.641	3.613	2.736	2.012
0.03	9.763	8.389	7.114	5.951	4.894	3.955	3.134	2.428	1.841
0.04	8.409	7.266	6.201	5.226	4.336	3.541	2.842	2.239	1.734
0.05	7.485	6.499	5.577	4.729	3.953	3.256	2.640	2.107	1.659
0.06	6.805	5.932	5.115	4.362	3.668	3.043	2.490	2.008	1.602
0.07	6.274	5.492	4.755	4.075	3.446	2.877	2.371	1.930	1.557
0.08	5.851	5.136	4.464	3.842	3.265	2.742	2.275	1.866	1.520
0.09	5.497	4.840	4.222	3.649	3.115	2.630	2.194	1.813	1.489
0.1	5.199	4.592	4.017	3.485	2.988	2.534	2.126	1.768	1.463
0.2	3.527	3.202	2.881	2.575	2.281	2.004	1.745	1.513	1.312
0.3	2.713	2.526	2.333	2.139	1.946	1.755	1.568	1.394	1.240
0.4	2.217	2.105	1.985	1.859	1.729	1.594	1.456	1.320	1.196
0.5	1.881	1.819	1.742	1.661	1.573	1.477	1.374	1.267	1.165
0.6	1.651	1.611	1.566	1.512	1.453	1.386	1.310	1.226	1.141
0.7	1.482	1.461	1.431	1.399	1.359	1.313	1.258	1.192	1.122
0.8	1.353	1.344	1.329	1.309	1.285	1.254	1.215	1.164	1.106
0.9	1.253	1.253	1.247	1.238	1.225	1.204	1.178	1.139	1.092
1.0	1.175	1.179	1.181	1.179	1.174	1.163	1.146	1.118	1.080
1.2	1.056	1.070	1.080	1.090	1.096	1.098	1.095	1.083	1.060
1.4	0.9730	0.9921	1.009	1.024	1.038	1.049	1.056	1.054	1.044
1.6	0.9116	0.9345	0.9558	0.9761	0.9948	1.011	1.025	1.031	1.030

[a] Mason, O'Hara, and Smith (1972); an accuracy of 0.1% is estimated.

TABLE I-8—*Continued.*

$\Omega^{(1,\,1)*}$

T^*	$a^* = 0$	$a^* = 0.1$	$a^* = 0.2$	$a^* = 0.3$	$a^* = 0.4$	$a^* = 0.5$	$a^* = 0.6$	$a^* = 0.7$	$a^* = 0.8$
1.8	0.8647	0.8899	0.9139	0.9382	0.9605	0.9812	0.9996	1.012	1.018
2.0	0.8278	0.8546	0.8808	0.9075	0.9333	0.9568	0.9789	0.9959	1.008
2.5	0.7620	0.7921	0.8221	0.8523	0.8829	0.9124	0.9409	0.9655	0.9877
3.0	0.7180	0.7503	0.7830	0.8159	0.8488	0.8821	0.9143	0.9441	0.9726
3.5	0.6866	0.7201	0.7544	0.7893	0.8245	0.8597	0.8953	0.9281	0.9611
4.0	0.6625	0.6972	0.7327	0.7689	0.8058	0.8427	0.8800	0.9157	0.9519
4.5	0.6434	0.6790	0.7154	0.7528	0.7909	0.8294	0.8681	0.9061	0.9445
5.0	0.6275	0.6639	0.7012	0.7394	0.7786	0.8183	0.8583	0.8981	0.9383
6.0	0.6027	0.6403	0.6789	0.7186	0.7593	0.8009	0.8432	0.8853	0.9285
8.0	0.5688	0.6079	0.6483	0.6901	0.7331	0.7773	0.8225	0.8683	0.9155
10.0	0.5455	0.5858	0.6274	0.6706	0.7153	0.7614	0.8086	0.8570	0.9066
20.0	0.4836	0.5269	0.5722	0.6195	0.6687	0.7199	0.7730	0.8279	0.8848
40.0	0.4311	0.4770	0.5253	0.5762	0.6295	0.6853	0.7436	0.8043	0.8676
60.0	0.4034	0.4505	0.5004	0.5532	0.6086	0.6669	0.7280	0.7920	0.8588
80.0	0.3848	0.4328	0.4837	0.5376	0.5946	0.6545	0.7175	0.7836	0.8528
100.0	0.3710	0.4195	0.4712	0.5260	0.5840	0.6452	0.7097	0.7774	0.8484
200.0	0.3311	0.3811	0.4347	0.4921	0.5531	0.6180	0.6866	0.7591	0.8355
400.0	0.2954	0.3464	0.4015	0.4610	0.5248	0.5929	0.6653	0.7422	0.8237
600.0	0.2763	0.3276	0.3835	0.4441	0.5092	0.5791	0.6536	0.7329	0.8172
800.0	0.2634	0.3149	0.3714	0.4326	0.4987	0.5697	0.6456	0.7265	0.8127
1000.0	0.2538	0.3055	0.3622	0.4239	0.4907	0.5626	0.6396	0.7217	0.8093

[a] Mason, O'Hara, and Smith (1972); an accuracy of 0.1% is estimated.

TABLE I-9. A^* for (12-4) core potentials[a]

A^*

T^*	$a^* = 0$	$a^* = 0.1$	$a^* = 0.2$	$a^* = 0.3$	$a^* = 0.4$	$a^* = 0.5$	$a^* = 0.6$	$a^* = 0.7$	$a^* = 0.8$
0	0.8713	0.8713	0.8713	0.8713	0.8713	0.8713	0.8713	0.8713	0.8713
0.01	0.9168	0.9314	0.9489	0.9697	0.9936	1.0204	1.0498	1.0784	1.0994
0.02	0.9268	0.9423	0.9604	0.9827	1.0075	1.0342	1.0626	1.0871	1.1009
0.03	0.9342	0.9497	0.9684	0.9906	1.0156	1.0421	1.0693	1.0910	1.1005
0.04	0.9403	0.9556	0.9749	0.9963	1.0212	1.0475	1.0734	1.0932	1.0997
0.05	0.9457	0.9607	0.9804	1.0008	1.0254	1.0515	1.0762	1.0945	1.0988
0.06	0.9501	0.9652	0.9852	1.0046	1.0287	1.0545	1.0783	1.0953	1.0979
0.07	0.9551	0.9690	0.9894	1.0078	1.0315	1.0569	1.0798	1.0958	1.0969
0.08	0.9581	0.9733	0.9929	1.0108	1.0339	1.0589	1.0809	1.0960	1.0961
0.09	0.9624	0.9765	0.9964	1.0135	1.0360	1.0606	1.0818	1.0961	1.0952
0.1	0.9668	0.9802	0.9999	1.0160	1.0379	1.0620	1.0825	1.0960	1.0944
0.2	1.0210	1.0249	1.0332	1.0417	1.0543	1.0704	1.0847	1.0931	1.0875
0.3	1.0684	1.0672	1.0672	1.0680	1.0712	1.0776	1.0851	1.0895	1.0822
0.4	1.0933	1.0909	1.0874	1.0852	1.0840	1.0844	1.0858	1.0864	1.0779
0.5	1.1077	1.1011	1.0985	1.0944	1.0913	1.0884	1.0859	1.0838	1.0744
0.6	1.1127	1.1073	1.1013	1.0976	1.0935	1.0894	1.0852	1.0813	1.0712
0.7	1.1110	1.1062	1.1028	1.0973	1.0933	1.0880	1.0834	1.0786	1.0682
0.8	1.1110	1.1046	1.1013	1.0960	1.0912	1.0862	1.0809	1.0757	1.06 7
0.9	1.1085	1.1023	1.0979	1.0933	1.0885	1.0833	1.0782	1.0727	1.0630
1.0	1.1046	1.1005	1.0949	1.0905	1.0865	1.0801	1.0746	1.0695	1.0607
1.2	1.1025	1.0955	1.0914	1.0849	1.0799	1.0746	1.0689	1.0632	1.0561
1.4	1.0992	1.0922	1.0863	1.0810	1.0752	1.0701	1.0636	1.0575	1.0519

[a] Mason, O'Hara, and Smith (1972). An accuracy of 0.2% is estimated.

TABLE 1-9—*Continued.*

T*	A*								
	a* = 0	a* = 0.1	a* = 0.2	a* = 0.3	a* = 0.4	a* = 0.5	a* = 0.6	a* = 0.7	a* = 0.8
1.6	1.0976	1.0899	1.0835	1.0774	1.0719	1.0653	1.0598	1.0526	1.0483
1.8	1.0966	1.0887	1.0822	1.0749	1.0692	1.0632	1.0566	1.0486	1.0451
2.0	1.0960	1.0882	1.0812	1.0736	1.0666	1.0603	1.0537	1.0455	1.0423
2.5	1.0967	1.0876	1.0797	1.0720	1.0640	1.0566	1.0489	1.0403	1.0368
3.0	1.0985	1.0887	1.0797	1.0711	1.0631	1.0547	1.0467	1.0375	1.0331
3.5	1.1000	1.0902	1.0806	1.0713	1.0625	1.0541	1.0450	1.0362	1.0305
4.0	1.1021	1.0917	1.0817	1.0719	1.0626	1.0538	1.0445	1.0354	1.0287
4.5	1.1041	1.0932	1.0828	1.0727	1.0628	1.0535	1.0442	1.0346	1.0273
5.0	1.1060	1.0947	1.0839	1.0735	1.0633	1.0536	0.0439	1.0342	1.0264
6.0	1.1093	1.0974	1.0860	1.0749	1.0643	1.0540	1.0438	1.0341	1.0252
8.0	1.1140	1.1016	1.0893	1.0774	1.0659	1.0549	1.0442	1.0338	1.0239
10.0	1.1174	1.1044	1.0917	1.0793	1.0673	1.0557	1.0446	1.0338	1.0235
20.0	1.1261	1.1116	1.0975	1.0838	1.0707	1.0580	1.0458	1.0341	1.0229
40.0	1.1319	1.1159	1.1007	1.0862	1.0723	1.0590	1.0463	1.0341	1.0224
60.0	1.1343	1.1176	1.1018	1.0869	1.0727	1.0591	1.0462	1.0339	1.0222
80.0	1.1357	1.1185	1.1024	1.0872	1.0727	1.0590	1.0460	1.0337	1.0219
100.0	1.1366	1.1191	1.1026	1.0873	1.0726	1.0588	1.0458	1.0334	1.0218
200.0	1.1388	1.1202	1.1030	1.0870	1.0720	1.0580	1.0449	1.0326	1.0213
400.0	1.1400	1.1205	1.1025	1.0861	1.0708	1.0567	1.0436	1.0315	1.0206
600.0	1.1404	1.1203	1.1020	1.0853	1.0699	1.0558	1.0428	1.0308	1.0201
800.0	1.1407	1.1201	1.1015	1.0846	1.0692	1.0552	1.0422	1.0303	1.0197
1000.0	1.1408	1.1199	1.1011	1.0841	1.0687	1.0546	1.0417	1.0299	1.0194
∞	1.1419	1.0000	1.0000	1.0000	1.0000	1.0000	1.0000	1.0000	1.0000

[a] Mason, O'Hara, and Smith (1972). An accuracy of 0.2% is estimated.

TABLE 1-10. $(6C^*-5)$ for (12·4) core potentials[a]

				$6C^* - 5$				
T^*	$a^* = 0$	$a^* = 0.1$	$a^* = 0.2$	$a^* = 0.3$	$a^* = 0.4$	$a^* = 0.5$	$a^* = 0.6$	$a^* = 0.7$	$a^* = 0.8$
0	0	0	0	0	0	0	0	0	0
0.01	-0.024	0.002	0.032	0.067	0.111	0.169	0.245	0.349	0.497
0.02	-0.031	0.002	0.039	0.082	0.135	0.201	0.285	0.396	0.546
0.03	-0.036	0.002	0.044	0.093	0.152	0.222	0.312	0.426	0.575
0.04	-0.040	0.000	0.048	0.102	0.165	0.240	0.332	0.447	0.596
0.05	-0.045	0.000	0.050	0.109	0.176	0.254	0.348	0.465	0.612
0.06	-0.048	-0.001	0.052	0.115	0.185	0.266	0.362	0.479	0.624
0.07	-0.048	-0.002	0.053	0.119	0.192	0.276	0.373	0.492	0.636
0.08	-0.054	-0.003	0.054	0.122	0.199	0.285	0.384	0.503	0.645
0.09	-0.057	-0.003	0.054	0.126	0.205	0.294	0.393	0.513	0.654
0.1	-0.063	-0.008	0.057	0.128	0.210	0.301	0.402	0.522	0.661
0.2	-0.216	-0.105	0.005	0.114	0.227	0.342	0.458	0.581	0.710
0.3	-0.367	-0.229	-0.090	0.054	0.198	0.341	0.483	0.612	0.738
0.4	-0.445	-0.301	-0.151	-0.002	0.160	0.324	0.486	0.628	0.756
0.5	-0.447	-0.324	-0.176	-0.023	0.137	0.308	0.481	0.636	0.768
0.6	-0.429	-0.295	-0.171	-0.021	0.136	0.300	0.475	0.639	0.778
0.7	-0.385	-0.268	-0.134	-0.004	0.144	0.303	0.472	0.639	0.784
0.8	-0.328	-0.220	-0.105	0.027	0.164	0.312	0.473	0.639	0.788
0.9	-0.276	-0.173	-0.058	0.057	0.188	0.327	0.479	0.640	0.791
1.0	-0.222	-0.118	-0.015	0.097	0.217	0.346	0.488	0.642	0.792
1.2	-0.108	-0.022	0.075	0.172	0.274	0.387	0.512	0.650	0.795
1.4	-0.014	0.064	0.150	0.240	0.333	0.429	0.538	0.662	0.798

[a] Mason, O'Hara, and Smith (1972); the third decimal is uncertain.

TABLE I-10. (continued)

					$6C^* - 5$				
T^*	$a^* = 0$	$a^* = 0.1$	$a^* = 0.2$	$a^* = 0.3$	$a^* = 0.4$	$a^* = 0.5$	$a^* = 0.6$	$a^* = 0.7$	$a^* = 0.8$
1.6	0.069	0.141	0.217	0.297	0.383	0.471	0.566	0.675	0.802
1.8	0.139	0.207	0.279	0.350	0.428	0.509	0.592	0.690	0.807
2.0	0.197	0.263	0.322	0.396	0.464	0.540	0.617	0.705	0.811
2.5	0.311	0.368	0.428	0.487	0.543	0.605	0.669	0.738	0.825
3.0	0.390	0.442	0.495	0.547	0.601	0.652	0.707	0.768	0.840
3.5	0.444	0.495	0.544	0.592	0.640	0.689	0.735	0.792	0.851
4.0	0.485	0.533	0.581	0.626	0.670	0.716	0.760	0.810	0.862
4.5	0.517	0.563	0.608	0.651	0.693	0.737	0.779	0.824	0.871
5	0.542	0.586	0.629	0.672	0.712	0.753	0.794	0.835	0.880
6	0.576	0.619	0.660	0.701	0.739	0.777	0.814	0.854	0.893
8	0.615	0.657	0.696	0.734	0.770	0.806	0.840	0.876	0.908
10	0.634	0.677	0.716	0.753	0.788	0.822	0.855	0.888	0.919
20	0.664	0.707	0.747	0.784	0.818	0.851	0.881	0.911	0.938
40	0.671	0.717	0.759	0.797	0.831	0.864	0.893	0.921	0.947
60	0.672	0.720	0.762	0.801	0.836	0.868	0.898	0.926	0.951
80	0.672	0.721	0.764	0.803	0.839	0.871	0.900	0.927	0.953
100	0.672	0.722	0.765	0.805	0.840	0.873	0.902	0.929	0.954
200	0.671	0.723	0.770	0.810	0.846	0.878	0.907	0.933	0.957
400	0.670	0.725	0.773	0.814	0.851	0.882	0.911	0.937	0.960
600	0.669	0.726	0.775	0.817	0.853	0.885	0.914	0.939	0.962
800	0.668	0.727	0.776	0.819	0.855	0.887	0.915	0.939	0.962
1000	0.668	0.728	0.777	0.820	0.857	0.888	0.916	0.941	0.963
∞	0.6667	1.0000	1.0000	1.0000	1.0000	1.0000	1.0000	1.0000	1.0000

[a] Mason, O'Hara, and Smith (1972); the third decimal is uncertain.

TABLE I-11. Summary of classical collision integrals available for other potentials

Potential	Reference
Inverse power, repulsive and attractive	Kihara, Taylor, and Hirschfelder (1960); Higgins and Smith (1968)
Exponential, repulsive	Monchick (1959); Higgins and Smith (1968)
Exponential, attractive	Munn, Mason, and Smith (1965)
Screened Coulomb, repulsive and attractive	Mason, Munn, and Smith (1967); Hahn, Mason, and Smith (1971)
Morse	Smith and Munn (1964); Samoilov and Tsitelauri (1964)
Exponential-6	Mason (1954); Mason and Rice (1954); Umanskii and Bogdanova (1968)
n-6	Smith, Mason, and Munn (1965); Dymond, Rigby, and Smith (1966); Klein and Smith (1968); Schramm (1968); Lin and Hsu (1969)
12-6-3	Monchick and Mason (1961)
12-6-5	Smith, Munn, and Mason (1967)
28-7	Smith, Mason, and Munn (1965)
(12-6) spherical core	Barker, Fock, and Smith (1964)
(12-6) spherical shell	DeRocco, Storvick, and Spurling (1968); Schramm (1969)
Square well	Holleran and Hulburt (1951); Brush and Lawrence (1963)

REFERENCES

Barker, J. A., W. Fock and F. Smith (1964), *Phys. Fluids* **7**, 897.

Brush, S. G., and J. D. Lawrence (1963), University of California Lawrence Radiation Laboratory Report **UCRL-7376**. Available from Office of Technical Services, U.S. Dept. of Commerce, Washington, D.C.

DeRocco, A. G., T. S. Storvick, and T. H. Spurling (1968), *J. Chem. Phys.* **48**, 997.

Dymond, J. H., M. Rigby, and E. B. Smith (1966), *Phys. Fluids* **9**, 1222.

Hahn, H., E. A. Mason, and F. J. Smith (1971), *Phys. Fluids* **14**, 278.

Hassé, H. R. (1926), *Phil. Mag.* **1**, 139.

―――― and W. R. Cook (1927), *Phil. Mag.* **3**, 977.

―――― (1929), *Proc. Roy. Soc. (London)* **A125**, 196.

―――― (1931), *Phil. Mag.* **12**, 554.

Higgins, L. D., and F. J. Smith (1968), *Mol. Phys.* **14**, 399.

Hirschfelder, J. O., R. B. Bird, and E. L. Spotz (1948), *J. Chem. Phys.* **16**, 968.

Holleran, E. M., and H. M. Hulburt (1951), *J. Chem. Phys.* **19**, 232.

Kihara, T., and M. Kotani (1943), *Proc. Phys.-Math. Soc. Japan* **25**, 602.

Kihara, T., M. H. Taylor, and J. O. Hirschfelder (1960), *Phys. Fluids* **3**, 715.

Klein, M., and F. J. Smith (1968), *J. Res. Natl. Bur. Stds. (U. S.)* **72A**, 359.

Lin, S. T., and H. W. Hsu (1969), *J. Chem. Engng. Data* **14**, 328.

Mason, E. A. (1954), *J. Chem. Phys.* **22**, 169.

—— Munn, R. J., and Smith, F. J. (1967), *Phys. Fluids* **10**, 1827.

Mason, E. A., and W. E. Rice (1954), *J. Chem. Phys.* **22**, 843.

Mason, E. A., and H. W. Schamp (1958), *Ann. Phys. (N. Y.)* 4, **233**.

Mason, E. A., H. O'Hara, and F. J. Smith (1972), *J. Phys.* **B5**, 169.

Monchick, L. (1959), *Phys. Fluids* **2**, 695.

——, and E. A. Mason (1961), *J. Chem. Phys.* **35**, 1676.

Munn, R. J., E. A. Mason, and F. J. Smith, (1965), *Phys. Fluids* **8**, 1103.

Samoilov, E. V., and N. N. Tsitelauri (1964), *High Temp. (English Transl.)* **2**, 509 [*Tepl. Vys. Temp.* **2**, 565 (1964)].

Schramm, B. (1968), *Ber. Bunsen Ges. Physik. Chemie* **72**, 609.

—— (1969), *Ber. Bunsen Ges. Physik. Chemie* **73**, 217.

Smith, F. J., E. A. Mason, and R. J. Munn (1965), *J. Chem. Phys.* **42**, 1334.

Smith, F. J., and R. J. Munn (1964), *J. Chem. Phys.* **41**, 3560.

——, and E. A. Mason (1967), *J. Chem. Phys.* **46**, 317.

Umanskii, A. S., and S. S. Bogdanova (1968), *High Temp. (English Transl.)* **6**, 518 [*Tepl. Vys. Temp.* **6**, 543 (1968)].

TABLES OF PROPERTIES

USEFUL IN THE ESTIMATION OF

ION-NEUTRAL INTERACTION ENERGIES

II-1   Polarizabilities for Atoms
II-2   Polarizabilities for Molecules
II-3   Polarizabilities for Atomic Ions
II-4   Permanent Dipole and Quadrupole Moments for Molecules
II-5   Equivalent Oscillator Numbers for Dispersion Energies
II-6   Ion-Atom Potential Parameters Derived from Mobility
       Measurements
II-7   Summary of Information Available on Short-Range Ion-Neutral
       Interactions

TABLE II-1. Polarizabilities for atoms

Atom	Dipole Polarizability $\alpha(\text{Å}^3)^a$	Quadrupole Polarizability $\alpha_q(\text{Å}^5)$
He	0.205	0.101^f
Ne	0.395	0.370^g
Ar	1.64	2.19^g
Kr	2.48	
Xe	4.04	
H	0.667	0.622^h
B	5.1	
C	1.5^b	
N	1.13	
O	0.77	
F	0.6	
Li	24.0	60.0^i
Na	24.2	74.8^i
K	41.3	212^i
Rb	43.6	261^i
Cs	53	441^i
Be	5.4	9.1^j
Mg	11.1^c	
Hg	5.02^d	
He(2^1S)	119^e	
He(2^3S)	46.8^e	

Comprehensive reviews on atomic polarizabilities have been given by Dalgarno (1962) and by Teachout and Pack (1971).

[a] Values from Teachout and Pack (1971) unless otherwise noted.
[b] Miller and Kelly (1972).
[c] Stwalley (1971).
[d] Stwalley and Kramer (1968).
[e] Victor, Dalgarno, and Taylor (1968).
[f] Davison (1966).
[g] Burns (1959).
[h] Exact value (Dalgarno, 1962).
[i] Sternheimer (1970).
[j] Dalgarno and McNamee (1961).

344

TABLE II-2. Polarizabilities for molecules

Molecule	Mean Dipole Polarizability $\bar{\alpha}(\text{Å}^3)$	Anisotropy $\kappa = \dfrac{\alpha_{\parallel} - \alpha_{\perp}}{\alpha_{\parallel} + 2\alpha_{\perp}}$
H_2	0.808^a	$0.124^{a,d}$
N_2	1.76^b	$0.176,^b\ 0.131^d$
O_2	1.60^b	$0.238,^b\ 0.229^d$
Cl_2	4.61^b	0.215^b
Br_2	6.46^c	
CO	1.95^b	0.167^b
NO	1.70^c	
HF	0.79^c	$0.10,^b\ 0.093^e$
HCl	2.58^c	0.094^b
HBr	3.52^c	0.084^b
HI	5.23^c	0.104^b
H_2O	1.45^c	
H_2S	3.67^c	0.05^b
CO_2	2.59^c	0.264^b
N_2O	2.92^c	0.310^b
SO_2	3.78^c	0.26^b
NO_2	3.02^c	
$(NO_2)_2$	6.64^c	
CH_4	2.56^c	0
CF_4	2.82^c	0
CCl_4	10.2^c	0
SF_6	4.48^c	0

[a] Victor and Dalgarno (1969).
[b] Hirschfelder, Curtiss, and Bird (1964), p. 950.
[c] Maryott and Buckley (1953).
[d] Langhoff, Gordon, and Karplus (1971)
[e] Muenter (1972).

TABLE II-3. Polarizabilities for atomic ions

Ion	Dipole Polarizability $\alpha(\text{Å}^3)^a$
Li^+	0.0285^b
Na^+	0.280^c
K^+	1.09^c
Rb^+	1.81^c
Cs^+	2.73^c
He^+	0.0417
Ne^+	0.21
H^+	0
O^+	0.49
He^{++}	0
Ne^{++}	0.15
Be^{++}	0.0076
Mg^{++}	0.082
Ca^{++}	0.73
H^-	30.5^b
O^-	3.2
F^-	0.76
Cl^-	3.0
Br^-	4.2
I^-	6.2

[a] Values from the comprehensive review of Dalgarno (1962) unless otherwise noted.
[b] Chung (1971).
[c] Buckingham (1937a).

TABLE II-4. Dipole and quadrupole moments

Molecule	Dipole Moment $(10^{-18}\text{esu})^a$	Quadrupole Moment $(10^{-26}\text{esu})^b$
H_2	0	$+0.662$
N_2	0	-1.52
O_2	0	-0.39
F_2	0	$+0.88$
Cl_2	0	$+6.14$
CO	0.112	-2.5
NO	0.153	-1.8
HF	1.82	$+2.6$
HCl	1.08	$+3.8$
HBr	0.82	$+4$
HI	0.44	$+6$
H_2O	1.85	$\sim \pm 1$
H_2S	0.97	
CO_2	0	-4.3
N_2O	0.167	-3.0
SO_2	1.63	± 4.4
NO_2	0.316	

[a] Nelson, Lide, and Maryott (1967).
[b] Stogryn and Stogryn (1966).

TABLE II-5. Equivalent oscillator numbers

Species	N	Species	N
H	0.824[a]	H_2	1.86[a]
		N_2	5.62[a]
He	1.45[a]	O_2	1.99[d]
Ne	3.74[a]	CH_4	7.73[a]
Ar	5.52[a]		
Kr	6.41[a]	He^+	0.824[e]
Xe	6.38[a]		
		Li^+	1.93[f]
Li	0.779[a]	Na^+	3.94[f]
Na	0.988[a]	K^+	5.40[f]
K	1.03[a]	Rb^+	5.32[f]
Rb	1.11[a]	Cs^+	7.49[f]
Cs	1.20[a]		
		H^-	1.13[g]
Hg	2.63[b]	F^-	4.0[h]
$He(2^1S)$	440[c]		
$He(2^3S)$	610[c]		

[a] Dalgarno (1967).
[b] Stwalley and Kramer (1968).
[c] Victor, Dalgarno, and Taylor (1968).
[d] Langhoff and Karplus (1970).
[e] Same value as for H atom.
[f] Buckingham (1937b).
[g] From $H^- + H$ and $H^- + He$ calculations of Dalgarno and Kingston (1959) and Davison (1966).
[h] Donath (1963).

TABLE II-6. Potential parameters derived
from mobility measurements[a]

System	γ	ε (eV)	r_m (Å)
Li$^+$-He	0.10	0.0474	2.22
Na$^+$-He	0.15	0.0403	2.35
Cs$^+$-He	0.42	0.0140	3.36
K$^+$-Ar	0.2	0.121	3.00
Rb$^+$-Kr	0.2	0.119	3.34
Cs$^+$-Xe	0.2	0.106	3.88

$$V(r) = \frac{\varepsilon}{2}\left[(1 + \gamma)\left(\frac{r_m}{r}\right)^{12} - 4\gamma\left(\frac{r_m}{r}\right)^6 - 3(1 - \gamma)\left(\frac{r_m}{r}\right)^4\right]$$

[a] Mason and Schamp (1958).

TABLE II-7. Information available on short-range ion-neutral interactions

System	Nature of Interaction[a]	References	
		Theoretical[b]	Experimental
H$^+$-H	c.t.	Bates et al. (1953) Dalgarno and Yadav (1953)	Fite et al. (1960) Fite et al. (1962)
H$^+$-He	val.	Mason and Vanderslice (1957) Peyerimhoff (1965) Wolniewicz (1965) Michels (1966) Rich et al. (1971)	Simons et al. (1943b) Chupka and Russell (1968) Doverspike et al. (1970) Champion et al. (1970)
H$^+$-Ne	val.	Peyerimhoff (1965) Bobbio et al. (1971) Rich et al. (1971)	Chupka and Russell (1968) Champion et al. (1970)
H$^+$-Ar	val.	Rich et al. (1971) Klingbeil (1972)	Chupka and Russell (1968) Herrero et al. (1969) Champion et al. (1970)
H$^+$-Kr	val.	Rich et al. (1971)	Champion et al. (1970)
H$^+$-H$_2$	val.	Mason and Vanderslice (1959b)	Simons et al. (1943a) Cramer (1961)
H$^+$-CH$_4$	rep.	Cloney and Vanderslice (1962)	Simons and Fryburg (1945)
H$^+$-C$_2$H$_6$	rep.	Cloney and Vanderslice (1962)	Simons and McAllister (1952)
H$^+$-CF$_4$	rep.	Cloney and Vanderslice (1962)	Simons and Garber (1953)
H$^+$-C$_2$F$_6$	rep.	Cloney and Vanderslice (1962)	Simons and Garber (1953)
D$^+$-D$_2$	val.		
He$^+$-H	rep.	Michels (1966)	Cramer and Marcus (1960)

He⁺-He	c.t.	Rapp and Francis (1962)	Cramer and Simons (1957)
		Olson and Mueller (1967)	Lorents and Aberth (1965)
		Gupta and Matsen (1967)	Mahadevan and Magnuson
		Hasted (1968)	(1968)
		Gilbert and Wahl (1971)	
He⁺-Ne	rep.	Smith et al. (1967)	Cramer (1958)
		Coffey et al. (1969)	Aberth and Lorents (1966)
		Olson and Smith (1971)	
He⁺-Ar	rep.	Smith et al. (1967)	Aberth and Lorents (1966)
		Smith et al. (1970)	
Ne⁺-He	rep.	Coffey et al. (1969)	Cramer (1958)
Ne⁺-Ne	c.t.	Mason and Vanderslice (1959a)	Ghosh and Sheridan (1957a)
		Rapp and Francis (1962)	Cramer (1958)
		Hasted (1968)	
		Gilbert and Wahl (1971)	
Ne⁺-Ar	rep.		Cramer (1959)
Ar⁺-He	rep.	Smith et al. (1970)	Chiang et al. (1971)
Ar⁺-Ne	rep.		Cramer (1959)
Ar⁺-Ar⁺	c.t.	Cloney et al. (1962)	Ghosh and Sheridan (1957a)
		Rapp and Francis (1962)	Cramer (1959)
		Hasted (1968)	Nichols and Witteborn (1966)
		Johnson (1970)	Aberth and Lorents (1966)
		Gilbert and Wahl (1971)	Mahadevan and Magnuson
			(1968)
Kr⁺-Kr	c.t.	Rapp and Francis (1962)	Ghosh and Sheridan (1957a)
		Hasted (1968)	
		Devoto (1969)	

TABLE II-7. *Continued*

System	Nature of Interaction[a]	References	
		Theoretical[b]	Experimental
Xe⁺-Xe	c.t.	Rapp and Francis (1962) Devoto (1969)	Ghosh and Sheridan (1957a)
Li⁺-H	rep.	Platas et al. (1959)	
Li⁺-He	rep.	Fischer (1968) Olson et al. (1970) Catlow et al. (1970) Krauss et al. (1971)	Zehr and Berry (1967) Aberth and Lorents (1969) Inouye and Kita (1972b)
Li⁺-Li	c.t.	Peek et al. (1968) Fischer and Kemmey (1969) Bardsley (1971)	Lorents et al. (1965) Perel et al. (1969) Aberth et al. (1970)
Li⁺-H₂	rep.	Lester (1970)	
Li⁺-N₂	rep.		Aberth and Lorents (1969)
Li⁺-O₂	rep.		Aberth and Lorents (1969)
Na⁺-He	rep.	Krauss et al. (1971)	
Na⁺-Na	c.t.	Sheldon (1963) Davies et al. (1965)	
K⁺-He	rep.		Inouye and Kita (1972a) Amdur et al. (1972)
K⁺-Ne	rep.		Inouye and Kita (1972a) Amdur et al. (1972)
K⁺-Ar	rep.	Sida (1957)	Inouye and Kita (1971, 1972a) Amdur et al. (1971, 1972)

Ion pair	Type	References	
K$^+$-Kr	rep.		Inouye and Kita (1972a)
K$^+$-K	c.t.	Davies et al. (1965)	Chkuaseli et al. (1961)
Rb$^+$-Rb	c.t.	Rapp and Francis (1962); Davies et al. (1965); Olson (1969)	Perel et al. (1965)
Cs$^+$-Cs	c.t.	Rapp and Francis (1962); Davies et al. (1965); Olson (1969)	Marino et al. (1962); Perel et al. (1965); Gentry et al. (1968); Perel et al. (1970)
Zn$^+$-Zn	c.t.	Palyukh and Savchin (1969)	Kushnir et al. (1966)
Cd$^+$-Cd	c.t.	Palyukh and Savchin (1969)	Dillon et al. (1955)
Hg$^+$-Hg	c.t.	Popescu Iovitsu and Ionescu-Pallas (1960); Rapp and Francis (1962); Palyukh and Savchin (1969)	Kushnir et al. (1959)
N$^+$-N	c.t.	Knof et al. (1964); Gilmore (1965); Stallcop (1971)	
O$^+$-O	c.t.	Knof et al. (1964)	Stebbings et al. (1964)
He^{++}-H	val.	Bates and Carson (1956)	
He^{++}-He	c.t.	Browne (1965); Dickinson (1968)	
Be^{++}-He	rep.	Hayes and Gole (1971)	Giffen and Berry (1971)
Be^{++}-Ne	rep.	Hayes and Gole (1971)	
Ca^{++}-Ne	rep.	Sida (1957)	

TABLE II-7. *Continued.*

System	Nature of Interaction[a]	References	
		Theoretical[b]	Experimental
H⁻-H	c.t.	Dalgarno and McDowell (1956)	Hummer et al. (1960)
		Bardsley et al. (1966)	
		Burke (1968)	
		Davidović and Janev (1969)	
H⁻-He	rep.	Mason and Vanderslice (1958a)	Bailey et al. (1957)
H⁻-Ne	rep.	Mason and Vanderslice (1958b)	Bailey et al. (1957)
H⁻-Ar	rep.	Mason and Vanderslice (1958b)	Bailey et al. (1957)
H⁻-H₂	rep.	Mason and Vanderslice (1958b)	Muschlitz et al. (1957)
O⁻-O	c.t.	Davidović and Janev (1971)	Snow et al. (1969)
O⁻-O₂	rep.		Muschlitz (1960)
Na⁻-Na	c.t.	Davidović and Janev (1969)	Bydin (1964)
K⁻-K	c.t.	Davidović and Janev (1969)	Bydin (1964)
Rb⁻-Rb	c.t.	Davidović and Janev (1969)	Bydin (1964)
Cs⁻-Cs	c.t.	Davidović and Janev (1969)	Bydin (1964)
F⁻-F	c.t.	Gilbert and Wahl (1971)	
Cl⁻-Cl	c.t.	Gilbert and Wahl (1971)	
H₂⁺-He	rep.	Mason and Vanderslice (1957)	Simons et al. (1943b)
H₂⁺-CH₄	rep.	Cloney and Vanderslice (1962)	Simons and Fryburg (1945)
H₂⁺-C₂H₆	rep.	Cloney and Vanderslice (1962)	Simons and McAllister (1952)

354

H_2^+-CF_4	rep.	Cloney and Vanderslice (1962)	Simons and Garber (1953)
H_2^+-C_2F_6	rep.	Cloney and Vanderslice (1962)	Simons and Garber (1953)
H_3^+-He	rep.	Mason and Vanderslice (1957)	Simons et al. (1943b)
N_2^+-N_2	c.t.		Stebbings et al. (1963)
			Nichols and Witteborn (1966)
			Savage and Witteborn (1968)
CO^+-CO	c.t.		Dillon et al. (1955)
			Ghosh and Sheridan (1957b)
NO^+-NO	c.t.		Dillon et al. (1955)
			Ghosh and Sheridan (1957b)
			Stebbings et al. (1963)
O_2^+-O_2	c.t.		Dillon et al. (1955)
			Ghosh and Sheridan (1957b)
			Stebbings et al. (1963)
Cl_2^+-Cl_2	c.t.		Dillon et al. (1955)
			Ghosh and Sheridan (1957b)
OH^--O_2	rep.(?)		Baker et al. (1962)
O_2^--O_2	c.t.		Muschlitz (1960)

[a] *rep.* = short-range repulsive force.

val. = short-range attractive valence force.

c.t. = resonant charge transfer.

[b] Either calculations or analysis of experimental data.

A review on elastic ion scattering has been given by Mason and Vanderslice (1962) and on charge transfer by Fite (1964) and by Hasted (1962, 1968).

REFERENCES

Aberth, W., O. Bernardini, D. Coffey, D. C. Lorents, and R. E. Olson, (1970), *Phys. Rev. Letters* **24**, 345.

Aberth, W., and D. C. Lorents (1966), *Phys. Rev.* **144**, 109.

——— (1969), *Phys. Rev.* **182**, 162.

Amdur, I., J. E. Jordan, K.-R. Chien, L. W.-M. Fung, R. L. Hance, E. Hulpke, and S. E. Johnson (1972), *J. Chem. Phys.* **57**, 2117.

Amdur, I., J. E. Jordan, L. W. Fung, R. L. Hance, E. Hulpke, and S. E. Johnson (1971), *Electronic and Atomic Collisions Abstracts, 7th Intern. Conf. Physics Electronic Atomic Collisions*, North-Holland, Amsterdam, p. 955.

Bailey, T. L., C. J. May, and E. E. Muschlitz (1957), *J. Chem. Phys.* **26**, 1446.

Baker, C. E., J. M. McGuire, and E. E. Muschlitz (1962), *J. Chem. Phys.* **37**, 2571.

Bardsley, J. N. (1971), *Phys. Rev.* **A3**, 1317 (1971).

———, A. Herzenberg, and F. Mandl (1966), *Proc. Phys. Soc. (London)* **89**, 305.

Bates, D. R., and T. R. Carson (1956), *Proc. Roy. Soc. (London)* **A234**, 207.

Bates, D. R., K. Ledsham, and A. L. Stewart (1953), *Phil. Trans. Roy. Soc. (London)* **A246**, 215.

Bobbio, S. M., W. G. Rich, L. D. Doverspike, and R. L. Champion (1971), *Phys. Rev.* **A4**, 957.

Browne, J. C. (1965), *J. Chem. Phys.* **42**, 1428.

Buckingham, R. A. (1937a), *Proc. Roy. Soc. (London)* **A160**, 94.

——— (1937b), *Proc. Roy. Soc. (London)* **A160**, 113.

Burke, P. G. (1968), *J. Phys.* **B1**, 586.

Burns, G. (1959), *J. Chem. Phys.* **31**, 1253.

Bydin, Yu. F. (1964), *Soviet Phys.-JETP (English transl.)* **19**, 1091 [*Zh. Eksperim. Teor. Fiz.* **46**, 1612 (1964)].

Catlow, G. W., M. R. C. McDowell, J. J. Kaufman, L. M. Sachs, and E. S. Chang (1970), *J. Phys.* **B3**, 833.

Champion, R. L., L. D. Doverspike, W. G. Rich, and S. M. Bobbio (1970), *Phys. Rev.* **A2**, 2327.

Chiang, M. H., E. A. Gislason, B. H. Mahan, C. W. Tsao, and A. S. Werner (1971), *J. Chem. Phys.* **55**, 3937.

Chkuaseli, D. V., V. D. Nikoleishvili, and A. I. Guldamashvili (1961), *Soviet Phys.-Tech. Phys. (English transl.)* **5**, 770 [*Zh. Tekh. Fiz.* **30**, 817 (1960)].

Chung, K. T. (1971), *Phys. Rev.* **A4**, 7.

Chupka, W. A., and M. E. Russell (1968), *J. Chem. Phys.* **49**, 5426.

Cloney, R. D., E. A. Mason, and J. T. Vanderslice (1962), *J. Chem. Phys.* **36**, 1103. The potentials reported are inadequate because a two-state rather than a four-state analysis was used, but the summarized charge transfer cross sections are unaffected and can be used directly in mobility calculations.

Cloney, R. D., and J. T. Vanderslice (1962), *J. Chem. Phys.* **36**, 1866.

Coffey, D., D. C. Lorents, and F. T. Smith (1969), *Phys. Rev.* **187**, 201.

Cramer, W. H. (1958), *J. Chem. Phys.* **28**, 688.

────── (1959), *J. Chem. Phys.* **30**, 641.

────── (1961), *J. Chem. Phys.* **35**, 836.

──────, and A. B. Marcus (1960), *J. Chem. Phys.* **32**, 186.

Cramer, W. H., and J. H. Simons (1957), *J. Chem. Phys.* **26**, 1272.

Dalgarno, A. (1962), *Adv. Phys.* **11**, 281.

────── (1967), *Adv. Chem. Phys.* **12**, 143.

──────, and A. E. Kingston (1959), *Proc. Phys. Soc. (London)* **73**, 455.

Dalgarno, A., and M. R. C. McDowell (1956), *Proc. Phys. Soc. (London)* **A69**, 615.

Dalgarno, A., and J. M. McNamee (1961), *J. Chem. Phys.* **35**, 1517.

Dalgarno, A., and H. N. Yadav (1953), *Proc. Phys. Soc. (London)* **A66**, 173.

Davidović, D. M., and R. K. Janev (1969), *Phys. Rev.* **186**, 89.

────── (1971), *Phys. Rev.* **A3**, 604.

Davies, R. H., E. A. Mason, and R. J. Munn (1965), *Phys. Fluids* **8**, 444.

Davison, W. D. (1966), *Proc. Phys. Soc. (London)* **87**, 133.

Devoto, R. S. (1969), *AIAA J.* **7**, 199.

Dickinson, A. S. (1968), *J. Phys.* **B1**, 395.

Dillon, J. A., W. F. Sheridan, H. D. Edwards, and S. N. Ghosh (1955), *J. Chem. Phys.* **23**, 776.

Donath, W. E. (1963), *J. Chem. Phys.* **39**, 2685.

Doverspike, L. D., R. L. Champion, S. M. Bobbio, and W. G. Rich (1970), *Phys. Rev. Letters* **25**, 909.

Fischer, C. R. (1968), *J. Chem. Phys.* **48**, 215.

──────, and P. J. Kemmey (1969), *Phys. Rev.* **186**, 272.

Fite, W. L. (1964), *Ann. Géophysique* **20**, 47.

──────, A. C. H. Smith, and R. F. Stebbings (1962), *Proc. Roy. Soc. (London)* **A268**, 527.

Fite, W. L., R. F. Stebbings, D. G. Hummer, and R. T. Brackmann (1960), *Phys. Rev.* **119**, 663.

Gentry, W. R., Y. Lee, and B. H. Mahan (1968), *J. Chem. Phys.* **49**, 1758.

Ghosh, S. N., and W. F. Sheridan (1957a), *J. Chem. Phys.* **26**, 480.

────── (1957b), *J. Chem. Phys.* **27**, 1436.

Giffen, W. C., and H. W. Berry (1971), *Phys. Rev.* **A3**, 635.

Gilbert, T. L., and A. C. Wahl (1971), *J. Chem. Phys.* **55**, 5247. ⇐

Gilmore, F. R. (1965), *J. Quant. Spectry. Radiat. Transfer* **5**, 369.

Gupta, B. K., and F. A. Matsen (1967), *J. Chem. Phys.* **47**, 4860.

Hasted, J. B. (1962), in *Atomic and Molecular Processes*, D. R. Bates, Ed., Chapter 18, Academic, New York.

────── (1968), *Adv. Atom. Mol. Phys.* **4**, 237.

Hayes, E. F., and J. L. Gole (1971), *J. Chem. Phys.* **55**, 5132.

Herrero, F. A., E. M. Nemeth, and T. L. Bailey (1969), *J. Chem. Phys.* **50**, 4591.

Hirschfelder, J. O., C. F. Curtiss, and R. B. Bird (1964), *Molecular Theory of Gases and Liquids*, Wiley, New York.

Hummer, D. G., R. F. Stebbings, W. L. Fite, and L. M. Branscomb (1960), *Phys. Rev.* **119**, 668.

Inouye, H., and S. Kita (1971), *Electronic and Atomic Collisions Abstracts, 7th Intern. Conf. Physics Electronic Atomic Collisions*, North-Holland, Amsterdam, p. 948.

——— (1972a), *J. Chem. Phys.* **56**, 4877.

——— (1972b), *J. Chem. Phys.* **57**, 1301.

Johnson, R. E. (1970), *J. Phys.* **B3**, 539.

Klingbeil, R. (1972), *J. Chem. Phys.* **57**, 1066.

Knof, H., E. A. Mason, and J. T. Vanderslice (1964), *J. Chem. Phys.* **40**, 3548.

Krauss, M., P. Maldonado, and A. C. Wahl (1971), *J. Chem. Phys.* **54**, 4944.

Kushnir, R. M., B. M. Palyukh, and L. A. Sena (1959), *Bull. Acad. Sci. USSR, Phys. Ser.* **23**, 995.

Kushnir, R. M., B. M. Palyukh, and L. S. Savchin (1966), *Soviet Phys.-Tech. Phys. (English transl.)* **10**, 1695 [*Zh. Tekh. Fiz.* **35**, 2212 (1965)].

Langhoff, P. W., R. G. Gordon, and M. Karplus (1971), *J. Chem. Phys.* **55**, 2126.

Langhoff, P. W., and M. Karplus (1970), in *The Padé Approximant in Theoretical Physics*, Academic, New York, Chap. 2.

Lester, W. A. (1970), *J. Chem. Phys.* **53**, 1511.

Lorents, D. C., and W. Aberth (1965), *Phys. Rev.* **139**, A1017.

Lorents, D. C., G. Black, and O. Heinz (1965), *Phys. Rev.* **137**, A1049.

Mahadevan, P., and G. D. Magnuson (1968), *Phys. Rev.* **171**, 103.

Marino, L. L., A. C. H. Smith, and E. Caplinger (1962), *Phys. Rev.* **128**, 2243.

Maryott, A. A., and F. Buckley (1953), U. S. National Bureau of Standards Circular **537**.

Mason, E. A., and H. W. Schamp (1958), *Ann. Phys. (N.Y.)* **4**, 233.

Mason, E. A., and J. T. Vanderslice (1957), *J. Chem. Phys.* **27**, 917.

——— (1958a), *J. Chem. Phys.* **28**, 253.

——— (1958b), *J. Chem. Phys.* **28**, 1070.

——— (1959a), *J. Chem. Phys.* **30**, 599. The potentials reported are inadequate because a two-state rather than a four-state analysis was used, but the summarized charge transfer cross sections are unaffected and can be used directly in mobility calculations.

——— (1959b), *Phys. Rev.* **114**, 497.

——— (1962), in *Atomic and Molecular Processes*, D. R. Bates, Ed., Chapter 17, Academic, New York.

Michels, H. H. (1966), *J. Chem. Phys.* **44**, 3834.

Miller, J. H., and H. P. Kelly (1972), *Phys. Rev.* **A5**, 516.

Muenter, J. S. (1972), *J. Chem. Phys.* **56**, 5409.

Muschlitz, E. E. (1960) in *Proc. 4th Intern. Conf. Ionization Phenomena Gases, Uppsala 1954*, N. R. Nilsson, Ed., North-Holland, Amsterdam, Vol. **1A**, p. 52.

———, T. L. Bailey, and J. H. Simons (1957), *J. Chem. Phys.* **26**, 711.

Nelson, R. D., Jr., D. R. Lide, Jr., and A. A. Maryott (1967), U.S. National Bureau of Standards **NSRDS-NBS 10**.

Nichols, B. J., and F. C. Witteborn (1966), *NASA Tech. Note* **D-3265**.

Olson, R. E. (1969), *Phys. Rev.* **187**, 153.

——— and C. R. Mueller (1967), *J. Chem. Phys.* **46**, 3810.

Olson, R. E., and F. T. Smith (1971), *Phys. Rev.* **A3**, 1607.

————, and C. R. Mueller (1970), *Phys. Rev.* **A1**, 27.

Palyukh, B. M., and L. S. Savchin (1969), *Ukrainian Phys. J.* (*English transl.*) **13**, 1194 [*Ukrainskii Fiz. Zh.* **13**, 1679 (1968)].

Peek, J. M., T. A. Green, J. Perel, and H. H. Michels (1968), *Phys. Rev. Letters* **20**, 1419.

Perel, J., H. L. Daley, J. M. Peek, and T. A. Green (1969), *Phys. Rev. Letters* **23**, 677.

Perel, J., H. L. Daley, and F. J. Smith (1970), *Phys. Rev.* **A1**, 1626.

Perel, J., R. H. Vernon, and H. L. Daley (1965), *Phys. Rev.* **138**, A937.

Peyerimhoff, S. (1965), *J. Chem. Phys.* **43**, 998.

Platas, O., R. P. Hurst, and F. A. Matsen (1959), *J. Chem. Phys.* **31**, 501.

Popescu Iovitsu, I., and N. Ionescu-Pallas (1960), *Soviet Phys.-Tech. Phys.* (*English transl.*) **4**, 781 [*Zh. Tekh. Fiz.* **29**, 866 (1959)].

Rapp, D., and W. E. Francis (1962), *J. Chem. Phys.* **37**, 2631.

Rich, W. G., S. M. Bobbio, R. L. Champion, and L. D. Doverspike (1971), *Phys. Rev.* **A4**, 2253.

Savage, H. F., and F. C. Witteborn (1968), *J. Chem. Phys.* **48**, 1872.

Sheldon, J. W. (1963), *J. Appl. Phys.* **34**, 444.

Sida, D. W. (1957), *Phil. Mag.* **2**, 761.

Simons, J. H., C. M. Fontana, E. E. Muschlitz, and S. R. Jackson (1943a), *J. Chem. Phys.* **11**, 307.

Simons, J. H., and G. C. Fryburg (1945), *J. Chem. Phys.* **13**, 216.

Simons, J. H., and C. S. Garber (1953), *J. Chem. Phys.* **21**, 689.

Simons, J. H., and S. A. McAllister (1952), *J. Chem. Phys.* **20**, 1431.

Simons, J. H., E. E. Muschlitz, and L. G. Unger (1943b), *J. Chem. Phys.* **11**, 322.

Smith, F. T., H. H. Fleischmann, and R. A. Young, (1970), *Phys. Rev.* **A2**, 379.

Smith, F. T., R. P. Marchi, W. Aberth, D. C. Lorents, and O. Heinz (1967), *Phys. Rev.* **161**, 31.

Snow, W. R., R. D. Rundel, and R. Geballe (1969), *Phys. Rev.* **178**, 228.

Stallcop, J. R. (1971), *J. Chem. Phys.* **54**, 2602.

Stebbings, R. F., A. C. H. Smith, and H. Ehrhardt (1964), in *Atomic Collision Processes*, M. R. C. McDowell, Ed., North-Holland, Amsterdam, p. 814.

Stebbings, R. F., B. R. Turner, and A. C. H. Smith (1963), *J. Chem. Phys.* **38**, 2277.

Sternheimer, R. M. (1970), *Phys. Rev.* **A1**, 321.

Stogryn, D. E., and A. P. Stogryn (1966), *Mol. Phys.* **11**, 371.

Stwalley, W. C. (1971), *J. Chem. Phys.* **54**, 4517.

————, and H. L. Kramer (1968), *J. Chem. Phys.* **49**, 5555.

Teachout, R. R., and R. T. Pack (1971), *Atomic Data* **3**, 195.

Victor, G. A., and A. Dalgarno (1969), *J. Chem. Phys.* **50**, 2535.

————, and A. J. Taylor (1968), *J. Phys.* **B1**, 13.

Wolniewicz, L. (1965), *J. Chem. Phys.* **43**, 1087.

Zehr, F. J., and H. W. Berry (1967), *Phys. Rev.* **159**, 13.

AUTHOR INDEX

Albritton, D. L., 52–58, 82, 100, 230, 232
Alievskiĭ, M. Ya., 157, 232
Amdur, I., 255, 258, 263, 264, 356
Arthurs, A. M., 146, 232

Bailey, T. L., 356–358
Baker, C. E., 356
Baltog, A., 281
Bardsley, J. N., 356
Barker, J. A., 238, 263, 341
Barnes, W. S., 52, 81–83, 230, 232
Bates, D. R., 129, 232, 356
Beaty, E. C., 43, 72–74, 81, 82, 231, 232,
    268–270, 272–275, 277
Bederson, B., 8, 27
Bernardini, O., 356
Bernstein, R. B., 247, 264
Berry, H. W., 357, 359
Beyer, R. A., 61, 82, 275, 286
Biondi, M. A., 32, 62–65, 83, 100, 112,
    114–117, 159, 232, 269, 272, 275, 277–
    281, 286, 291, 306–311
Bird, R. B., 1, 27, 121, 128, 138, 140, 161,
    233, 237, 248–253, 264, 341, 357
Black, G., 358
Blanc, A., 134, 159, 232
Bobbio, S. M., 356, 357, 359
Bogdanova, S. S., 342
Boyd, A. H., 129, 232
Brackmann, R. T., 357
Bradbury, N. E., 30, 60, 82
Branscomb, L. M., 357
Brown, H. L., 62–65, 82, 83, 269
Brown, S. C., 114–116, 149, 234
Browne, J. C., 231, 232, 270, 356
Brush, S. G., 341
Buckingham, R. A., 243, 263, 356
Buckley, F., 358
Burch, D. S., 57, 82
Burhop, E. H. S., 9, 27, 32, 33, 83, 258,
    259, 265

Burke, P. G., 356
Burns, G., 356
Bydin, Yu. F., 356
Byers-Brown, W., 255, 263

Caplinger, E., 358
Carson, T. R., 356
Catlow, G. W., 262, 356
Center, R. E., 275
Čermák, V., 32, 46, 83, 100, 116
Champion, R. L., 356, 357, 359
Chang, E. S., 262, 356
Chanin, L. M., 159, 232, 269, 270, 272,
    275, 277–281, 306, 308, 310–312
Chantry, P. J., 279
Chapman, S., 121, 122, 126, 136, 138, 140,
    141, 161, 170, 187, 189, 194, 232
Chen, C. L., 269
Chen, F. F., 107, 116
Chiang, M. H., 356
Chien, K. R., 356
Chkuaseli, D. V., 356
Chung, K. T., 356
Chupka, W. A., 356
Cloney, R. D., 356, 357
Coffey, D., 356, 357
Colgate, S. O., 255, 263
Colonna-Romano, L. M., 275
Cook, W. R., 142, 146, 232, 237, 264, 341
Copsey, M. J., 109, 111, 117
Courville, G. E., 272, 306, 307, 309, 310
Cowling, T. G., 121, 122, 125, 126, 136,
    138, 140, 141, 161, 170, 187, 189, 194,
    232
Cramer, W. H., 263, 357
Crane, H. R., 305
Crank, J., 1, 14, 27
Creaser, R. P., 58–62, 82
Cromey, P. R., 109, 111, 117
Crompton, R. W., 5, 8, 10, 27, 29, 56, 82,
    83, 92, 98, 272, 286

Curtiss, C. F., 1, 27, 121, 128, 138, 140, 161, 233, 247–253, 264, 357

Dahler, J. S., 157, 233, 234
Daley, H. L., 359
Dalgarno, A., 32, 46, 83, 100, 116, 128, 132, 143, 146, 228, 231, 232, 252, 257, 263, 264, 270, 357, 359
Davidović, D. M., 357
Davies, P. G., 286
Davies, R. H., 357
Davison, W. D., 357
Dean, A. G., 111, 113, 117, 272, 275, 277
de Bethune, A. J., 133, 234
de Boer, J., 157, 234, 237, 264
De Rocco, A. G., 238, 264, 341
Devoto, R. S., 357
Dickinson, A. S., 156, 221, 223, 232, 308, 357
Dillon, J. A., 357
Donath, W. E., 357
Doverspike, L. D., 356, 357, 359
Kukowicz, J. K., 263, 264
Dutton, J., 94–96, 98, 286, 316
Dymond, J. H., 238, 264, 341

Edelson, D., 44, 52, 68–72, 75, 80, 81, 83, 84, 275
Edwards, H. D., 357
Ehrhardt, H., 359
Einstein, A., 2, 27
Elford, M. T., 5, 6, 27, 29, 56, 58–62, 82, 272, 282, 286
Eliason, M. A., 130, 233

Fahr, H., 32, 49, 83, 212, 213, 232
Falconer, W. E., 68–72, 84, 275
Fehsenfeld, F. C., 100, 116
Ferguson, E. E., 32, 46, 83, 100, 116
Fick, A., 1, 27
Firsov, O. B., 129, 225, 232, 259, 264
Fischer, C. R., 260, 264, 357
Fite, W. L., 259, 264, 357
Flannery, M. R., 6, 8, 27, 28, 37, 45, 61, 84, 286, 299, 302, 317, 325, 326
Fleischmann, H. H., 359
Fleming, I. A., 93, 94, 98, 208, 209, 232, 282, 286, 316, 318
Fock, W., 238, 263, 341
Fontana, C. M., 359

Ford, G. W., 168, 170, 232
Ford, K. W., 239, 240, 246, 264
Fouracre, R. A., 111, 117
Fox, J. W., 243, 263
Francis, W. E., 259, 265, 359
Franklin, J. L., 46, 83, 100, 116
Frenkel, S. P., 132, 232
Friedman, L., 32, 46, 83, 100, 116
Frommhold, L., 80, 83, 114–116
Fryburg, G. C., 359
Fung, L. W., 356

Gal, E., 243, 263
Garber, C. S., 359
Gascoigne, J., 56, 82
Gatland, I. R., 6, 27, 28, 37, 45, 48, 61, 76–92, 98, 275, 276, 286, 299–302, 317, 318, 325, 326
Geballe, R., 57, 82, 359
Geltman, S., 231, 232
Gentry, W. R., 357
Gerber, R. A., 268, 269
Ghosh, S. N., 357
Giffen, W. C., 357
Gilbert, T. L., 357
Gilbody, H. B., 9, 27, 32, 33, 83
Gilmore, F. R., 260, 264, 357
Gislason, E. A., 356
Glueck, A. R., 238, 264
Gole, J. L., 357
Goodall, C. V., 113, 117
Gordon, R. G., 260, 264, 358
Graham, E., 6, 28, 37, 45, 46, 48, 61, 84, 275, 276, 282, 286, 317, 325, 326
Gray, A., 78, 83
Gray, D. R., 96–98, 323
Gray, E. P., 113, 116
Green, T. A., 358
Greene, F. T., 31, 84
Grew, K. E., 161, 232
Guldamashvili, A. I., 356
Gupta, B. K., 213, 232, 357
Gurnee, E. F., 259, 264

Hagena, O. F., 30, 83
Hahn, H., 133, 135, 168, 170, 187, 202, 205, 206, 208, 209, 232, 233, 341
Hance, R. L., 356
Harmer, D. S., 52, 83
Hassé, H. R., 142, 146, 232, 237, 264, 341

Hasted, J. B., 9, 27, 33, 83, 258, 259, 264, 357
Hayes, E. F., 357
Heiche, G., 146, 215, 217, 218, 220–223, 226, 228, 232
Heimerl, J. M., 32, 62–65, 83
Heinz, O., 358, 359
Herrero, F. A., 357
Herschbach, D. R., 252, 264
Herzberg, G., 214, 232
Herzenberg, A., 356
Higgins, L. D., 129, 195, 226, 233, 341
Hirschfelder, J. O., 1, 27, 121, 128, 130, 138, 140, 161, 195, 226, 233, 237, 248–253, 264, 341, 342, 357
Hoffman, D. K., 157, 233
Holleran, E. M., 237, 264, 341
Holstein, T., 159, 210, 233, 311
Hornbeck, J. A., 74, 75, 83, 271
Hoselitz, K., 149, 224, 233, 262, 264, 312
Howells, P., 94–96, 98
Hsu, H. W., 342
Hulburt, H. M., 237, 264, 341
Hulpke, E., 356
Hummer, D. G., 357
Hurst, R. P., 359
Huxley, L. G. H., 5, 8, 10, 27, 29, 83, 92, 98

Ibbs, T. L., 161, 232
Iman-Rahajoe, S., 247, 264
Inouye, H., 358
Ionescu-Pallas, N., 359
Itean, E. C., 238, 264

Jackson, S. R., 359
James, D. R., 6, 28, 37, 45, 48, 61, 84, 275, 276, 286, 299–302, 317, 325, 326
Janev, R. K., 357
Jeans, J. H., 125, 233
Johnsen, R., 32, 62–65, 83, 269, 286, 291
Johnson, R. E., 358
Johnson, S. E., 356
Jordan, J. E., 255, 258, 263, 264, 356

Kagan, Yu., 157, 233
Karplus, M., 358
Kasner, W. H., 114, 116
Kaufman, J. J., 262, 263, 356
Keller, G. E., 52, 61, 81–83, 275, 286
Kelly, H. P., 358

Kemmey, P. J., 357
Kennard, E. H., 129, 145, 233
Kennedy, M., 244, 264
Kerr, D. E., 113, 116
Kestner, N. R., 248, 254, 256, 265
Kieffer, L. J., 8, 27
Kihara, T., 165, 167, 168, 170, 175, 176, 185, 187, 194, 195, 226, 233, 237, 264, 342
Kim, Y. S., 260, 264
Kingston, A. E., 357
Kirkwood, J. G., 249, 252, 264, 265
Kita, S., 358
Klein, M., 342
Klingbeil, R., 358
Knof, H., 358
Kotani, M., 237, 264, 342
Kovar, F. R., 281
Kramer, H. L., 252, 264, 359
Kramers, H. A., 126, 233
Krauss, M., 358
Kregel, M. D., 81, 83
Kumar, K., 165, 233
Kishnir, R. M., 358

Landau, L. D., 156, 219, 233
Langer, R. E., 244, 264
Langevin, P., 132, 146, 233, 237, 264
Langhoff, P. W., 358
Laser, R. D., 84, 296–298, 324
Lawrence, J. D., 341
Ledsham, K., 356
Lee, Y., 281, 357
Lester, W. A., 358
Lide, D. R., Jr., 358
Lifshitz, E. M., 156, 219, 233
Lin, S. T., 342
Lineberger, W. C., 100–107, 116
Lippincott, E. R., 256, 265
Livingston, P. M., 246, 264
Llewellyn Jones, F., 93, 94–96, 98, 286, 316
Loeb, L. B., 8, 27, 30, 50, 83
Lorents, D. C., 356–359
Lowke, J. J., 190, 233, 234

McAfee, K. B., 43, 44, 52, 57, 75, 80, 83, 275
McAllister, S. A., 359
McCormack, F. J., 235
McCoy, D. G., 58, 84

McDaniel, E. W., 2, 4, 6, 8, 9, 15, 25, 27, 28, 32, 34, 36, 37, 39, 43–46, 48, 52–58, 61, 76, 81–92, 98–100, 112–114, 116, 149, 230, 232, 233, 258, 259, 265, 275, 276, 282–302, 305, 316–322, 324–326

McDowell, M. R. C., 44, 83, 128, 232, 257, 262–264, 356, 357

McFarland, M., 100

McGuire, J. M., 356

McKnight, L. G., 43, 44, 57, 65–68, 80, 81, 83, 275, 286, 291

McLachlan, N. W., 22, 27

McNamee, J. M., 357

MacRoberts, T. M., 78, 83

Madson, J. M., 270, 275

Magee, J. L., 259, 264

Magnuson, G. P., 358

Mahadevan, P., 358

Mahan, B. H., 281, 356, 357

Maldonado, P., 358

Mandl, F., 356

Marchi, R. P., 359

Marcus, A. B., 357

Margenau, H., 143, 233, 248–250, 254–256, 265

Marino, L. L., 358

Märk, T. D., 272

Marrero, T. R., 118, 233

Martin, D. W., 36, 37, 39, 43–46, 52–58, 76, 81–92, 98, 149, 230, 232, 233, 282–298, 316–322, 324

Maryott, A. A., 358

Mason, E. A., 45, 84, 118, 124, 127, 133, 135, 136, 140–150, 154, 157, 160, 161, 165, 168–171, 174, 187, 202, 205–213, 218, 220–223, 226, 228, 230–249, 254–265, 319, 341, 342, 356–358

Massey, H. S. W., 8, 9, 27, 30, 32, 33, 50, 84, 216, 219, 225, 233, 234, 258, 259, 265

Matcha, R. L., 255, 265

Mathews, G. B., 78, 83

Matsen, F. A., 213, 232, 357, 359

Maxwell, J. C., 132, 233, 237, 265

May, C. J., 356

Mehr, F. J., 114, 116, 117

Michels, H. H., 358, 359

Miller, J. H., 358

Miller, T. M., 36, 37, 43, 52–58, 82, 84–92, 98, 149, 230, 232, 233, 282–285, 316

Milne, T. A., 31, 84

Mohr, C. B. O., 216, 225, 233

Moiseiwitsch, B. L., 9, 27

Monchick, L., 127, 131, 144, 157, 233, 234, 237, 238, 242, 247, 256, 259, 260, 265, 342

Morrison, J. A., 44, 68, 80, 81, 83

Moseley, J. T., 36, 37, 39, 43–45, 76, 77, 83–92, 98, 149, 233, 282–290, 316–319

Mott, N. F., 219, 234

Mueller, C. R., 358, 359

Muenter, J. S., 358

Müller, K. G., 32, 49, 83, 212, 213, 232

Munn, R. J., 154, 157, 161, 233, 234, 238, 240, 243, 244, 246, 247, 265, 342, 357

Munson, R. J., 278, 280

Musa, G., 281

Muschlitz, E. E., 356, 358, 359

Mustată, I., 281

Năstase, L., 281

Nelson, R. D., 358

Nemeth, E. M., 357

Nernst, W., 2, 28

Nesbet, R. K., 255, 265

Neynaber, R. H., 33, 84

Nichols, B. J., 358

Niculescu, N. D., 281

Nielsen, R. A., 30, 60, 82

Niloleishvili, V. D., 356

Obert, W., 30, 83

O'Hara, H., 150, 233, 237, 265, 342

Olson, R. E., 356, 358, 359

Orient, O. J., 155, 223, 231, 234, 262, 265, 269, 308, 309

Oskam, H. J., 100, 113, 114, 116, 268–270, 272, 275

Pack, R. T., 359

Palyukh, B. M., 358, 359

Parker, J. H., Jr., 190, 233, 234

Parkes, D. A., 56, 84

Patterson, P. L., 43, 44, 74, 81, 82, 84, 150, 151, 223, 224, 231, 234, 262, 265, 268–270, 272, 273, 303, 304, 308, 309

Pearce, A. F., 307

Peek, J. M., 359

Perel, J., 359

Persson, K. B., 115, 116, 149, 234

Peyerimhoff, S., 156, 234, 359
Phelps, A. V., 8, 9, 28, 29
Philbrick, J., 114, 117
Platas, O., 359
Plumb, I. C., 116
Popescu, A., 281
Popescu Iovitsu, I., 359
Power, J. D., 255
Present, R. D., 132, 133, 234
Puckett, L. J., 58, 84, 100–107, 117

Raether, M., 269
Rapp, D., 259, 265, 359
Rayleigh, Lord, 128, 187, 234
Rees, J. A., 93, 94, 96–98, 208, 209, 232, 282, 286, 316, 318, 323
Rees, W. D., 94–96, 98, 316
Rice, W. E., 342
Rich, W. G., 356, 357, 359
Rigby, M., 238, 264, 341
Robson, R. E., 165, 233
Rundel, R. D., 359
Russell, M. E., 356

Sachs, L. M., 262, 263, 356
Samoilov, E. V., 238, 265, 342
Sandler, S. I., 157, 159, 160, 234
Sather, N. F., 157, 234
Sauter, G. F., 268, 269
Savage, H. F., 359
Savchin, L. S., 358, 359
Sawina, J. M., 83, 275, 286, 291
Schamp, H. W., 124, 142–149, 168, 170, 171, 174, 213, 223, 231, 233, 237, 249, 254, 257, 265, 342, 358
Schmeltekopf, A. L., 100, 116
Schramm, B., 342
Schottky, W., 25, 28
Schummers, J. H., 6, 28, 37, 45, 46, 48, 61, 77, 81, 82, 84, 286, 291–302, 317, 320–322, 324–326
Sena, L. A., 358
She, R. S. C., 157, 234
Sheldon, J. W., 359
Sheridan, W. F., 357
Sida, D. W., 260, 265, 359
Simons, J. H., 263, 357–359
Sipler, D., 43, 44, 57, 68, 75, 80, 81, 83, 275
Skullerud, H. R., 94, 98, 190, 192, 193, 201, 207, 212, 234

Slater, J. C., 252, 265
Smirnov, B. M., 188, 212, 213, 234
Smith, A. C. H., 357–359
Smith, D., 107–113, 116, 117, 272, 275, 277
Smith, E. B., 238, 264, 341
Smith, F., 238, 263, 341
Smith, F. J., 129, 150, 154, 161, 195, 215, 217, 226, 233, 237, 238, 240, 243, 244, 246, 247, 256, 264, 265, 341, 342, 359
Smith, F. T., 357–359
Smith, R. A., 216, 233
Smith, S. J., 9, 27
Snider, R. F., 157, 234
Snow, W. R., 359
Snuggs, R. M., 39, 43–46, 70, 84–86, 98, 286–295, 318–322
Spotz, E. L., 237, 264, 341
Spurling, T. H., 238, 264, 341
Stallcop, J. R., 359
Stebbings, R. F., 357, 359
Steele, D., 256, 265
Steen, R. D., 281
Stefan, J., 132, 234
Sternheimer, R. M., 359
Stewart, A. L., 356
Stogryn, A. P., 359
Stogryn, D. E., 359
Storvick, T. S., 238, 264, 341
Stwalley, W. C., 359
Sullivan, M. R., 81, 83
Svehla, R. A., 238, 264

Taylor, A. J., 359
Taylor, M. H., 195, 226, 233, 237, 264, 342
Taxman, N., 157, 234
Teachout, R. R., 359
Teague, M. W., 58, 84
Thomas, E. W., 9, 28
Thomson, G. M., 6, 28, 37, 45, 48, 61, 84, 275, 276, 286, 299–302, 317, 325, 326
Thomson, G. P., 8, 28
Thomson, J. J., 8, 28
Townsend, J. S., 2, 28
Tsao, C. W., 356
Tsitelauri, N. N., 238, 265, 342
Turner, B. R., 359
Tunnicliffe, R. J., 93, 94, 98, 208, 209, 232, 282, 286, 316, 318
Tyndall, A. M., 30, 72, 84, 257, 265, 269, 272, 275, 277–280, 282, 286

Uhlenbeck, G. E., 157, 234
Umanskii, A. S., 342
Unger, L. G., 359

Vanderslice, J. T., 149, 230, 233, 256, 260, 265, 356–358
Van Kranendonk, J., 237, 264
Varney, R. N., 74, 84, 234, 277–280
Varshni, Y. P., 256, 265
Veatch, G. E., 269
Vernon, R. H., 359
Victor, G. A., 359
Volz, D. J., 46, 81, 84, 291–298, 320–322, 324

Wahl, A. C., 357, 358
Waldmann, L., 157, 234
Wang Chang, C. S., 157, 234
Wannier, G. H., 4, 7, 8, 28, 32, 84, 120–128, 187–190, 194, 196–198, 201, 202, 208, 235, 314, 315
Weber, G. G., 260, 265
Weller, C. S., 114, 115, 117

Werner, A. S., 356
Whealton, J. H., 45, 81, 82, 84, 135, 136, 165, 208, 209, 235, 319
Wheeler, J. A., 239, 240, 246, 264
Williams, A., 128, 143, 232, 257, 264
Williams, E. M., 94–96, 98, 316
Wilson, J. N., 252, 265
Witteborn, F. C., 358, 359
Wolniewicz, L., 156, 235, 359
Woo, S. B., 81, 82, 84
Wood, H. T., 152, 235

Yadav, H. N., 357
Yos, J. M., 210, 211, 233
Young, C. E., 68–72, 84, 275
Young, R. A., 359
Yun, K. S., 157, 234

Zehr, F. J., 359
Zener, C., 255, 265
Zhdanov, V. M., 157, 232
Zwanzig, R., 118, 235

SUBJECT INDEX

Transport data on specific ion-molecule systems are referenced separately only if they appear outside of Chapter 7, which contains the main collection of experimental data.

Accuracy of measurements, diffusion coefficients, 44, 46, 88, 92, 94, 96, 98, 100, 107, 313, 314, 316
mobilities and drift velocities, 35–37, 41, 43, 44, 267, 268, 313
reaction rate coefficients, 46, 49, 92, 105
Additional residence time method, 65
Afterglow, accuracy of measurements, 100, 107
Afterglow, definition, 99
diffusion controlled, 113
flowing, 100
recombination controlled, 113
stationary, 100–116
Ambipolar diffusion, definition, 25
ion-ion, 99, 105–107
Ambipolar diffusion coefficient, 26, 27
Ambipolar diffusion equation, 26
Angle of deflection, as function of impact parameter, 238–239
ion-neutral potential and, 127, 237
orbiting collisions, 128–129, 238–239
rainbow scattering, 239–240
relation, to cross section, 127, 133, 153, 237
to phase shifts, 153, 244, 245
small-angle approximation, 129, 225
Anisotropy of polarizability, 248–249, 345
Approximate cross sections, capture, 128
charge-transfer, 132, 224–228
effective-diameter, 129–131
impact-parameter, 224–227
orbiting, 128–130
random-angle, 128–129
random-phase, 224–229
Attraction, ion-neutral, 146, 147, 248, 251–254
Arrival time spectra, nonreacting ions, 36–40

Attenuation method, for measuring, reaction rate coefficients, 88–92
transverse diffusion coefficients, 85–92
Average energy of ions in gas in electric field, 32, 49, 50

Beam experiments, limitations on, 32, 33
Beam scattering and ion-neutral interactions, 257–258, 350–355
Blanc's law, deviations at high fields, 134–136, 208–209
at low fields, 159–161
experimental tests of, 304–306
for multicomponent mixtures, 133–135, 159
momentum-transfer theory, 133–134, 208–209
Boltzmann equation, Chapman-Enskog solution, 136–140; see also Chapman-Enskog kinetic theory
high-field solutions, 187–201
Kihara solution, 165–176
low-field solution, 136–140
medium-field solution, 165–176
modification for inelastic collisions, 156–158
relation to other theories, 144–145
Bose-Einstein statistics, 215–217
Boundary conditions for diffusion, 14, 15
Bradbury-Nielsen method, 60, 61
Burnett functions, 168

Chapman-Enskog kinetic theory, connection with elementary theories, 144–145
convergence of approximations, 140–144
general method, 136–140
inelastic collisions, 156–158
limitations, 140

multicomponent diffusion, 159
quantum effects, 151–156, 244–247
Charge-dipole interaction, 129, 253
Charge exchange, *see* Charge transfer
Charge-induced dipole interaction, 129, 146, 248, 254
Charge-induced quadrupole interaction, 129, 143, 248, 253, 254
Charge-quadrupole interaction, 129, 253
Charge transfer, cross sections, 132, 224–229, 263
    effect on mobility and diffusion, 42, 45, 86, 132, 209–211, 247–248, 315, 317, 318
    quantum-mechanical description, 214–219
    reaction, definition, 29
    relation to ion-neutral potential, 213
    semi-classical description, 210–213
Chemical reaction, definition, 29
Chemical reactions, effects on measurements with drift tubes, 39–44
Classical calculation of mobility, 237–242
Clustering of molecules about ions, 61, 62
Collision integral, definition, 139, 157–158
    for high fields, 205
    for inelastic collisions, 157–158
    quantum effects, 155–156, 243, 247
    tabulations, 331–341
    temperature dependence, 145–151
Collisions, mean free time between, 120–121, 124
Convergence of formulas, 140–144, 176–182
Cross sections, angle of deflection, 127, 133, 153, 237
    approximate, *see* Approximate cross sections
    calculation from interaction potentials, 237–247
    charge-transfer, 132, 210–211, 216, 218–219, 221, 222, 226–228, 258–259, 263; *see also* Charge transfer
    differential, 127, 151, 152, 157–158, 216
    diffusion, 127, 152, 153, 158
    inelastic collisions, 157–158
    momentum-transfer, 127, 133; *see also* Cross Sections, diffusion
    nuclear spin effects, 217–219, 222
    phase shifts, 152, 216–217, 243

quantum effects, 153–156, 246–247
statistics, 215–219
tabulations, 328–330
transport, definition, 139, 157–158
Cs⁺, cross sections in Cs, 222, 227
    mobility, in, Cs, 223–224
    in Xe, 223–224

de Boer parameter, 154, 243
Debye length, 25
Delta-function model, *see* Interaction energy
Diffusion, Blanc's law, *see* Blanc's law
    Chapman-Enskog theory, *see* Chapman-Enskog kinetic theory
    composition dependence, 136, 159–161, 208–209
    cross sections, *see* Cross section, diffusion
    definition, 1
    density dependence, 119
    during drift in electric field, 12, 13
    Einstein relation, *see* Einstein relation
    elementary theory, 125–126, 136
    field dependence, 119–120, 126, 192, 194, 196–197, 207, 208, 314–326
    higher approximations, 140–144
    multicomponent mixtures, 136, 159, 208
    Nernst-Townsend relation, *see* Einstein relation
    relation to mobility, *see* Einstein relation
    temperature dependence, 145–151
    thermal, 161–165
    unbounded, 10–12
Diffusion coefficient (scalar), definition, 1
Diffusion coefficient (tensor), 3
Diffusion coefficients, uses of data, 8
Diffusion cooling, 112
Diffusion data—ions in, argon, 319, 326
    carbon monoxide, 318, 325, 326
    hydrogen, 316
    nitric oxide, 318, 324
    nitrogen, 317–319, 326
    oxygen, 318, 320–323
Diffusion equation, solution for various geometries, 16–24
    time-dependent, 14
    time-independent, 14
Diffusion length, 17, 19, 21, 24
Dipole-dipole dispersion interaction, 143, 252, 253, 348
Dipole-dipole interaction, 253

Dipole-induced dipole interaction, 251
Dipole moment, 251, 253, 347
Dipole, polarizability, *see* Polarizability
Dipole-quadrupole dispersion interaction, 252
Dipole-quadrupole interaction, 253
Dispersion energy, 143, 252, 253, 348
Dissociation of ions during sampling, 57
Distrubiton function, 136–138, 166–167, 188–191, 199, 201; *see also* Boltzmann equation
Drift tubes, descriptions of, 50–75
general conditions of operation, 31, 32
general description, 30
Drift tube mass spectrometer, general description, 31
Drift velocity, definition, 2
measurements, general considerations, 35–44
see also Mobility

Einstein relation, 1, 118–119, 139, 196, 208, 314
relation of ND to K_O, 316
Elastic scattering, ion beams, 257–258
Electron-ion recombination, studies of, 113–115
Electrons, transport of, 9, 10
Electrostatic energy, 253, 347
Elementary theories, Free-flight, *see* Free-flight theory
Free-Path, *see* Free-flight theory
mean-free-flight, *see* Free-flight theory
mean-free-path, *see* Free-flight theory
momentum-transfer, *see* Momentum-transfer theory
End effects in drift tubes, 35, 36
Energy, partitioning of ion, 32, 122, 123, 134, 135, 184–187, 195, 201–204
Energy of interaction, *see* Interaction energy
Equation of continuity, 13
E/N, significance of, 3, 4
relation to E/p, 4
E/p, significance of, 4
Exchange, charge, *see* Charge transfer
electron, *see* Charge transfer
ion, *see* Ion transfer
Exponential potential, 130–131, 254–256

Fermi-Dirac statistics, 215–217
Fick's first law, 1
Fick's second law, *see* Diffusion equation, time-dependent
Field energy of ions, 4, 5; *see also* Energy, partitioning of ion
Flux density, definition, 1
Forces, *see* Interaction energy
Four-gauze electrical shutter method, 72
Free diffusion coefficient, 25
Free-flight theory, mobility and diffusion, 120–127, 212
relation to other theories, 144–145
Free-path theory, *see* Free-flight theory

Glory scattering, 239, 240, 245
Gray-Kerr criteria for afterglows, 113

H^+ interaction, with H_2, 148–149, 350
with He, 148, 155–156, 263, 350
H^+ mobility, in H_2, 148–149
in He, 148, 155–156, 263
H_3^+ mobility in H_2, 229–230
H_3^+ proton transfer in H_2, 229
He^+ cross-sections in He, 220–221, 227
He^+ interaction with He, 213–214, 262, 351
He^+ mobility in He, 223, 262–263
He_2^+ ion transfer in He, 231
He_2^+ mobility in He, 231
High-field conditions, 5

Identical particles, 215–217
Impact-parameter approximations, 224–227
Impurities in gases, effects of, 10, 51, 52
Induction energy, *see* Polarization energy
Inelastic collisions, 156–158
Interaction energy, attraction, *see* Attraction, ion-neutral
combination rules, 252, 255–256
delta-function model, 260
determination of, 257–260
dispersion, *see* Dispersion energy
electrostatic, *see* Electrostatic energy
induction, *see* Polarization energy
intermediate-range, theory, 256
long-range, theory, 248–253
polarization, *see* Polarization energy
repulsion, *see* Repulsion, ion-neutral
short-range, theory, 254–256, 259–260, 350–355

statistical model, 260
van der Waals, *see* Dispersion energy
Intermediate-range forces, *see* Interaction energy
Inverse-power potential, 128, 130, 131, 142, 255
Ion-dipole interaction, 129, 253
Ion exchange, *see* Ion transfer
Ion-induced dipole interaction, 129, 146, 248, 254
Ion-induced quadrupole interaction, 129, 143, 248, 253, 254
Ion-ion interactions, 6–8
Ion-molecule reaction, definition, 29
Ion-neutral interactions, *see* Interaction energy
Ion-quadrupole interaction, 129, 253
Ion sampling from gas mixtures, 65
Ions, states of excitation, 34
Intermediate-field conditions, 4
Ion transfer, 229–231

JWKB approximation, 153–155, 225, 243–247

K^+ mobility, in $H_2 + N_2$ mixtures, 208–209
in $N_2 + O_2$ mixtures, 209
Kihara theory, 165–176, 194
Kinetic theory, *see* Boltzman equation; Chapman-Enskog kinetic theory

Landau-Zener effect, 219
Langmuir probes, 107–110, 113
Legendre polynomials, 168, 199, 202
Li^+, interaction with He, 213, 214, 261
mobility in He, 149, 223, 261–262
Lifetime, of ions against collisions with walls, 12
Linear extrapolation distance, 15
London forces, *see* Dispersion energy
Long-range forces, *see* Interaction energy
Lorentz model, 141, 189–193
Lorentzian mixture, *see* Lorentz model
Low-field conditions, criterion for, 5
definition, 3
Longitudinal diffusion coefficient (definition), 3, 4
Longitudinal diffusion coefficient measurements, general considerations, 44–46

Magnetic field, effect on ionic motion, 4
Mass discrimination during ion sampling, 56
Maxwell model, 141, 193–198
Maxwellian molecules, *see* Maxwell model
Mean free path, *see* Free-flight theory
Mean free time, *see* Free-flight theory
Merging beams experiments, 33, 46
Microwave techniques, 114–116
Mixtures, diffusion in, *see* Diffusion: Blanc's law, multicomponent mixtures
ion energy in, 201–204
mobility in, *see* Mobility: Blanc's law, multicomponent mixtures
Mobility, Blanc's law, *see* Blanc's law
Chapman-Enskog theory, *see* Chapman-Enskog kinetic theory
Composition dependence, *see* Blanc's law
cross sections, *see* Diffusion cross sections, momentum-transfer cross sections
definition, 2
dependence, on gas temperature, 5
on molecular number density, 5, 6, 119–120
Einstein relation, *see* Einstein relation
elementary theory, *see* Elementary theories
field dependence, 119–124, 174–175, 181, 182–184, 204–207, 222
higher approximations, 140–144, 176–182
Kihara theory, *see* Kihara theory
multicomponent mixtures, *see* Blanc's law
Nernst-Townsend relation, *see* Einstein relation
quantum effects, 151–156, 182, 214–222
relation to diffusion, *see* Einstein relation
temperature dependence, 145–151, 182–184, 222–224
Mobility data—ions in mixtures of gases, helium-neon mixtures, 304, 306, 307
oxygen with other gases, 304, 305
ions in pure gases and vapors, argon, 274–276
carbon monoxide, 299–302
deuterium, 283–285
helium, 268–271
hydrogen, 282, 283, 285
krypton, 277, 278
metal vapors, 281
neon, 272, 273
nitric oxide, 296–298

nitrogen, 286–290
oxygen, 291–295
sulfur hexafluoride, 303, 304
xenon, 279, 280
variation with gas temperature, argon, 311,
    312
  helium, 306–309
  hydrogen, 311–313
  krypton, 311, 312
  neon, 308, 310
  xenon, 311
Mobility determinations, accuracy of, 35
Mobility, uses of data, 8
Momentum-transfer cross sections, see
    Cross sections, momentum-transfer
Momentum-transfer theory, mixtures, 133–
    136, 208–209
  mobility and diffusion, 132–136
  relation to other theories, 144–145
Monte Carlo calculations, 201
Morse potential, 256
Multipole moments, see Dipole moment;
    Quadrupole moment

Nernst-Townsend relation, see Einstein
    relation
Noncentral interactions, 248, 252–253
    see also Inelastic collisions
Nuclear spin, effect on mobility, 215–219

Orbiting, 128, 130, 238–241, 244–245
Orbiting, resonances, 153–155, 220
Ordinary diffusion coefficient, 25
Oscillator model for dispersion energy, 252,
    348
Oscillator model, for quadrupole
    polarizability, 250–251
Overlap energy, 254

Partial waves, 151–152
Peesistence of velocities, 125, 126, 192, 212
Phase shift, as function, of energy, 244
    of l, 245
  ion-neutral potential, 153, 242–243
  JWKB approximation, see JWKB approxi-
    mation
  relation, to angle of deflection, 153, 244,
    245
    to cross section, 152, 216–217, 243
    to orbiting, 244–245

small-angle approximation, 225
Plasma, definition, 99
Poisson's equation, 7
Polar molecules, see Dipole moment
Polarizability, dipole, 146, 248–252, 344–
    346
  quadrupole, 248–251, 344
Polarization, approximation, 224, 226, 229
  energy, 143, 146, 248–251, 253, 254,
    344–346
  limit for mobility, 146, 226
Potential functions, cross sections and colli-
    sion integrals for, 327–341
  field dependence of mobility, 183
  temperature dependence of mobility, 145–
    151, 183
  see also Interaction energy
Primary ions, definition, 33
Protons, see H^+

Quadrupole, moment, 253, 347
Quadrupole, polarizability, see Polarizability
Quantum-mechanical calculations–mobility,
    151–156, 242–247
  interaction energies, 213–214, 259
  effects on mobility, 151–156, 182, 214–
    222
  theory of scattering, charge-transfer, 214–
    217, 247–248
  elastic, 151–156, 242–244
Quasi-Lorentzian gas, 141, 187; see also
    Rayleigh model

Rainbow scattering, 239–240
Rayleigh model, 187–189
Reaction rate coefficients, determination
    from arrival time spectra, 46–50
Reduced mobility, definition, 6
Reduced pressure, definition, 4
Repulsion, ion-neutral, 130, 146–147, 150,
    254–256
Resonance, see Charge transfer; Orbiting
Resonant charge transfer, see Charge
    transfer
Reversible field drift tube, 62–65

Scattering, see also Cross sections
Scattering, amplitude, 152, 215–216
  angle, see Angle of deflection
  classical theory of, 127–128, 237–240

elastic, 127, 257–258
glory, *see* Glory scattering
inelastic, 156–158
ion-beam, 257–258, 259
orbiting, *see* Orbiting
quantum theory of, *see* Quantum-mechanical theory of scattering
rainbow, *see* Rainbow scattering
Screening constants, 249
Semi-classical theory of, charge transfer, 210–212
scattering, 153, 210–211, 224–227, 243–246; *see also* JWKB approximation
Short-range forces, *see* Interaction energy
Solutions of ionic transport equations, 75–82
Sonine polynomials, 139, 168
Space charge effects, 7, 24, 25, 37, 39, 45
Spin, effect on mobility, *see* Nuclear spin
Spiraling, *see* Orbiting
Standard conditions of pressure and temperature, 6
Statistical model, *see* Interaction energy
Symmetry effects, 215–217

Thermal diffusion, 161–165
Thermal energy, 121, 123, 125, 134, 135, 137, 188, 191, 204

Thermionic emmiters of ions, 34
Townsend (Td), definition, 5
Townsend method for measuring transverse diffusion coefficients, 92–98
Transfer, *see* Charge transfer; Ion transfer
Transport, coefficients, calculations of, *see* Chapman-Enskog kinetic theory
collision integrals, 139, 157–158, 169, 170, 172–173, 176–177, 205, 237, 241, 247, 331–341
cross sections, 139, 152–154, 156, 158, 205, 215–222, 237, 241, 243, 246–247, 328–330
Transverse diffusion coefficient, definition, 3
measurement of, 85–98
Tyndall gate, 55, 66, 69, 72

Valence forces, 254–256; *see also* Interaction energy; Overlap energy
van der Waals forces, *see* Dispersion energy

Wannier diffusion equations, 314, 315, 319
Wannier theory, 194–202, 314, 315, 319
Wavelength, de Broglie, 154, 242, 243
Whealton-Mason diffusion theory, 208, 319
WKB approximation, *see* JWKB approximation

111860